THE INTERNATIONAL SERIES OF
MONOGRAPHS ON CHEMISTRY

THE INTERNATIONAL SERIES OF
MONOGRAPHS ON CHEMISTRY

---

1 J. D. Lambert: *Vibrational and rotational relaxation in gases*
2 N. G. Parsonage and L. A. K. Staveley: *Disorder in crystals*
3 G. C. Maitland, M. Rigby, E. B. Smith, and W. A. Wakeham: *Intermolecular forces: their origin and determination*
4 W. G. Richards, H. P. Trivedi, and D. L. Cooper: *Spin–orbit coupling in molecules*
5 C. F. Cullis and M. M. Hirschler: The *combustion of organic polymers*
6 R. T. Bailey, A. M. North, and R. A. Pethrick: *Molecular motion in high polymers*
7 Atta-ur-Rahman and A. Basha: *Biosynthesis of indole alkaloids*
8 J. S. Rowlinson and B. Widom: *Molecular theory of capillarity*
9 C. G. Gray and K. E. Gubbins: *Theory of molecular fluids, volume 1: Fundamentals*
10 C. G. Gray and K. E. Gubbins: *Theory of molecular fluids, volume 2: Applications* (in preparation)
11 S. Wilson: *Electron correlations in molecules*
12 E. Haslam: *Metabolites and metabolism: a commentary on secondary metabolism*
13 G. R. Fleming: *Chemical applications of ultrafast spectroscopy*
14 R. R. Ernst, G. Bodenhausen, and A. Wokaun: *Principles of nuclear magnetic resonance in one and two dimensions*
15 M. Goldman: *Quantum description of high-resolution NMR in liquids*
16 R. G. Parr and W. Yang: *Density-functional theory of atoms and molecules*
17 J. C. Vickerman, A. Brown, and N. M. Reed (editors): *Secondary ion mass spectrometry: principles and applications*
18 F. R. McCourt, J. Beenakker, W. E. Kohler, and I. Kuscer: *Nonequilibrium phenomena in polyatomic gases, volume 1: Dilute gases*
19 F. R. McCourt, J. Beenakker, W. E. Kohler, and I. Kuscer: *Nonequilibrium phenomena in polyatomic gases, volume 2: Cross-sections, scattering, and rarefied gases*
20 T. Mukaiyama: *Challenges in synthetic organic chemistry*
21 P. Gray and S. K. Scott: *Chemical oscillations and instabilities: non-linear chemical kinetics*
22 R. F. W. Bader: *Atoms in molecules: a quantum theory*
23 J. H. Jones: *The chemical synthesis of peptides*
24 S. K. Scott: *Chemical chaos*
25 M. S. Child: *Semiclassical mechanics with molecular applications*
26 D. T. Sawyer: *Oxygen chemistry*
27 P. A. Cox: *Transition metal oxides: an introduction to their electronic structure and properties*
28 B. R. Brown: *The organic chemistry of aliphatic nitrogen compounds*
29 Y. Yamaguchi, Y. Osumara, J. D. Goddard, and H. F. Schaeffer: *A new dimension to quantum chemistry: analytic derivative methods in* ab initio *molecular electronic structure theory*
30 P. W. Fowler and D. E. Manolopoulos: *An atlas of fullerenes*

Transition Metal Oxides

An Introduction to their Electronic Structure and Properties

P. A. COX

Fellow of New College and
Lecturer in Inorganic Chemistry
University of Oxford

CLARENDON PRESS · OXFORD

1995

Oxford University Press, Walton Street, Oxford OX2 6DP
Oxford New York
Athens Auckland Bangkok Bombay
Calcutta Cape Town Dar es Salaam Delhi
Florence Hong Kong Istanbul Karachi
Kuala Lumpur Madras Madrid Melbourne
Mexico City Nairobi Paris Singapore
Taipei Tokyo Toronto
and associated companies in
Berlin Ibadan

Oxford is a trade mark of Oxford University Press

Published in the United States by
Oxford University Press Inc., New York

First published 1992
First published in paperback 1995

A catalogue record for this book is available from the British Library

Library of Congress Cataloging in Publication Data
Cox, P. A.
Transition metal oxides : an introduction to their electronic structure and properties / P. A. Cox.
p. cm.—(The International series of monographs on chemistry : 27)
Includes bibliographical references and index.
I. Transition metal oxides. I. Title. II. Series.
QD172.T6C68 1992 546.6—DC20 91–38222
ISBN 0 19 855925 9 (Pbk)

Printed in Great Britain by
Biddles Ltd., Guildford and King's Lynn

PREFACE

Transition metal oxides form a series of compounds with a uniquely wide range of electronic properties. Some of these—the magnetism of lodestone and the colours of gems and minerals containing transition metals—have been known since antiquity. Other properties, especially the 'high-temperature' superconductivity of mixed oxides containing copper, have been discovered only recently. Serious attempts to characterize and understand the electronic structure of transition metal oxides began in the late 1930s, but as the discovery of high-temperature superconductors illustrates, these compounds continue to surprise and baffle us in many ways.

The present book is *not* primarily about superconductors, although it is true that the current interest in these compounds provided a strong motivation for writing it. The aim is to describe the range of electronic properties found in transition metal oxides—including magnetic, optical, and spectroscopic properties as well as electrical conductivity—and to discuss in a fairly critical way the various models that have been proposed to interpret them. I have made no attempt at anything approaching a complete survey, which would be quite impossible in such a limited space. Rather, the book is intended as an *introduction* to chemists and physicists who wish to get some idea of what the field is about. I have tried to select both compounds and theories so as to provide a reasonably balanced account, although the choice must inevitably reflect my own interests. Only bulk properties are treated here: surfaces of both transition and non-transition metal oxides will form the subject matter of a forthcoming book in collaboration with Vic Henrich.

The level of the book is intended to be suitable for graduate students and other researchers with a background in solid-state chemistry or physics, or in materials science. Because of this diverse background—which reflects the interdisciplinary nature of the subject—I have included some introductory material. Chapter 1 describes some chemical and structural concepts that may not be familiar to physicists. Chapter 2 explains the different models that have been used to interpret electronic properties; it seems particularly essential to do this, as the theories derive from the very different traditions of physical and chemical thinking about electronic structure. My aim has been to describe the physical basis of the different models, rather than the mathematical details. If theorists are unhappy about the lack of

Hamiltonians and of computational details, then I offer no apologies, as their kind of meat is unfortunately the experimentalists' (and especially the chemists') poison: but they will find references that I hope will provide them with suitable nourishment. In the same way, the discussion of experimental results concentrates on what is being measured, rather than how the experiments are done: again there are plenty of other places where these details can be found.

My interest in transition metal oxides goes back several years, and has been greatly stimulated by collaboration and discussion with many other people in this field. I must mention especially John Goodenough, whose own contributions to this area are paramount, as well as Tony Orchard, Russ Egdell, Andrew Hamnett, Peter Dickens, Tony Cheetham, Marshall Stoneham, and Vic Henrich. The immediate stimulus for this book came from a stay in the United States. In the Fall Semester of 1988 I gave a course of lectures at Cornell University, which formed the basis of the present account. I would like to thank Roald Hoffmann and Frank DiSalvo for the invitation to Cornell, and to many other people there who made my stay so enjoyable. I am also grateful to Vic Henrich and his colleagues for an invitation to Yale, where I tried out some of my ideas, and had many interesting discussions.

A number of people read the first draft of all or part of the book, and I would like to thank especially Marshall Stoneham, Jeremy Kemp, and Wendy Flavell for their helpful and sometimes critical comments. Christine Palmer's assistance in drawing the diagrams was invaluable. Finally, I must thank Christine, Stephen, Andrew, and Emma once again for their encouragement and for their patience during my absences, both physical and mental.

Oxford P.A.C.
June 1991

CONTENTS

1 Introduction 1

1.1 Scope and plan of the book 1

1.2 Chemical aspects 3

1.2.1 Oxidation state and electron configuration 3

1.2.2 Thermodynamic stability and phase diagrams 10

1.3 Structural principles 15

1.3.1 Metal coordination geometries 15

1.3.2 Structural families 20

1.3.3 Defects and the accommodation of non-stoichiometry 25

1.4 Electronic classification 29

References 32

2 Models of electronic structure 36

2.1 Ionic models 37

2.1.1 Crystal field theory 37

2.1.2 Charge transfer and band gaps 47

2.2 Cluster models 53

2.2.1 The molecular orbital (MO) method 54

2.2.2 Configuration interaction models 60

2.3 Band theory 64

2.3.1 Band structure and chemical bonding 64

2.3.2 The properties of conduction electrons 72

2.3.3 Magnetic band structures 75

2.3.4 Peierls instabilities and Fermi-surface nesting 78

2.4 Intermediate models 81

2.4.1 The Hubbard model 82

2.4.2 Excitons 87

2.4.3 Impurity states 89

2.4.4 Anderson localization 91

2.4.5 Polarons 92

References 95

3 Insulating oxides 100

3.1 d^0 compounds 100

3.1.1 Band gaps and spectroscopic transitions 101

3.1.2 Dielectric and non-linear properties 108

3.2 Other closed-shell oxides 115
3.3 Transition metal impurities 118
3.3.1 Spectroscopic properties of d^n states 118
3.3.2 Charge transfer and impurity-state energies 122
3.3.3 Interaction between impurities 129
3.4 Magnetic insulators 131
3.4.1 Survey of properties 132
3.4.2 The nature of the band gap 136
3.4.3 Magnetic ordering 142
3.4.4 Origin of exchange interactions 148
References 153

4 Defects and semiconduction 157

4.1 Electronic carrier properties 157
4.1.1 Free carriers in thermal equilibrium 157
4.1.2 Transport properties 160
4.2 The point-defect model 169
4.2.1 The Kröger–Vink notation 169
4.2.2 The law of mass action 172
4.2.3 Defect mobility and diffusion 178
4.2.4 Limitations of point-defect theories 182
4.3 Carrier binding energies and spectroscopy 184
4.3.1 Spectroscopic studies of free and bound carriers 184
4.3.2 Interpretation of defect ionization energies 190
4.4 Transition to the metallic state 192
4.4.1 Properties of heavily doped oxides 192
4.4.2 Models of the transition 196
References 200

5 Metallic oxides 204

5.1 Simple metals 204
5.1.1 Band structure and Fermi surfaces 205
5.1.2 Transport properties 210
5.1.3 Optical and spectroscopic properties 213
5.2 Electron correlation and magnetic anomalies 219
5.2.1 Band magnetism 220
5.2.2 Mixed valency and double exchange 221
5.2.3 The 'Mott transition' 225
5.2.4 Chemical trends in the Mott transition 232
5.3 Lattice interactions 234
5.3.1 Fermi-surface instabilities 234

5.3.2 Metal–metal bonding 237
5.3.3 Mixed-valency compounds: charge ordering and disproportionation 242
5.4 Superconductivity 249
5.4.1 Characteristics of 'high-T_c' superconductors 250
5.4.2 Electronic structure of doped copper oxides 260
5.4.3 Problems in understanding high-T_c superconductivity 265
References 273

Compound Index 277

Subject Index 279

1

INTRODUCTION

1.1 Scope and plan of the book

The three *transition series* form the short groups of elements in the periodic table, interposed between the longer *main groups* (see Fig. 1.1). The occurrence of these elements, as well as the extraordinarily varied chemical and physical properties of their compounds, is a feature of the progressive filling of shells of *d* orbitals across each series.[1,2] The precise boundaries of the transition series vary slightly in different definitions. In oxides, as in other compounds, the properties associated with the *d* shells are most apparent with the elements in the shaded region of Fig. 1.1. All the compounds discussed in this book, therefore, contain an element between Ti and Cu in the 3*d* series, between Zr and Ag in the 4*d* series, or between Hf and Au in the 5*d* series.

Binary oxides contain a metallic element and oxygen, as in TiO_2, Fe_3O_4, or NiO. The range of such compounds is already quite large, but it is greatly extended by considering ternary and yet more complex compounds, where additional metallic elements are present. These may themselves be transition elements, as in Fe_2CoO_4, but more frequently in the compounds discussed here the additional elements will be from the pre-transition or post-transition groups, for example in $LaNiO_3$ and $PbTiO_3$.

A few transition metal oxides, such as OsO_4, are volatile compounds consisting of discrete molecules.[2] By far the majority, however, are solid under normal conditions of temperature and pressure, and it is the properties of these solid compounds which form the main subject of the book. A notable characteristic is the enormous range of electronic properties found.[3] Transition metal oxides may be good insulators (TiO_2), semiconductors ($Fe_{0.9}O$), metals (ReO_3), and, of course, superconductors ($YBa_2Cu_3O_7$). Many compounds show transitions from a metallic to a non-metallic state as a function of temperature (VO_2), pressure (V_2O_3), or composition (Na_xWO_3). Along with these variations in electrical conductivity go wide differences in other physical properties related to electronic structure. The optical and magnetic behaviour in particular form the basis for many important applications. This range of properties also poses many difficult problems of scientific understanding. The main aim of the book is give examples of the

H																	He
Li	Be											B	C	N	O	F	Ne
Na	Mg											Al	Si	P	S	Cl	Ar
K	Ca	Sc	Ti	V	Cr	Mn	Fe	Co	Ni	Cu	Zn	Ga	Ge	As	Se	Br	Kr
Rb	Sr	Y	Zr	Nb	Mo	Tc	Ru	Rh	Pd	Ag	Cd	In	Sn	Sb	Te	I	Xe
Cs	Ba	La[1]	Hf	Ta	W	Re	Os	Ir	Pt	Au	Hg	Tl	Pb	Bi	Po	At	Rn
Fr	Ra	Ac[2]															

[1]Lanthanides	Ce	Pr	Nd	Pm	Sm	Eu	Gd	Tb	Dy	Ho	Er	Tm	Yb	Lu
[2]Actinides	Th	Pa	U	Np	Pu	Am	Cm	Bk	Cf	Es	Fm	Md	No	Lw

FIG. 1.1 The Periodic Table of the Elements. Compounds discussed in this book are oxides containing one or more of the elements shown in the shaded area.

different kinds of electronic properties found in transition metal oxides, and to discuss their interpretation. Applications will be mentioned at appropriate points, although these are not treated in detail.

This chapter is introductory in nature, and discusses some features which form an essential background to the main theme. Chemists have, of course, many important roles to play in studying these materials, not least in the areas of synthesis and characterization.[4–6] It would be inappropriate here to describe the experimental aspects involved. But chemical *thinking* about transition metal oxides, using concepts evolved from studying many different types of compound, is essential for the understanding of electronic properties. Section 1.2.1 describes some chemical concepts that will be used widely throughout the book. It is also important to have some idea of the variables that control the stability of different compounds, and this problem is discussed from the point of view of chemical thermodynamics in Section 1.2.2. The subsequent sections (1.3) describe some of the important features of the crystal structures of transition metal oxides, which are again essential for an understanding of electronic properties. The chapter concludes with an attempt to classify the compounds according to their electronic properties, making use of the chemical classification discussed earlier. Before looking in detail at the different types of behaviour, in Chapters 3 to 5, it is also necessary to appreciate the equally wide range of theoretical models of electronic structure that have been applied to this field. These form the subject of Chapter 2.

1.2 Chemical aspects

The following sections introduce, firstly, some of the chemical concepts and language that are essential for thinking about the electronic structure of transition metal oxides and, secondly, some elementary ideas of chemical thermodynamics that are important in understanding the conditions under which these compounds can be studied.

1.2.1 Oxidation state and electron configuration

The first step in the chemical characterization of any compound is to establish its precise elemental composition or **stoichiometry**, which is expressed in a chemical formula such as TiO_2 or $Na_{0.7}WO_3$. The composition can, in principle, be determined by chemical analysis for the various elements present. Apart from the difficulties and inaccuracies that may be inherent in the analytical procedures, it is important to be aware of various other problems. In the first place, one must be sure that a polycrystalline sample is a genuinely *single-phase* compound and not a mixture of solid phases with different compositions. Newly prepared samples are usually checked by X-ray powder diffraction; if the desired compound has a simple structure, or one that is known already and listed, for example, in the powder-diffraction file,[7] the X-ray lines should show whether extraneous phases are present. With unknown compounds, especially if their structure is complicated, it is desirable to examine polycrystalline samples by electron microscopy; the analytical facilities available on modern microscopes enable rough compositions of different crystallites to be measured and checked for uniformity.[8]

Even with single crystals, or with polycrystalline samples confirmed as a single phase, problems with composition remain. As the example of $Na_{0.7}WO_3$ shows, many transition metal oxides show the phenomenon of **non-stoichiometry**, where elements are not present in simple integral proportions.[9,10] Variations of stoichiometry are responsible for some of the complexity in structure and properties of oxides. Establishing the precise composition is an essential prerequisite for meaningful interpretations of electronic properties, and can place severe demands on the methods used for synthesis and analysis of these compounds. Within a series with closely related compositions, it is also important to know whether the composition can be varied continuously within a single phase, or whether there is a series of **line phases** each with fixed stoichiometry. The sodium–tungsten bronze series Na_xWO_3 shows genuine non-stoichiometry, with x being a continuous variable up to about 0.9, although there are small structural changes at lower x values.[11] On the other hand, the molybdenum bronzes $K_{0.30}MoO_3$ and $K_{0.33}MoO_3$ are quite distinct in their structure and electronic properties, and should probably be regarded as compounds with fixed stoichiometry, $K_3Mo_{10}O_{30}$ for example.[12] Only careful studies, where variation of the preparative conditions is combined with structural and electronic investigations of each composition, can distinguish these two kinds of behaviour.

The oxygen content is a particular problem with many compounds, as this element is normally not analysed for directly, and is easily gained or lost during the course of a synthesis. This quickly became apparent in work on high-T_c superconductors, when it was found that superconducting behaviour is very dependent on the precise preparative conditions, and especially on the environment in which samples were cooled and annealed after synthesis. Thus in the compounds $YBa_2Cu_3O_x$ compositions with $x < 6.4$ are non-metallic, whereas superconductivity appears as x increases towards 7.[13] Again, one must make the distinction between genuine non-stoichiometry and series of line phases with closely related stoichiometry, as for example in W_nO_{3n-1} and W_nO_{3n-2} with integral values of *n*.[14] The structural features of this type of line phase will be described briefly later (Section 1.3.3); the distinction between line phases and genuinely non-stoichiometric ones is also discussed from a thermodynamic point of view in Section 1.2.2.

The many disputes which still exist about quite fundamental properties of some oxide systems, such as metallic versus non-metallic properties, or the existence of magnetic moments on transition metal atoms, suggest that different research groups are frequently not studying the same compound, and in particular may have failed to ascertain the oxygen content correctly.

Knowledge of the composition generally leads to an assignment of **oxidation states** to each element present. The chemists' oxidation state is most simply thought of as a formal, or notional, ionic charge.[1] The utility of the *ionic model*, where every atom is imagined to have lost or gained an integral number of electrons, will be explored in Chapter 2. The oxidation-state concept is more general and not restricted to compounds where the ionic model gives a 'good' description. In normal cases, every oxygen atom is assigned a -2 charge, and metal atoms then take up appropriately balancing charges. Thus we have Ni^{2+} and Ti^{4+} in NiO and TiO_2. In ternary and more complex cases the positive charge must be distributed somehow between more than one element. This causes no difficulty with elements that are known to take only one oxidation state in oxide environments, for example Li^+, Sr^{2+}, and La^{3+}. It is clear then that $LiNbO_3$ has Nb^{5+} and $LaFeO_3$ has Fe^{3+}. Ambiguities may arise when charge has to be partitioned between elements, especially transition metals, which show variability in oxidation state. Plausible assignments in ilmenite, $FeTiO_3$, could be Fe^{3+} and Ti^{3+}, or Fe^{2+} and Ti^{4+}. In this case a comparison with the stabilities of oxidation states in aqueous solution (a common oxide environment!) suggests that the latter is more likely, but one might not want to trust such an analogy. It is not evident *a priori* that this kind of distinction is even meaningful. We shall see, however, that different oxidation states of an element do often give rise to characteristically different structural and electronic properties, so that while the concept may have occasional limitations, it is surprisingly powerful in interpreting the electronic structure of a wide range of transition metal oxides. Techniques that have been used to assign oxidation states under different circumstances include structural studies (especially of metal-oxygen distances); magnetic measurements including neutron diffraction and the resonance techniques of ESR and NMR; optical absorption or emission spectroscopy; Mössbauer spectroscopy; and core-level spectroscopies such as XPS. Their uses (and sometimes abuses) will be referred to later.[15]

Another ambiguity arises in **mixed-valency**† compounds, such as Fe_3O_4.[16,17] The average oxidation state is $Fe^{2.67+}$. Does this mean that each iron atom has a fractional oxidation state equal to 2.67, or should one rather think of two Fe^{3+} ions to every one Fe^{2+}? Again, only the structure and physical properties can show whether such a distinction is meaningful. In a case such as Fe_3O_4, electrical conductivity is clearly a relevant property. Equal but fractional oxidation states for the iron would imply mobility of electrons between atoms and thus high conductivity. This is in fact found above 120 K, although we shall see that it is more appropriate to think of one Fe^{3+} and two $Fe^{2.5+}$ rather than three $Fe^{2.67+}$ in this case. Below this temperature in Fe_3O_4 a transition intervenes (the 'Verwey' transition), in which electrons, and presumably oxidation states, become localized or trapped, so that it is better to imagine integral Fe^{2+} and Fe^{3+} states[19] (see Section 5.3.3).

Mixed valency, in this sense just used for the case of Fe_3O_4, is a very common feature of transition metal oxides, and is often associated with non-stoichiometry. In Na_xWO_3 for example, the oxidation state of tungsten is between 5 and 6; as with Fe_3O_4 only structural and other measurements can show whether it is reasonable to assign oxidation states to individual atoms. Mixed valency is frequently associated with semiconducting or metallic properties (discussed in Chapters 4 and 5) and appears to be an essential feature of the copper oxide superconductors.[20]

It has been assumed so far that oxygen is to be regarded invariably as O^{2-}. This is not always the case. The paramagnetic O^- ion has been recognized as a defect in some oxides.[21] Better known as stable species are the peroxide (O_2^{2-}) and superoxide (O_2^-) ions, found in solid salts of alkali metals and in aqueous solution. The O–O distances (148 and 129 pm respectively, compared with 121 pm in O_2 and at least double this value for the normal distance between oxide ions in crystals) clearly indicate the presence of a bond between oxygen atoms.[2] Peroxide compounds of transition metals are not as common as disulphides containing S_2^{2-} (such as FeS_2) but some are known for the earlier transition elements in high oxidation states.[22] For example K_3CrO_8 does not contain the improbable Cr^{13+}, as its structure shows that Cr is coordinated

†The definition of *mixed valency* by chemists is distinct from that which has become common in the physics literature. There it is used to describe compounds such as TmSe in which the electronic configuration of Tm apparently fluctuates between two states.[18] The chemical oxidation state of thulium is Tm^{2+}, irrespective of this fluctuation. In this book the term 'mixed valency' will be used in its chemical sense of mixed or fractional oxidation state.

by oxygen atoms in pairs, each with an O–O distance of 141 pm characteristic of O_2^{2-}; the more reasonable Cr^{5+} is confirmed by the Cr–O distance, and by magnetic measurements which show one unpaired electron per chromium as expected.

Peroxides should be easy to identify if the structure of a compound is well enough known. But there are other situations where it has been claimed that an unconventional assignment of charges represents the electronic structure better. This is the case particularly with 'high-T_c' copper oxide superconductors.[20] The change in composition from $YBa_2Cu_3O_{6.5}$ to $YBa_2Cu_3O_7$ suggests the oxidation of some copper from Cu^{2+} to Cu^{3+}, as the oxidation states Ba^{2+} and Y^{3+} are essentially invariant. However, many electronic measurements suggest that electrons are being removed from states more closely associated with oxygen than copper atomic orbitals: hence the suggestion that the oxidation should be regarded as a conversion of some O^{2-} to O^-. This raises the question, not often confronted by chemists who assign oxidation states to elements, of how much these are intended to relate to 'real' charges present on atoms in the solid. The problem will be discussed again at the appropriate point (see Section 5.4.2). For the moment, we shall take the conventional chemical view that oxygen is to be regarded as O^{2-} unless there is structural evidence for peroxide formation. Although we may often use the ionic model as a first approximation in thinking about electronic structure, we must recognize that a chemical oxidation state is not meant to represent a real ionic charge, but is more of a formal book-keeping device. Of course the reason for using such a concept must lie in its interpretive power, which we shall see through most of the book is impressive; we should not be surprised, however, if such a man-made classification occasionally breaks down and loses some of its utility.

Figure 1.2 shows the range of oxidation states found in oxides containing elements of the three transition series.[2,7] In some cases, marked *T* in the Figures, oxidation states not apparently stable in simple binary oxides can be found in ternary and mixed-valency compounds. Sometimes, an especially high (or low) oxidation state may be of marginal stability, imperfectly characterized, or exist only in a mixed-valency situation. For example Fe^{6+} is found in $BaFeO_4$, but this compound probably always has some oxygen deficit, implying the presence of some iron in a lower oxidation state.

The trends apparent in Fig. 1.2 are for the most part characteristic of the elements concerned, and reflected in other aspects of their chemistry. Early elements in each series show progressively higher oxidation states on passing to the right, until a point is reached where high oxidation states become less stable, and the series ends with low ones. The trend

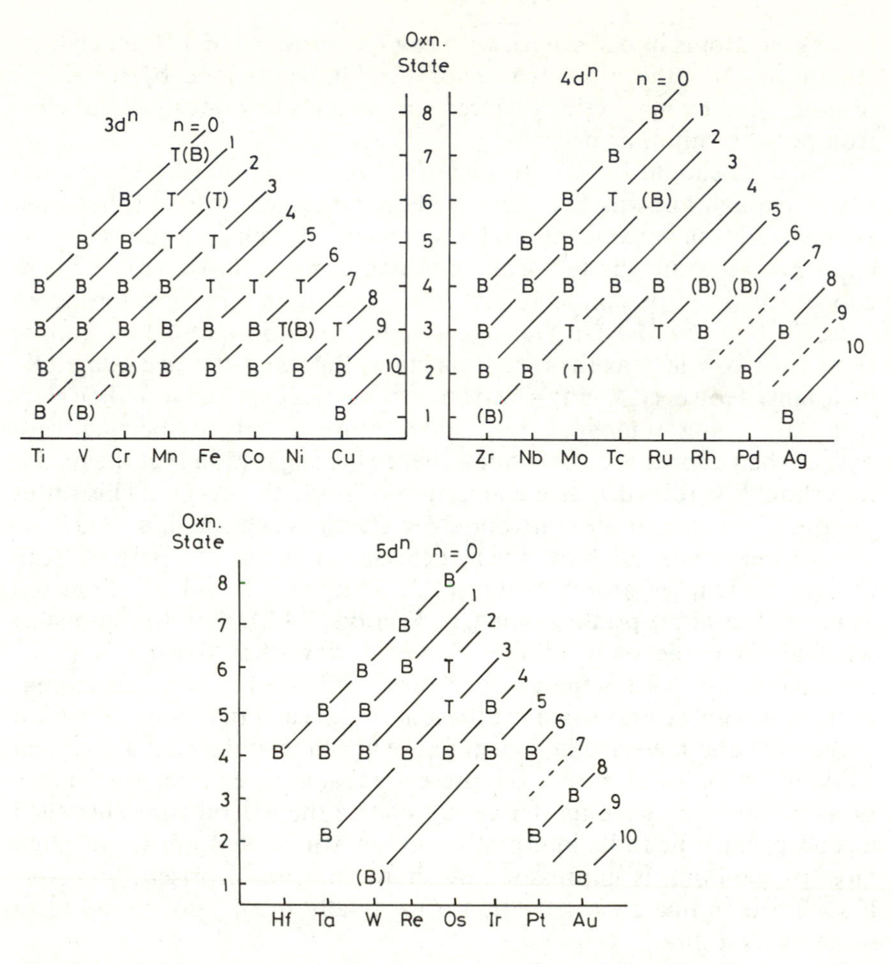

FIG. 1.2 Oxidation states and electron configurations found in transition metal oxides of (a) the $3d$, (b) the $4d$, and (c) the $5d$ series. The diagonal lines show the d^n configuration. B indicates oxidation states found in well characterized binary oxides and T indicates those found only in ternary or more complex systems. (B) and (T) represent doubtful or poorly characterized systems, or ones with mixed valency.

to higher oxidation states persists longer in the $4d$ and $5d$ series than in the $3d$ one. These trends can be interpreted, qualitatively at least, in terms of the atomic structure of the elements; for example, the 'collapse' to low oxidation states is related to an increase of atomic ionization energies in the later elements of each period.[23]

The ionic model suggests—rightly or wrongly—that the electronic properties of a solid should be related closely to those of the notional ions that make it up. At the simplest level, attention focuses on the electron configuration, that is the occupancy of atomic orbitals. Electron configurations of the neutral transition metal atoms are quite complicated, with both *d* and *s* shells involved: e.g. V $3d^3 4s^2$, Cr $3d^5 4s^1$ (here, and in all subsequent examples, the filled argon core is omitted).[23] However, it is always the outer *s* electrons that are the first to be ionized, so that when elements are in the positive oxidation states found in oxides the relevant configurations are of the type d^n. The values of n for the oxidation states found in transition metal oxides are shown as the diagonal lines in Fig. 1.2. These values are quite important for interpreting the electronic properties of oxides, even though the ionic model arguments on which they are based are not always justified. In fact, relaxation of the ionic assumption allows us to think more in terms of 'bonding' and 'antibonding' orbitals in a solid, and the arguments presented in Chapter 2 show that the value of n in 'd^n' corresponds to the number of electrons remaining when all metal–oxygen bonding orbitals have been filled. All d^0 oxides are diamagnetic insulators; it is the additional electrons where $0 < n < 10$ that give rise to the range of electronic and magnetic properties displayed by transition metal oxides.

Figure 1.2 shows that the maximum oxidation state for early transition metals corresponds to d^0, as expected from simple arguments (since the energy required to ionize further electrons would be prohibitively large). This is one simple example of the use of the ideas of oxidation state and electron configuration. One or two other trends are apparent in the Figure. It seems that for the later elements of the 4*d* and 5*d* series, even n values are preferred over odd ones. The fact that no oxides appear to be known with the electron configurations $4d^7$, $4d^9$, $5d^7$, or $5d^9$ is quite noteworthy, and expresses a trend that is apparent in other aspects of the chemistry of the so-called *platinum-group* elements (Ru, Rh, Pd, Os, Ir, Pt).[2]

One further elaboration of the electron configuration idea will be useful for our discussion of crystal structures later in this chapter. This is concerned with the number of unpaired electrons, and is important for the magnetic properties of transition metal compounds. In the ground states of the free d^n ions, as many electrons as possible are left unpaired, in accordance with Hund's first rule.[24] Thus the number of unpaired electrons is equal to n up to the half-filled shell at $n = 5$. Beyond that point electrons must be paired up in *d* orbitals, so that the number of unpaired electrons now falls with increasing n, reaching zero at the filled shell with $n = 10$. Observations on transition metal compounds show that in some cases the transition metal atom has the same number of unpaired electrons as in the free ion with corresponding

n.[1,2,23] For example Mn^{2+} in MnO has five unpaired electrons as expected for $3d^5$. These are known as **high-spin** configurations, and are generally characteristic of the 3*d* series compounds. In the 4*d* and 5*d* series, **low-spin** compounds are more common, where the number of unpaired electrons is less than in the free ion. This clearly indicates stronger interaction of the 'ion' with its surroundings. Although a majority of 3*d* series oxides show the high-spin configuration there are a few exceptions. These occur with later 3*d* elements in high oxidation states. The best known is Co^{3+}, but Ni^{3+} and Cu^{3+} oxides also have low-spin configurations.[20,25]

1.2.2 Thermodynamic stability and phase diagrams

Any study of a solid compound must take account of the conditions under which it is stable. Chemical thermodynamics shows that this depends on the **Gibbs free-energy change**

$$\Delta G = \Delta H - T\Delta S \tag{1.1}$$

for reactions involved in formation or decomposition.[26,27] Consider for example the possible decomposition of an oxide MO_m to one with less oxygen, MO_{m-n}:

$$MO_m = MO_{m-n} + n/2\,O_2. \tag{1.2}$$

At a given temperature, the enthalpy change ΔH is almost independent of the oxygen pressure p_{O_2}. On the other hand, the entropy change ΔS is dominated by the production of gas molecules in this reaction, and varies with p_{O_2} according to

$$\Delta S = \Delta S^0 - (nR/2)\ln p_{O_2} \tag{1.3}$$

where R is the gas constant (8.31 J K^{-1} mol^{-1}). From (1.1) therefore:

$$\Delta G = \Delta G^0 + (nRT/2)\ln p_{O_2}. \tag{1.4}$$

The condition for (1.2) to proceed in a forward direction is that $\Delta G < 0$, and this will happen, according to (1.4), when the oxygen pressure is below a value p_d determined by the standard free energy change ΔG^0:

$$\ln p_d = -2\Delta G^0/nRT. \tag{1.5}$$

Any oxide can therefore decompose to one where the metal is in a lower oxidation state (or to the metallic element itself) if the ambient oxygen pressure falls below the appropriate value p_d, although in the case of many oxides this value can be exceedingly low.

The other variable under experimental control is the temperature. According to the **Gibbs–Helmholtz equation**:[26]

$$d(\Delta G/T)/dT = -\Delta H/T^2 \tag{1.6}$$

and so

$$d(\ln p_d)/dT = 2\,\Delta H/nRT^2. \tag{1.7}$$

ΔH for reaction (1.2) must be positive if the compound MO_m is stable under any conditions. Increasing the temperature therefore increases the decomposition pressure, so that higher oxygen pressures are necessary to prevent decomposition. At sufficiently high temperatures decomposition will occur whatever the ambient oxygen pressure.

Information on the thermodynamic stability of different compounds is conveniently displayed in the form of **phase diagrams**.[4,28] The example in Fig. 1.3 shows the binary oxides CoO and Co_3O_4, together with elemental cobalt.[29] The lines in the diagram show the decomposition pressures for reactions of the type considered above; the fact that they give approximate straight lines in this plot of log p_{O_2} against $1/T$ is expected from integrating equation (1.7), with the assumption that the enthalpy change ΔH does not depend strongly on temperature.

With compounds that are significantly non-stoichiometric in oxygen content, the type of phase diagram just shown has an important limitation, as it fails to show the stoichiometry range possible for each phase.

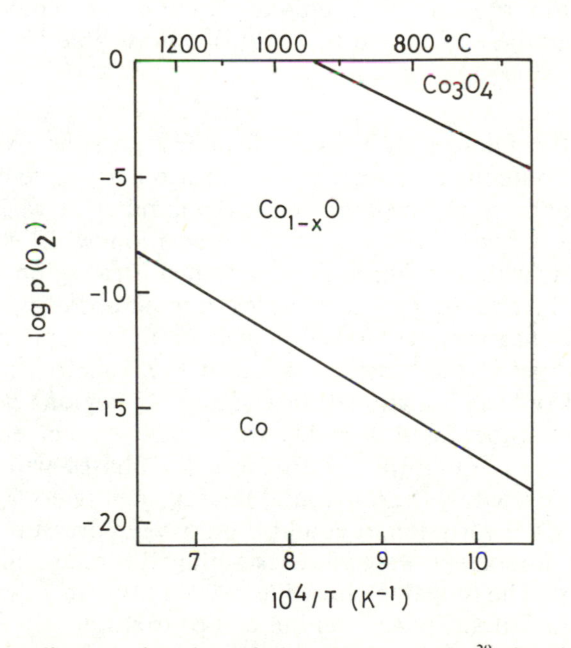

FIG. 1.3 Phase diagram for the cobalt–oxygen system,[29] showing the stable solids as a function of temperature and oxygen pressure.

Even in the cobalt case such non-stoichiometry is significant, but a much higher degree of non-stoichiometry is shown for example in the Fe–O binary system. In this case an alternative plot, using overall composition and temperature as the principal variables, is more useful, as the diagram of Fig. 1.4 illustrates.[30] The latter form of diagram makes an important distinction between regions with two (solid) phases, for example $Fe_3O_4 + Fe_2O_3$, and those with a single solid phase of variable composition, such as the 'wüstite' region between $Fe_{0.85}O$ and $Fe_{0.96}O$ at 800°C. **Line phases** are those with a very narrow range of stoichiometry, such as Fe_2O_3. In many cases this range may broaden out at higher temperatures, giving a single-phase region of variable composition.

Figure 1.4 allows one to predict which phases will be present at given temperature, in a system with known overall Fe/O composition; however, it is also important to know the value of p_{O_2} present in equilibrium with such a solid, as it is generally by controlling this variable that the composition is fixed. Such diagrams do not always show p_{O_2} values, but in Fig. 1.4 they are shown by the broken lines. It can be seen that there is a difference between one- and two-phase regions. The sloping lines in the former show that oxygen pressure and temperature can both be varied while the phase is stable; on the other hand in two-phase regions the lines are horizontal, showing that the equilibrium decomposition pressure is a unique function of temperature. This difference can be understood by using the **phase rule**.[4,26]

$$F + P = C + 2. \tag{1.8}$$

Here P is the number of phases (including gaseous oxygen), C the number of components (equal to two in a binary system), and F the number of degrees of freedom: that is the number of variables such as temperature, oxygen pressure, and composition which may be fixed independently while equilibrium is maintained. In a region with one solid phase, $P = 2$, and so $F = 2$. Thus the temperature and p_{O_2} may be chosen, but the phase composition will be a function of these. The sloping p_{O_2} lines in the 'wüstite' region of Fig. 1.4, for example, allow one to read off this functional dependence. In regions with two solid phases, on the other hand, $P = 3$ so that $F = 1$. At a given temperature the p_{O_2} value in equilibrium with the solid phases will therefore be fixed, as illustrated by the horizontal lines in such regions in the figure.

Measuring the variation of equilibrium oxygen pressure with composition is an important way of determining the phase diagram of an oxide system. The range of accessible oxygen pressures can be extended by making use of gas-phase equilibria. For example, the reactions

$$CO_2 = CO + \tfrac{1}{2}O_2 \tag{1.9}$$

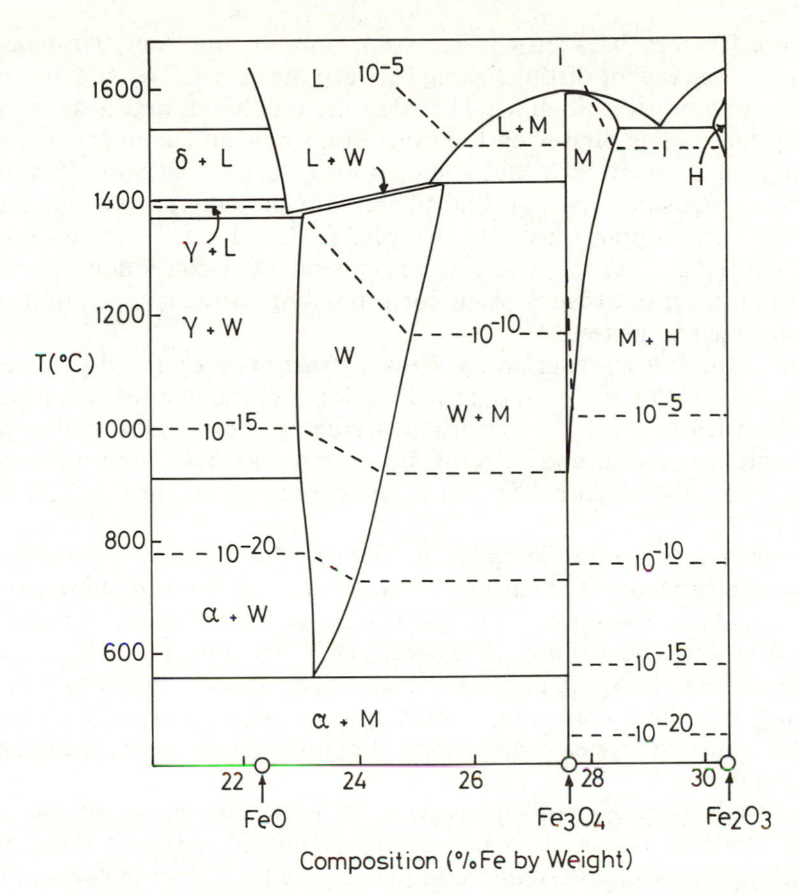

FIG. 1.4 Phase diagram for the iron–oxygen system,[30] with stable phases shown against overall Fe/O composition and temperature, and the equilibrium oxygen pressure shown by broken lines. The phases are labelled α, γ, δ (phases of metallic Fe), L (liquid), W (wüstite), M (magnetite), and H (haematite). Note the distinction between *single-phase* regions (such as the non-stoichiometric wüstite phase) and *two-phase* regions.

and

$$H_2O = H_2 + \tfrac{1}{2}O_2 \qquad (1.10)$$

both have equilibrium p_{O_2} values that can be controlled by making known mixtures of CO_2 and CO, or of H_2O and H_2.[28] In this way measurements can be made with effective p_{O_2} values of 10^{-25} atmospheres or below, which would be quite impossible with pure oxygen.

The difference between the p_{O_2} behaviour in one- and two-phase regions is one way of distinguishing between the alternative possibilities of a single non-stoichiometric phase on the one hand, and a series of closely spaced line phases on the other. In a non-stoichiometric phase the oxygen pressure at a given temperature varies continuously with overall composition. Where line phases are present, however, intermediate compositions must give two phases. In this case, therefore p_{O_2} cannot vary at a given temperature; instead of a continuous curve, we obtain a series of steps, each corresponding to the composition of a particular line phase.[14]

Structural studies, generally by X-ray diffraction, are normally carried out at the same time, and should help to identify the number of phases present. Clearly this is an important prerequisite of serious studies of electronic properties, and many misleading results have been reported on 'compounds' which are in fact an indeterminate mixture of several phases.

In ternary and more complicated systems the number of variables becomes larger, and it is correspondingly more difficult to show them on a single phase diagram. This interpretation of triangular and other such diagrams is described elsewhere.[4] The important point from the perspective of electronic properties is the same as with binary systems; that not only the overall composition, but the phases present in a given sample, must be properly characterized before results can be meaningfully interpreted.

Although thermodynamic phase stability is an essential consideration in the study of oxides, it is important to remember that it refers to *equilibrium* conditions. Actual attainment of equilibrium in solid-state reactions normally depends on the diffusion of elements within the bulk of a material. This can be very slow, especially at low temperatures. Clearly this fact is important in the actual determination of a phase diagram; but it also implies that it may be possible to obtain and study phases that are not in equilibrium. Thus the Fe–O diagram in Fig. 1.4 shows that the 'FeO' wüstite phase disappears below 570°C. However, by rapidly quenching this phase formed at high temperatures, decomposition to Fe and Fe_3O_4 can be prevented, and although wüstite is rapidly oxidized in air, it can be studied at room temperature.[31] Sometimes thermodynamically unstable compounds can be prepared by low-temperature techniques, for example by ion-exchange or electrochemical methods. Thus it is possible to make compositions approaching CoO_2, by oxidation of $LiCoO_2$.[32] Not only can thermodynamically unstable compositions be made, but also unstable structures of known compounds. For example, an unusual form of TiO_2 has recently been made by low-temperature synthesis;[33] another type of example is the preservation of a high-temperature, disordered phase of $SrFeO_{2.5}$ by rapid quenching to low temperatures, where a more ordered structure is the thermodynamically stable one.[34] These structural differences may have an important influence on electronic properties, and show that all aspects of the preparation of a compound can be significant.

1.3 Structural principles

In the determination of a crystal structure, one starts by identifying the lattice type and unit cell, and only at a later stage are the detailed positions of atoms, their bond lengths and angles, and so on, filled in.[4] In contrast to this 'top-down' approach, the opposite 'bottom-up' approach is a more helpful way of *understanding* crystal structures.[35] To a large extent, it is the local bonding interactions of atoms that control what structure is stable; we shall see that these near-neighbour interactions between atoms also have the most important effect on electronic properties. We shall start this very brief survey therefore by looking at the local coordination geometry of atoms, and then discuss the crystal structures which result from fitting these units together.

1.3.1 Metal coordination geometries

The most straightforward way to interpret the crystal structures of metal oxides is in terms of the geometrical requirements of the metal atoms. Transition metal oxides are predominantly 'ionic' in their structural features: that is the closest interatomic distances are between metal and oxygen, and it is this metal–oxygen bonding that provides stability to the structure. The most significant geometrical factor, therefore, is the way in which a metal atom is surrounded by oxygen. Different aspects of this are the *bond lengths*, and the *coordination number* and *geometry*.

Interatomic distances are normally rationalized using *ionic radii*. The idea that distances can be expressed as the sum of atomic parameters in this way is not obvious, and empirically can only be justified as a first approximation.[1] Even when it works, the division of an observed distance into two ionic radii is based on more-or-less arbitrary assumptions. Nevertheless, such ionic radii are widely used in connection with oxide structures, most modern values being based on the tabulations of Shannon and Prewitt.[36] A selection for transition elements in common oxidation states is displayed in Fig. 1.5.

As can be seen, the ionic radius is a strong function of oxidation state, decreasing as electrons are removed. Other major trends apparent

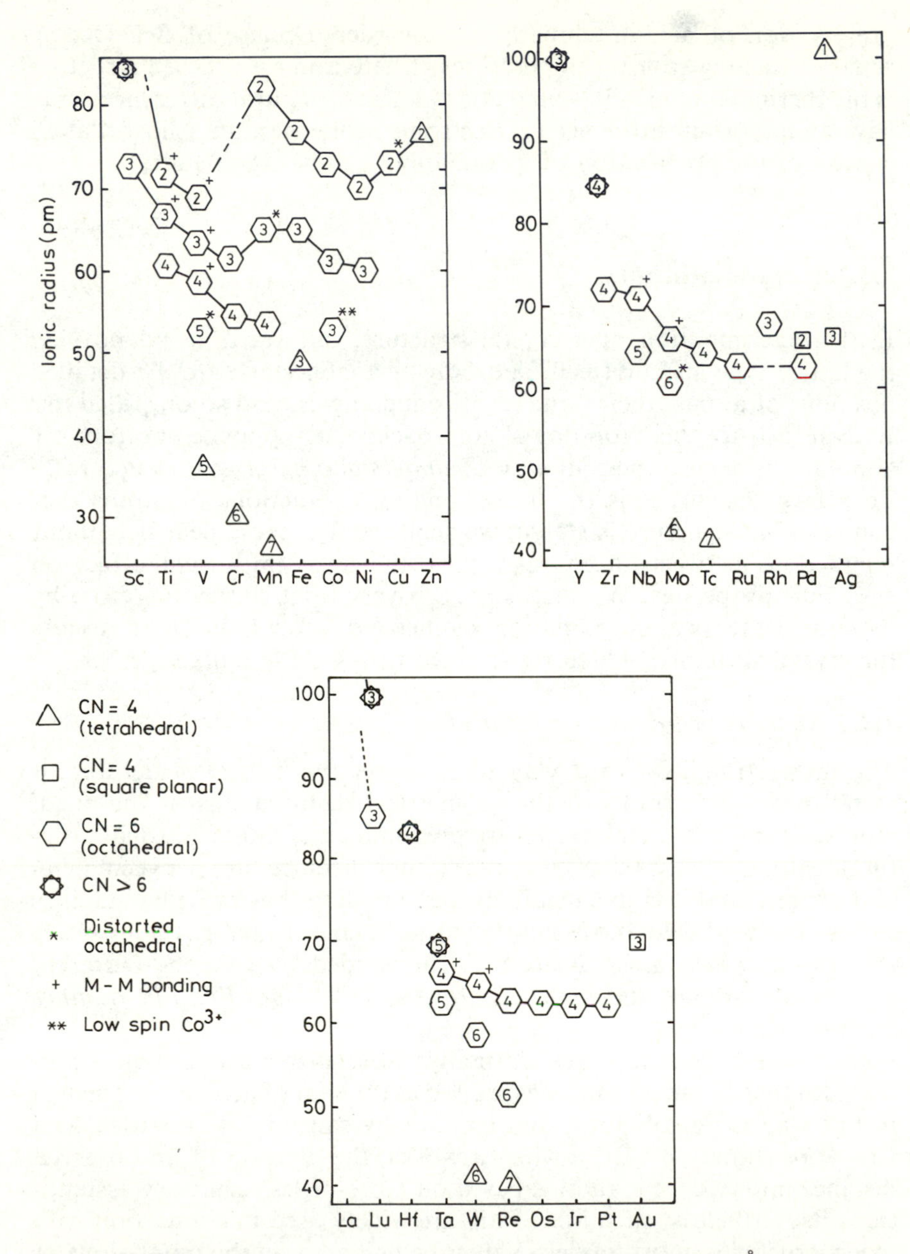

FIG. 1.5 Ionic radii for transition metals in oxides (100 pm = 1 Å). The values are taken from Shannon and Prewitt,[36] and based on an octahedral oxygen radius of 126 pm. The number in each symbol is the oxidation state; the symbols plotted show the coordination number and geometry appropriate to that radius. Coordination geometries are shown more explicitly in Fig. 1.6.

are the increase in size down each group, and the (not always regular) contraction of ions with increasing atomic number in each period. Both features are a result of atomic-structure effects. The contraction across a period comes from the increasing nuclear charge, which is only partially compensated by shielding from the added electrons in the *d* shell.[23] The same effect gives rise to an increase in the atomic ionization energies, which is important in understanding some of the trends in electronic structure of oxides. The irregularities in the contraction, which are most apparent in the high-spin ions of the 3*d* series, are a consequence of crystal-field effects, explained in Section 2.1.1.

The increase in size from 3*d* to 4*d* is a natural consequence of an increase in the principal quantum number. The change of radius between the 4*d* and 5*d* series is much less than between 3*d* and 4*d*, and is hardly perceptible for elements early in the two series. This contrast is a result of the filling of the 4*f* shell between lanthanum and hafnium, at the beginning of the 5*d* series. The nuclear charge increases by 14 units, the increase being only partially shielded by the added 4*f* electrons. The resulting *lanthanide contraction* is an important feature of the 4*f* elements themselves, as the change in size along the series is frequently the principal cause of chemical differentiation between elements.[2] It also has the result that the 5*d* elements shortly following the lanthanides, particularly hafnium and tantalum, are very similar in size and in chemistry to the corresponding 4*d* elements.

Figure 1.5 also indicates the type of coordination geometry found most frequently for each element in its different oxidation states. Common coordination numbers range from four (tetrahedral, occasionally square planar) to eight (cubic), although the earlier (and therefore larger) lanthanides may show even higher numbers. The coordination geometries corresponding to the different symbols in Fig. 1.5 are illustrated in Fig. 1.6. It is the 6-fold octahedral coordination that is commonest, but it is frequently not completely regular; the two distorted variants shown are discussed below. The list of geometries shown is not by any means complete: for example, 5-coordinate square pyramidal coordination is sometimes shown by Cu^{2+}, and occasionally by other elements.[35]

There is evidently a strong correlation between coordination number and ionic radius. Tetrahedral coordination is the rule for metal ions with radii less than 50 pm, octahedral for radii between 50 and 80 pm, and larger coordination numbers for radii above 80 pm. In the most extreme (hard sphere) version of the ionic model, the coordination number is related to the relative sizes of anion and cation by simple geometrical considerations. The so-called **radius-ratio rules** predict structural changes as a function of increasing cation size, but they are

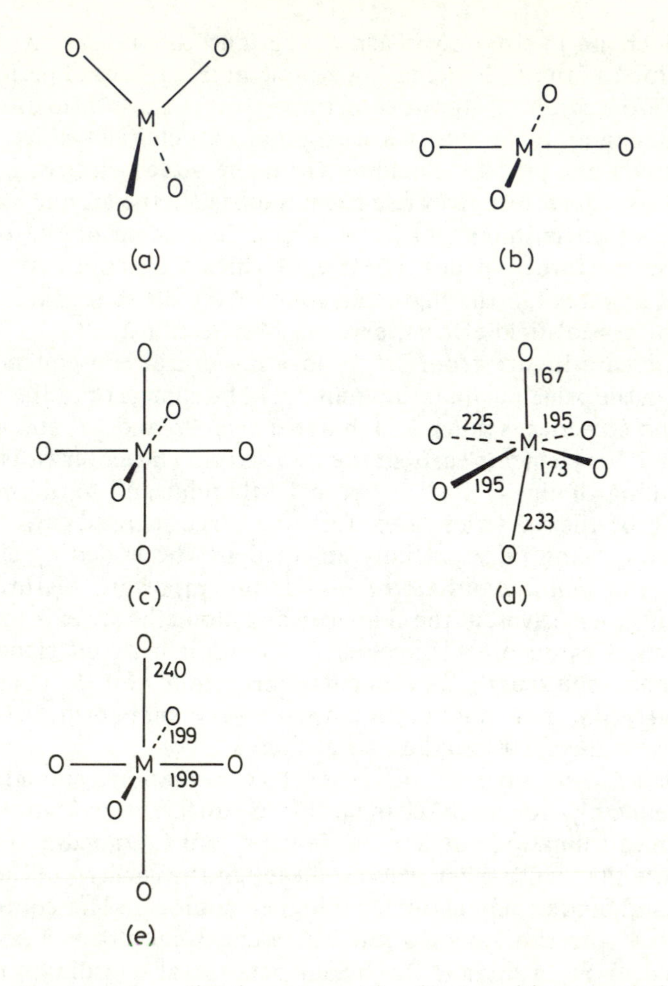

Fig. 1.6 Common coordination geometries for transition metals in oxides. (a) tetrahedral; (b) square planar; (c) octahedral; (d) distorted octahedral: Mo in MoO_3;[37] (e) distorted octahedral: Cu in $La_{1.85}Sr_{0.15}CuO_4$[38] (metal–oxygen distances in pm).

not quantitatively reliable, even in a series of compounds as strongly ionic as the alkali halides.[1,4] There may be various reasons for this, the most obvious being that ions are not hard spheres, and the apparent ionic radii result from a balance between attractive and repulsive forces. One important consequence of this can be seen in Fig. 1.5: ionic radii themselves depend on coordination number. For example Fe^{3+} has a radius of 65 pm in octahedral and 49 pm in tetrahedral coordination.

This change can be understood in purely ionic terms, because the balance of attractive (predominantly long-range Coulomb) and repulsive (near-neighbour overlap) forces changes with coordination number.[23] However, the trend could also be interpreted in terms of metal–oxygen covalent bonds. The same number of orbitals shared between fewer bonds will make each bond stronger, and therefore shorter. Whichever explanation is preferred, one can obviously not expect simple 'hard sphere' arguments like the radius-ratio rules to work very well.

In addition to the regular 4–, 6–, and higher-fold geometries expected from ionic forces with no directional character, there are a number of less regular coordinations found in specific cases. Two versions of distorted octahedral geometry are shown in Fig. 1.6.[37,38] Certain d^0 ions, especially V^{5+} and Mo^{6+}, often adopt a coordination where one or two M–O bonds are shortened relative to the others in the octahedron. From the point of view of covalent bonding, these could be regarded as double bonds, giving groups such as $Mo(=O)_2^{2+}$ which are also found as constituents of molecular complexes such as $[MoO_2(\text{8-hydroxyquinolinate})_2]$.[2] An alternative ionic explanation is suggested by the radii of these ions, which are close to the borderline of stability between octahedral and lower coordination numbers; thus one can imagine that these ions are 'too small' to fit comfortably in a regular octahedral site.[39]

Another irregularity is that which gives a lengthening of two opposed bonds, thus lowering the symmetry to tetragonal. It is nearly always found with two specific electron configurations, d^4 (high-spin only) and d^9. Examples are Mn^{3+} in $LaMnO_3$ and Cu^{2+} in La_2CuO_4.[38] These distortions are related to crystal-field effects, and as discussed in Section 2.1.1 are normally attributed to *Jahn–Teller* distortions of a regular octahedron. The square planar geometry can be regarded as an extreme limit of tetragonal distortion, and is especially found with a low-spin d^8 configuration, as in PdO.[4,35]

The assumption implicitly made above is that only metal–oxygen bonding is important. This is not true in some oxides which show evidence of direct **metal–metal bonding.**[40] Sometimes this leads to the formation of pairs (as in VO_2, discussed in Section 5.3.2) or larger clusters of metal atoms, with M–M distances comparable to those found in the elements. In other cases, such as $NaMo_4O_8$[41] the metallic bonding comes to dominate the structure totally, leading to chains with directly bonded molybdenum octahedra. Metal–metal bonding is a feature of oxides of the early transition elements, in rather low oxidation states. The d^1 and (in the 4d and 5d series) d^2 and d^3 electron configurations seem to be especially susceptible. The interesting relation between metal–metal bonding and electronic properties is discussed in Section 5.3.2.

1.3.2 Structural families

The complexity and variety shown by transition metal oxide structures cannot be adequately described in such a brief space.[35] Nevertheless it is useful to look at some of the commoner (and simpler) structural types found, which will be referred to frequently in later chapters. It is important to have some idea of the ways in which metal–oxygen units connect together in typical crystal structures, as these connections have an important influence on electronic properties. At the same time a discussion of simple structures will introduce some of the language of structural chemistry, which will be useful when more complex structures are considered at appropriate places later.

Figure 1.7 shows a selection of important binary structures, drawn so as to emphasize the near-neighbour metal–oxygen interactions. Some of these structures also occur in modified forms, e.g. distorted, with

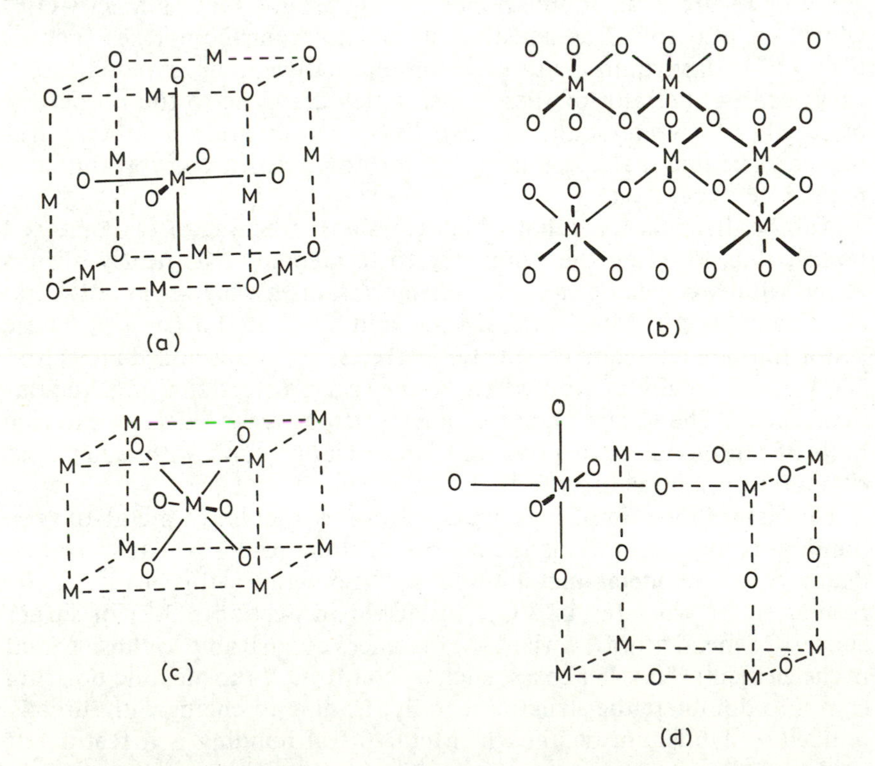

FIG. 1.7 Some important binary structures, drawn so as to emphasize the octahedral metal–oxygen coordination. (a) Rock-salt structure of MO composition; (b) corundum (M_2O_3); (c) rutile (MO_2); (d) rhenium trioxide (MO_3). Crystallographic unit cells are shown except in (b).

defects, or with substitution of some of the metallic element by another.

The **rock-salt** or sodium chloride structure is that of the monoxides MnO, CoO, NiO. 'FeO', as we have seen above, is always iron-deficient, and has defects (see also below). TiO_x and VO_x are also non-stoichiometric phases with defect versions of the rock-salt structure containing large numbers of lattice vacancies. Another variant occurs with NbO. Substitution of ions is possible; thus many compounds of the form AMO_2, where A is an alkali metal such as Li, have structures where half the transition metal atoms M are replaced by A, frequently in alternate layers.

The **corundum** structure (shown as only a part of its 30-atom unit cell) is common for 3*d* oxides M_2O_3. The relative closeness of pairs of metal atoms allows metal–metal bonding in some cases, e.g. in Ti_2O_3, which has a low-temperature phase where the hexagonal *c* axis and the Ti–Ti distance is anomolously low.[42] Ternary variants formed by ordered substitution of the corundum structure are also important. For example ilmenite, $FeTiO_3$, and lithium niobate, $LiNbO_3$, are formed by substitution of Al–Al pairs with Fe–Ti and Li–Nb respectively.

Rutile is the most important crystal form for TiO_2, and is a common structure for dioxides (except for those of the large Zr^{4+} and Hf^{4+} ions). Distorted variants can occur (with V, Nb, and Mo) in which M–M bonding again seems to be important, as metal atoms come together in pairs.

The **rhenium trioxide** structure is not itself common, being found in oxides only with the parent compound ReO_3 and (in distorted variants) by WO_3. It is interesting for its own sake as a beautifully simple structure, but its chief importance is that it forms the basis for the **perovskite** structure, one of the most important ternaries (see below).

In all of these structures the metal atom is octahedrally coordinated, although the symmetry is not perfect in either corundum or rutile, and even rock-salt compounds frequently show slight distortions from perfect symmetry. This is indeed the commonest coordination for binary transition metal oxides of all series. The structures clearly differ in the way in which neighbouring octahedral groups are connected, and this is illustrated in an alternative representation of some structures in Fig. 1.8. In this illustration atoms are not shown explicitly. Each octahedron must be imagined to have a metal atom at the centre, and an oxygen at each of the six corners. The connections between octahedra may be characterized as **corner, edge-**, or **face-sharing**. Thus rock-salt has edge-sharing contacts, ReO_3 corner-sharing, and corundum face-sharing. Rutile has a mixture of corner- and edge-sharing, the latter occurring in chains along the unique tetragonal *c* axis of the structure. Clearly such edge-sharing contacts entail a closer approach of metal

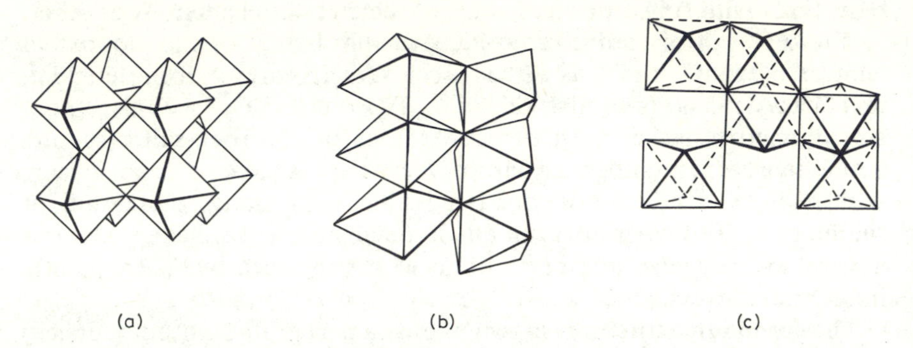

FIG. 1.8 Some binary structures of Fig. 1.7 drawn so as to show the connectivity of metal–oxygen octahedra. Individual atoms are not shown; each octahedron should be imagined to have a metal atom at the centre, and oxygens at each vertex. (a) Rhenium trioxide; (b) rutile; (c) corundum.

atoms than in corner-sharing, and there is some evidence in rutile compounds for stronger interactions between metal along the *c* axis. Closest of all would be the contacts given in a face-sharing structure. Such contacts are not favourable for ionic compounds, and it is notable that the **nickel arsenide** structure, in which extensive face-sharing takes place in chains along the hexagonal axis, is not found in oxides. As noted, such sharing does take place in corundum, but only between pairs of atoms which are otherwise rather isolated. The absence of metal atoms in sites either side of the pair allows considerable structural freedom, and in most cases distortion of the octahedral coordination allows metal atoms to move further apart. The consequences of this for the properties of $LiNbO_3$ will be discussed in Chapter 3. Occasionally, as with Ti_2O_3, the natural proximity of metal atoms is made use of to form a bond.

The representation in Fig. 1.8 is a very convenient short-hand notation for structures, and will be used at some later points in this book. Figure 1.9(a) and (b) show representations of two simple ternary structures, where transition metals and oxygen are shown in the same way, and a larger ternary metal by spheres. The **perovskite** structure is another one of those apparently simple structures beloved of elementary courses. It is indeed common for compounds of the formula AMO_3, where A must be larger in size than a typical transition metal ion, as it occupies a site with 12-fold oxygen coordination. Typically, A is a pre-transition metal such as K (as in $KTaO_3$), Sr (in $SrRuO_3$), or La (in $LaVO_3$). Occupation of the 12-fold site (which makes the difference between perovskite and ReO_3 structures) may be only partial, as in

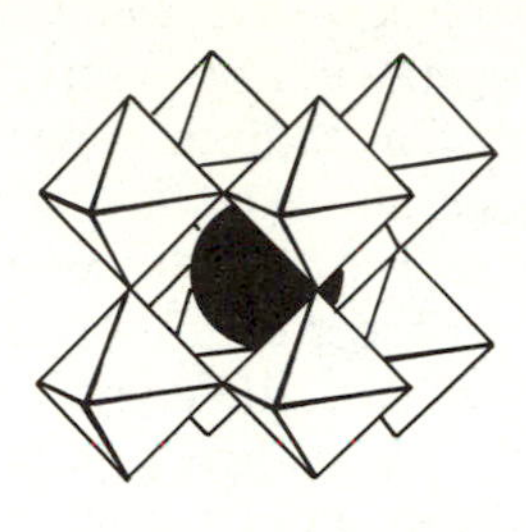
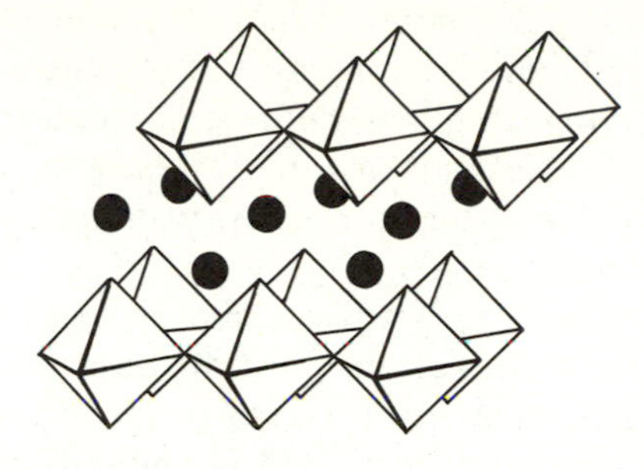

(a) (b)

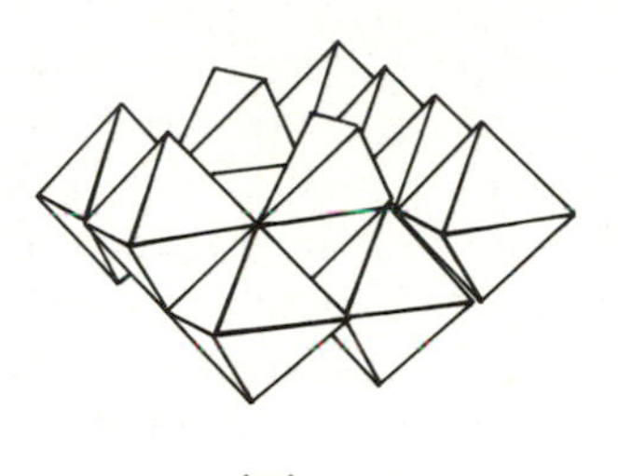

(c)

Fig. 1.9 Some ternary structures: (a) perovskite (AMO_3); (b) layer perovskite, or K_2NiF_4 structure; (c) spinel AB_2O_4 structure. Metal–oxygen octahedra (and tetrahedra in (c)) are shown as in Fig. 1.8, and the larger ternary element by spheres.

Na_xWO_3. Occasionally post-transition metal atoms such as Pb are found, as in $PbTiO_3$. The number of perovskite compounds which adopt the ideal cubic form is, however, quite small, and a great variety of distortions can be found. Sometimes these may be attributable to ionic sizes, which are frequently incompatible with the ideal cubic form;[43] in other cases specific geometrical preferences of metal ions may come into play, as with some d^0 cases discussed in Chapter 3. Related to perovskite is the **K_2NiF_4** structure; as its alternative name, the **layer perovskite** structure suggests, it is composed of layers of octahedra which corner-share in two dimensions only, these being interspersed with the larger ion. An oxide example is the superconductor $La_{1.85}Sr_{0.85}CuO_4$: the corner-sharing copper–oxygen layers, with coordination generally very far from a regular octahedron, are typical of many more complex 'high-T_c' superconductors.[38]

The network of corner-sharing octahedra is characteristic of many transition metal oxide **bronzes**.[44,45] This term normally designates an originally d^0 binary oxide in which another electropositive element—generally an alkali metal, but occasionally hydrogen—has been inserted to varying (frequently non-stoichiometric) extents. The sodium–tungsten bronze series Na_xWO_3 has already been mentioned above as having the perovskite structure. As with many bronzes there are in fact a variety of structures occurring for different x values. The corner-sharing framework can be vary varied, and give rise to interstices with different sizes, so that different cations can be accommodated. Many bronzes show interesting electronic properties, and will be discussed later in connection with semiconducting and metallic oxides (Chapters 4 and 5).

Two more complex but important structures are **spinel** and **garnet**, named after the parent minerals $MgAl_2O_4$ and $Ca_3Al_2Si_3O_{12}$ respectively. The spinel structure is found in binaries such as Fe_3O_4 and Co_3O_4, where clearly there is a mixed oxidation state (2+ and 3+) present, and in many ternaries, including a range of **ferrite** compounds with technologically important magnetic properties.[4,46] Figure 1.9 (c) illustrates the way in which both octahedral and tetrahedral coordination of metal atoms is found; in each formula unit of spinel, there is one tetrahedral metal, and two are in octahedral sites. The distribution of metal ions among these sites has been much studied; it frequently shows some disorder.[10,47] The two extreme limits are the **normal spinel** characteristic of $MgAl_2O_4$ itself, and the **inverse spinel** form of Fe_3O_4. In the normal spinel the 3+ ions occupy the octahedral sites, a situation denoted $Mg^{2+}[Al_2^{3+}]O_4$, with square brackets showing the octahedral ions. The same notation gives $Fe^{3+}[Fe^{2+}Fe^{3+}]O_4$ indicating that in the inverse spinel the 2+ ions go into octahedral sites, displacing half the 3+ ions into the tetrahedral positions. In general, the site occupancy can be characterized by a parameter γ, equal to the fraction of 2+ ions occupying octahedral sites; the limiting values are thus 0 (for the normal structure) and 1 (for an inverse case). Values of γ found for a variety of binary and ternary spinels of general formula $A^{2+}(B^{3+})_2O_4$ are given in Table 1.1.[4] Certain features which stand out, such as the relative preference of Fe^{3+} for tetrahedral coordination, may be related to ion sizes more than any other factor,[48] although ligand-field stabilization energies (see Section 2.1.1) have been invoked to explain them,[47] and the strong preference of Co^{3+} and Cr^{3+} for octahedral sites is particularly notable.

The garnet structure has a complex unit cell, and is not drawn here. Again, various coordination sites exist, and in a typical transition metal example, yttrium–iron garnet (YIG, $Y_3Fe_5O_{12}$), the Fe^{3+} ions have tetrahedral and octahedral coordination (two and three per formula unit respectively) while the large Y^{3+} occupies a larger eight-coordinate site. Garnets containing transition metals and lanthanides are also

Table 1.1 *Cation distributions in spinels. The values give the proportion of A^{2+} ions on octahedral sites in $A^{2+}(B^{3+})_2O_4$ (from ref. 4)*

	A^{2+}						
B^{3+}	Mg^{2+}	Mn^{2+}	Fe^{2+}	Co^{2+}	Ni^{2+}	Cu^{2+}	Zn^{2+}
Al^{3+}	0	0.3	0	0	0.75	0.4	0
Cr^{3+}	0	0	0	0	0	0	0
Fe^{3+}	0.9	0.2	1	1	1	1	0
Mn^{3+}	0	0	0.67	0	1	0	0
Co^{3+}	–	–	–	0	–	–	–

used as magnetic materials, and yttrium–aluminium garnet (YAG, $Y_3Al_5O_{12}$) doped with lanthanides is an important phosphor material, employed in solid-state lasers and in television screens.[4]

1.3.3 Defects and the accommodation of non-stoichiometry

The importance of non-stoichiometry in transition metal oxides has already been mentioned. Obviously it must involve some departure from the perfectly regular atomic arrangements that make up an ideal crystal. Elementary thermodynamic considerations show that all real crystals must have defects;[21] but in many transition metal compounds, the oxides in particular, the concentration of these can be very large, with important consequences for many physical properties.[10]

The most elementary defects are the **lattice vacancy**, where an atom is absent from a normal lattice site, and an **interstitial atom** in an 'irregular' site not normally occupied. Binary compounds can maintain their stoichiometry by having balanced combinations of such defects, for example the **Frenkel** (vacancy plus interstitial of the same ion) or **Schottky** (vacancies of both ions) types. Non-stoichiometry can arise from an imbalance in the defect types, and often one kind may predominate. For example a metal deficiency as in $M_{1-x}O$ may be accommodated either by metal vacancies or by oxygen interstitials; the former is more likely, as in most structures there is insufficient space for large oxygen interstitial atoms. Oxygen deficiency, for example in MO_{2-x}, may result from oxygen vacancies or metal interstitials. Distinguishing between different possible defect types is often difficult. When the deviation from stoichiometry is significant, various physical measurements may help the most simple of which is the density, which may be compared with a value computed from the unit-cell volume on the basis of no defects. Then vacancies will make the observed density less, and interstitials will make it greater, than this ideal estimate. Diffraction studies (especially with neutrons, making use of profile-refinement techniques) and high-resolution transmission electron microscopy are also useful.[14] Unfortunately, as we shall see below, when the defect concentration is high enough for such techniques to be really useful, the structural chemistry becomes more complicated. For very low deviations from stoichiometry, when the picture of isolated point defects is at least approximately valid, less direct methods of investigation are required. As discussed in Chapter 4, studies of defect mobility and some spectroscopic methods can help. Theoretical calculations of defect structure and energetics, most often with sophisticated versions of the ionic model, are also used.[21,49]

The simple point-defect model is appealing, but there is a great deal of evidence that it is inadequate to describe anything other than very small departures from perfect stoichiometry. Point defects interact strongly with one another, through long-range Coulomb interactions coming from their charged nature, and from elastic interactions caused by the local distortions they entail. In some cases, direct chemical bonding between defects may even occur. These forces result in the clustering or ordering or defects except when their concentration is very low.

One system in which defect clustering has been studied extensively is wüstite, $Fe_{1-x}O$, where the basic structure is rock-salt. The density shows that Fe vacancies are present, but a variety of techniques, such as magnetic measurements and diffraction, also suggest the occupation of tetrahedral sites. We have already noted the relative preference of Fe^{3+} for such a geometry, and that in the Fe_3O_4 phase formed by further oxidation of wüstite, half the Fe^{3+} is indeed in the tetrahedral sites. The model developed to explain the observations on wüstite is shown in Fig. 1.10.[50] Neighbouring Fe vacancies make it more favourable for Fe^{3+} to move into a tetrahedral interstice. This favourable interaction in turn causes the clustering of the vacancies, and the development of larger and larger clusters. The phase diagram (see Fig. 1.4 (b)) shows that the wüstite and magnetite phases never meet, but nevertheless it is possible to imagine that the defect clusters in wüstite are similar to small parts of the inverse-spinel structure.

The model just described for wüstite is one of defect clusters which, although they may grow quite large, remain *isolated* from one another. A different kind of possibility is that some partial or complete *ordering* of defects occurs throughout the structure. This type of ordering is common when the defect concentration is high. One example is in the 'monoxides' TiO_x and VO_x. In fact these phases have a range of compositions which extend to both metal- and oxygen-rich sides. They are based on the rock-salt structure, but with a high proportion of vacancies

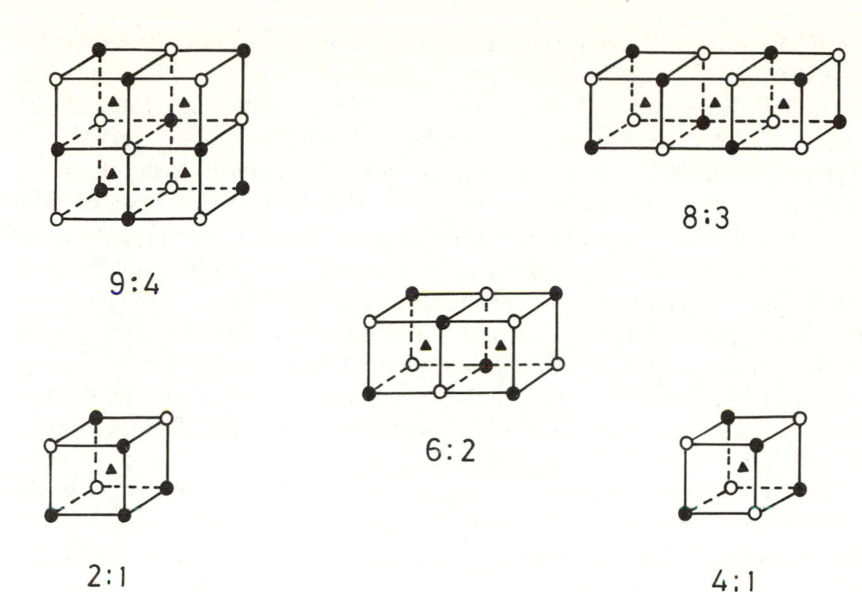

FIG. 1.10 Some of the defect clusters proposed for $Fe_{1-x}O$. The symbols show: (○) Fe^{2+} vacancy at an octahedral site of the rock-salt lattice, and (▲) Fe^{3+} interstitial. There is presumed to be a range of cluster sizes present, larger ones predominating at higher deviations from stoichiometry.[50]

in both sites. Thus apparently stoichiometric TiO has about 15 per cent of Ti and O sites vacant. At high temperatures the vacancies have no long-range order within the structure, but below 900 K there is an ordered phase in which the vacant sites (one sixth of the total) become ordered.[35] Another example is the 'defect perovskite' $SrFeO_{2.5}$.[34] One sixth of oxygen sites from the ideal perovskite lattice are vacant, again in an apparently random way at high temperatures, but ordered in the **brownmillerite** structure below 1100 K. In this case, as in many others, the random arrangement may be obtained at lower temperatures by rapid *quenching* of the high-temperature form. There are other variations possible in this system; for example the oxygen content may be increased according to the preparative conditions, the pure defect-free perovskite composition $SrFeO_3$ being approached by treatment with oxygen at high pressures. The ordering of the brownmillerite form is only possible with $SrFeO_{2.5}$ and a variety of distorted perovskite variants is found for intermediate compositions. This by no means untypical example shows how not only the composition, but also the structure arising, may be a sensitive function of preparative conditions; and clearly the physical properties of the resulting phases may be expected to be different.

A quite different kind of defect clustering occurs with reduced d^0 oxides of Ti, V, Nb, Mo, and W. In these compounds, point defects are largely eliminated by the formation of **crystallographic shear phases.**[14] A picture of how this happens is shown in Fig. 1.11 (which is schematic and is not meant to represent a *mechanism* for crystallographic shear (CS) formation). By aligning oxygen vacancies in a plane (shown in section on the left of the figure) it is possible to 'shear' the structure in such a way that the vacancies are eliminated. In the case shown, the original structure is that of ReO_3 with all corner-sharing metal–oxygen octahedra (see above); in the CS phase the elimination of oxygen vacancies entails some edge sharing. This formation of groups of more closely spaced metal atoms is one characteristic of CS phases, and various factors have been considered to explain their existence. From an ionic point of view, it would seem to be electrostatically unfavourable, but ionic-model calculations have emphasized the importance of polarization in mitigating the extra repulsion.[49] The d^0 oxides that form CS phases on reduction all have high static dielectric constants.[51] This

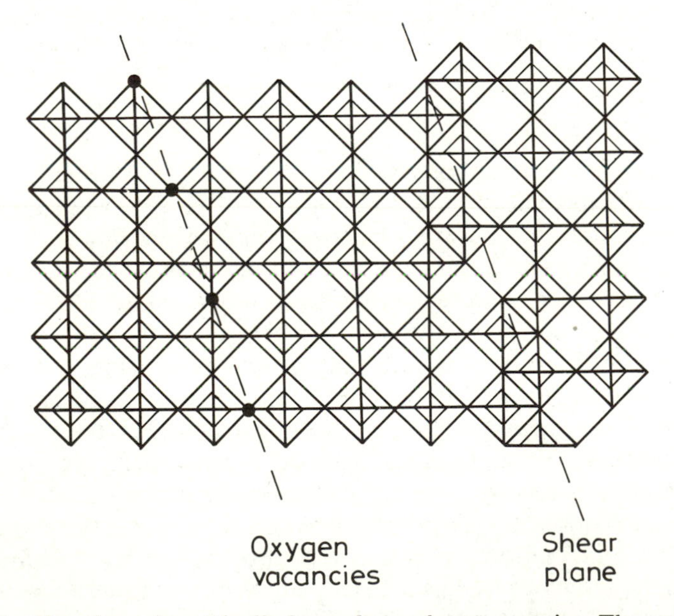

FIG. 1.11 Showing schematically how *shear planes* can arise. The ordering of oxygen vacancies in a plane in the ReO_3 lattice (left) allows the shearing of the structure (right) to eliminate the point defects. Groups of more closely spaced metal atoms form edge-sharing octahedra.

results from a high polarizability of the lattice, and may be related to the facile distortions of metal–oxygen octahedra containing these ions (see Section 3.1.3). In the ionic viewpoint, such distortion is crucial in lowering the repulsion between the cations brought together by CS formation. However, there is another very different possibility. All the metals that form CS phases show some signs of metal–metal bonding in their reduced oxides. This may be present in the edge-shared groupings of the reduced phases.

The most powerful structural technique for studying CS phases is electron microscopy.[14,52] Conventional diffraction methods are often less useful because of the very large crystallographic unit cells involved. One general feature of these materials is the regular ordering of the CS planes. The existence of 'homologous series' of phases (with general composition W_nO_{3n-1} for example) can thus be understood in terms of the same basic CS structure, but with different spacings between the shear planes. As n increases the planes get further apart, and the forces which lead to their ordering presumably become weaker. Successive values of n also correspond to more similar compositions. Thus eventually it may be impossible, either from structural measurements or from the thermodynamic considerations discussed in Section 1.2.2, to distinguish such a series of line phases from a range of continuously varying stoichiometry.

1.4 Electronic classification

The extremely wide range of electronic behaviour of transition metal oxides obviously creates difficulties, not just of understanding and interpretation, but also of classification.[3,53] Whatever the scheme employed, there would always be compounds that failed somehow to fit in. A reasonably logical method, however, is to keep to the 'elementary' division of solids into insulators, semiconductors, and metals.[54] This is the scheme employed in Chapters 3, 4, and 5. Before proceeding to these more descriptive accounts, Chapter 2 gives a comparison of some of the varied theoretical models that have been used to interpret the electronic structure of transition metal oxides.

A more detailed classification of electronic properties is given below; some of the features described can be understood by reference to Fig. 1.12.

d^0 insulators are stoichiometric oxides with the d^0 electron configuration. Not only are they good insulators with quite large band gaps (e.g. 3 eV in TiO_2), but they show other properties expected of insulators: they have no optical absorptions at photon energies less than the band

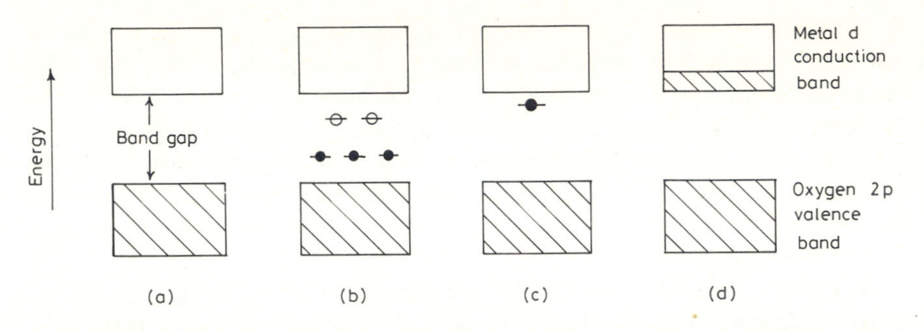

FIG. 1.12 Qualitative electron energy-level diagrams for transition metal oxides. (a) The bands of a d^0 compound, with a gap between the 'oxygen 2*p*' valence band and the empty 'metal *d*' conduction band; (b) the localized *d* levels appropriate to a transition metal impurity, and to a magnetic insulator; (c) a donor level associated with semiconduction in a non-stoichiometric oxide; and (d) the partially filled conduction band of a metallic oxide.

gap, and are diamagnetic with no unpaired electrons. Examples are TiO_2, $KTaO_3$, and $CaWO_4$. Their properties can be understood according to the qualitative energy-level diagram of Fig. 1.12 (a). The band gap is between a filled band of bonding orbitals, with predominantly oxygen 2*p* atomic character, and an empty 'metal *d*' band of antibonding orbitals. The evidence for this type of band structure will be discussed in more detail in Chapters 2 and 3. Many d^0 insulators are susceptible to loss of oxygen, which gives rise to semiconducting properties as discussed below.

Other closed-shell insulators arise when all or part of the metal *d* band is full. Most obvious is the d^{10} case such as Cu_2O, but other examples arise, since the *d* band is split by chemical-bonding effects. Crystal field arguments show how low-spin d^6 ions with octahedral coordination (for example $LaRhO_3$ with $4d^6$) and d^8 square planar ions (as in PdO) can give closed-shell insulating ground states (see Section 3.2). Metal–metal bonding can also split the *d* band, and some d^1 oxides such as Ti_2O_3 and VO_2 have non-metallic forms at low temperatures. The gaps are small, however, and metal–metal bonding in these compounds weakens with increasing temperature, which is associated with transitions to a metallic state. For this reason, Ti_2O_3 and some related examples, are treated in Chapter 5 as metallic oxides with lattice interactions that interfere with the electronic structure at low temperatures.

d^n **impurities**: transition metal impurities are common in insulating oxides, both of the d^0 transition metal type, and of non-transition metals (e.g. MgO and Al_2O_3). Their magnetic and spectroscopic properties (which give rise, for example, to many of the desirable optical properties of gemstones) are typical for isolated transition metal ions in a chemical environment, and similar to those found in complexes such as the hexa-aquo ions $M(H_2O)_6^{n+}$ common in aqueous solution. They are largely due to the partially occupied metal *d* levels, forming localized states within the band gap of the host solid (see Fig. 1.12 (b) and Section 3.3).

Magnetic insulators form a more surprising class of compound from the point of view of conventional band models (Section 3.4). These may be pure oxides with d^n ions such as Ni^{2+} (in NiO) or Cr^{3+} (in Cr_2O_3). Their properties are surprisingly similar to those of oxides containing the same ions as impurities, except that the concentration of the 'impurity' is much higher. Thus the localized 'impurity-like' *d* levels of Fig. 1.12 (b) still give the simplest picture of the electronic structure. Magnetic properties indicate the unpaired electrons expected for the appropriate d^n configuration, although weak interactions between magnetic moments give rise to co-operative phenomena, most commonly antiferromagnetic ordering, at low temperatures. *d–d* (crystal-field) transitions give rise to optical absorptions, again with only quite weak perturbations from those expected for isolated ions. At the same time, the pure compounds are good insulators with band gaps that may be as large as those in some d^0 compounds. Like the latter, however, the magnetic insulators are susceptible to non-stoichiometry, and this may give rise to semiconductive properties and strong optical absorptions. For example NiO is green and highly insulating when pure, but easily takes up excess oxygen to become black and semiconducting. Many ferrites and garnets with important magnetic applications are magnetic insulators, for example yttrium–iron garnet (YIG), $Y_3Fe_5O_8$.

Semiconducting properties are, in principle, a feature of all the 'insulating' oxides, but in most cases the band gaps are large enough that intrinsic semiconduction is hard to observe. In practical terms, semiconduction is characteristic of defective or non-stoichiometric oxides. These include reduced d^0 compounds such as TiO_{2-x}, non-stoichiometric samples of magnetic insulators such as $Ni_{1-x}O$, as well as oxides with deliberate or inadvertant doping by another element, as in $Li_xMn_{1-x}O$. The relationship between semiconduction and defect structure and chemistry is complex and imperfectly understood; a discussion of this question forms a major part of Chapter 4. At a qualitative level, Fig. 1.12 (c) shows how an electron introduced by chemical reduction can be held in a defect level with an energy slightly below the empty conduction band: thus one has an *n*-type semiconductor.

Transitions to the metallic state occur as the doping level is increased in many oxide semiconductors, as with more familiar materials such as silicon. Examples are in the sodium–tungsten bronzes Na_xWO_3, and some superconducting oxides such as $La_{2-x}Sr_xCuO_4$; the former is an *n*-type semiconductor for $x < 0.3$, the latter a *p*-type semiconductor for $x < 0.07$. The simplest way to understand such transitions is to imagine that defect levels associated with different impurity atoms start to overlap and form a band, but many different interactions may be at work, as discussed in Section 4.4

Simple metals in the sense normally understood in solid-state physics, that is solids where a free-electron model works well, are not found among transition metal oxides. Band-structure effects are always important; there are, however, some compounds which, when these are properly taken into account, seem to follow quite well the 'textbook' properties of a metal (Section 5.1). These are generally 4*d* and 5*d*, rather than 3*d* compounds, examples being ReO_3, Na_xWO_3 (where $0.3 < x < 0.9$) and RuO_2. Figure 1.12 (d) is appropriate here, showing a partially filled conduction band. Unfortunately for band-theorists, however, this class of compound is not large among transition metal oxides.

Anomalous magnetic properties are a very common sign of deviation from 'simple' metallic behaviour, due to electron-correlation effects not treated in band theory (Section 5.2). They include enhanced magnetic susceptibilities (e.g. in $LaNiO_3$) and cases where metallic oxides show magnetic ordering (for example, ferromagnetism in CrO_2). There are also many cases not easily interpreted with any simple band model: for example the very remarkable compound V_2O_3 shows a temperature-dependent transition between a metal and a magnetic insulator.

Lattice interactions (or *electron–phonon coupling*) encompass various types of interaction between conduction electrons and the structure or vibrational modes of a metallic solid. A number of different effects in oxides can give rise to metal/non-metal transitions as a function of temperature. These include 'low-dimensional' (Peierls distortion) phenomena in $K_{0.3}MoO_3$, metal–metal bonding in Ti_2O_3 and VO_2, and local distortions of the M–O bond lengths, associated with ordering the oxidation states in Fe_3O_4. These interactions are discussed in Section 5.3.

Superconductors include not just the 'high-T_c' copper oxides, but a number of other compounds such as NbO and $SrTiO_{3-x}$. According to the BCS theory, superconductivity is itself caused by electron–lattice interactions. This may not be the case with the high-T_c compounds, although at the time of writing the mechanism of superconductivity remains unknown (see Section 5.4).

References

1. Shriver, D.F., Atkins, P.W., and Langford, C.H. (1990). *Inorganic chemistry*. Oxford University Press.
2. Greenwood, N.N. and Earnshaw, A. (1984). *Chemistry of the elements*. Pergamon, Oxford.
3. Goodenough, J.B. (1971). *Prog. Solid State Chem.,* **5**, 145.
4. West, A.R. (1984). *Solid state chemistry and its applications*. Wiley, Chichester.
5. Cheetham, A.K. and Day, P. (eds) (1987). *Solid state chemistry: techniques*. Oxford University Press.
6. Hagenmuller, P. (ed.) (1972). *Preparative methods in solid state chemistry*. Academic, New York.
7. *X-ray powder data file* (1960). American Society for Testing Materials, Philadelphia.
8. Loretto, M.H. (1984). *Electron beam analysis of materials*. Chapman and Hall, London.
9. Greenwood, N.N. (1968). *Ionic crystals, lattice defects, and non-stoichiometry*. Butterworths, London.
10. Sorensen, O.T. (ed.) (1981). *Non-stoichiometric oxides*. Academic, New York.
11. Dickens, P.G. and Whittingham, M.S. (1968). *Quart. Rev.* **22**, 30.
12. Schlenker, C., Dumas, J., Escribe-Filippini, C., Guyot, H., Marcus, J., and Fourcaudet, G. (1985). *Phil. Mag. B,* **52**, 643.
13. Cava, R.J., Batlogg, B., Rabe, K.M., Rietman, E.A., Gallagher, P.K. and Repp, Jr., L.W. (1988). *Physica C,* **156**, 523.
14. Tilley, R.J.D. (1987). *Defect crystal chemistry*. Blackie, Glasgow.
15. See, for example, refs 16 and 17.
16. Robin, M.B. and Day, P. (1967). *Adv. Inorg. Chem. Radiochem.,* **10**, 247.
17. Brown, D.B. (ed.), (1980). *Mixed-valence compounds*. D. Reidel, Dordrecht.
18. Varma, C.M. (1976). *Rev. Mod. Phys.,* **48**, 218.
19. Goodenough, J.B. in ref. 17, p. 413.
20. Bednorz, J.G. and Muller, K.A. (eds) (1990). *Superconductivity*, Springer Series in Solid State Sciences, Vol. 90. Springer, Berlin.
21. Hayes, W. and Stoneham, A.M. (1985). *Defects and defect processes in non-metallic solids*. Wiley, New York.
22. Connor, J.A. and Ebsworth, E.A.V. (1964). *Adv. Inorg. Chem. Radiochem.* **6**, 249.
23. Phillips, C.S.G. and Williams, R.J.P. (1965) *Inorganic chemistry*. Clarendon Press, Oxford.
24. Gerloch, M. (1986) *Orbitals, terms, and states*. Wiley, Chichester.
25. Kemp, J.P., Cox, P.A., and Hodby, J.W. (1990). *J. Phys.: Condensed Matter,* **2**, 6699.
26. Atkins, P.W. (1990). *Physical chemistry*, (4th edn). Oxford University Press.

27. Johnson, D.A. (1982). *Some thermodynamic aspects of inorganic chemistry*, (2nd edn). Cambridge University Press.
28. For an extensive compilation of literature references, see Wisniak, J. (1981). *Phase diagrams: a literature source book*. Elsevier, Amsterdam.
29. Revoleschi, A. and Dhalenne, G. (1984). In *Basic properties of binary oxides*, (ed. A. Dominguez-Rodriguez, J. Carstaing, and R. Marquez). University of Seville.
30. Darken, L.S. and Gurry, R.W. (1946). *J. Am. Chem. Soc.*, **67**, 1398; (1946) Ibid., **68**, 798.
31. Cheetham, A.K. (1981). Structural studies on nonstoichiometric oxides using X-ray and neutron diffraction. In ref. 10.
32. Wizansky, A.R., Rauch, P.E., and DiSalvo, F. (1989). *J. Solid State Chem.*, **81**, 203.
33. Tournaix, M., Marchand, R., and Brohan, L. (1986). *Prog. Solid State Chem.*, **17**, 33.
34. Takeda, Y., Kanno, K., Takada, T., Yamamoto, O., Takano, M., Nakayama, N., and Bando, Y. (1986). *J. Solid State Chem.*, **63**, 237.
35. Wells, A.F. (1984) *Structural inorganic chemistry*, (5th edn), Oxford University Press.
36. Shannon, R.D. and Prewitt, C.T. (1969). *Acta Cryst.* B, **25**, 925; (1970). Ibid., **26**,1046.
37. Kihlborg, L. (1963). *Ark. Kemi*, **21**, 357.
38. Day, P. (1988). In *High temperature superconducting materials*, (ed. W.E. Hatfield, and J.H. Miller), Dekker, New York.
39. See for example Orgel, L.E. (1958). *Disc. Faraday Soc.*, **26**, 138; and ref. 47.
40. For proceedings of a recent symposium, see *J. Solid State Chem.* **57** (1985).
41. Torardi, C. and McCarley, R.E. (1979). *J. Am. Chem. Soc.*, **101**, 3963.
42. Rice, C.E. and Robinson, W.R. (1977). *Acta Cryst. B*, **33**, 1342.
43. Goodenough, J.B. and Longo, J.M. (1970). Crystallographic and magnetic properties of perovksite and perovskite related compounds. In *Magnetic and other properties of oxides and related compounds*, (ed. K.-H. Hellwege.) Landolt-Bornstein New Series III, Vol. 4. Springer, Berlin.
44. Dickens, P.G. and Wiseman, P.J.W. (1975). Oxide bronzes. In *Solid state chemistry*, MTP International Review of Science, Inorganic Chemistry Series 2, Vol. 10. Butterworth, London.
45. Hagenmuller, P. (1973). Vanadium Bronzes and related Compounds. In *Comprehensive inorganic chemistry*, (ed. A.F. Trotman-Dickenson) Vol. 4, p. 541, Pergamon, Oxford.
46. Trebble, R.S. and Graik, D.J. (1979). *Magnetic materials*. Wiley, New York.
47. Dunitz, J.D. and Orgel, L.E. (1960) *Adv. Inorg. Chem. Radiochem.*, **2**, 1.
48. Burdett, J.K., Price, G.D. and Price, S.L. (1982). *J. Am. Chem. Soc.*, **104**, 92.
49. Catlow, C.R.A. (1981). Defect clustering in nonstoichiometric oxides. In ref. 10.

50. Catlow, C.R.A. and Fender, B.E.F. (1975). *J. Phys. C: Solid State Phys.*, **8**, 3267.
51. Tilley, R.J.D. (1977). *Nature*, **269**, 229.
52. Eyring, LeRoy (1981). Structure, defects and nonstoichiometry in oxides. In ref. 10.
53. Hamnett, A. and Goodenough, J.B. (1984). Binary transition metal oxides. In *Physics of non-tetrahedrally bonded compounds*, (ed. O. Madelung), Landolt-Bornstein New Series III, Vol. 17. Springer, Berlin.
54. Cox, P.A. (1977). *The electronic structure and chemistry of solids*. Oxford University Press.

2

MODELS OF ELECTRONIC STRUCTURE

At the present time there exists no unified model for interpreting the electronic properties of transition metal oxides. The very great diversity of properties make it very unlikely in fact that such a 'super-theory' could be developed, or that it would ever be very useful. This chapter gives an outline of some of the models which have been found helpful, with a few examples for illustration. A more critical discussion of the validity of the different theories will come in the succeeding chapters, where the experimental properties that need to be explained will be described more fully.

The simplest way of classifying theories of the electronic structure of solids is to distinguish between those that take a *local* view, and those that describe *extended* or *delocalized* electron states. An extreme version of the first kind is the **ionic model**, which concentrates on the properties of individual ions, presumed to have integral charges given by the oxidation states of the different elements present. Some of the obvious deficiencies of this picture can be remedied by **cluster models**, where the electronic interactions within a small group of atoms are treated more explicitly. As one would expect, these approaches are most useful for non-metallic solids. The contrasting picture is that provided by **band theory**, which calculates the wavefunctions for electrons in a periodic lattice. The band model is most obviously applicable to metallic solids; however, because of the general popularity of this theory in solid-state physics, and especially with the current availability of powerful computer programs for calculating band structures, it has been applied very widely. Some of the controversies surrounding its applicability will be discussed later.

Between the fully localized and delocalized pictures stands another class of models, which attempt in some sense to interpolate between them, by describing the forces that cause electronic states to be more localized than predicted by band theory. Among these forces are the repulsion between electrons, the interaction of band electrons with lattice vibrations, and the influence of defects and disorder in the solid. These **intermediate models** will be discussed at the end of the chapter.

All the models discussed here are capable of considerable mathematical sophistication, and can make detailed numerical predictions. We shall look at the results of some of these calculations, but the mathematical and computational methods required will not be described. Our emphasis rather will be on the *physical content* of the different models, and a comparison of their predictions with experiment. The more technical aspects of how the calculations are done can be found in some of the references provided.

2.1 Ionic models

The ionic model is not intended as a serious theory of electronic structure, and for many purposes gives far too naïve a picture of the electron distribution in a solid.[1-3] Yet it is useful as a scheme for rationalizing many of the structures and properties of solids, and as we shall see later it also forms the basis for some useful predictions about ground and excited-state energy levels.[4,5] It was suggested in Chapter 1 that the electron configuration based on the ionic model is a useful basis for the simple classification of transition metal compounds. In a d^0 compound, the insulating band gap is predicted to be between a filled valence band, made up of oxygen $2p$ levels (as the O^{2-} ion has the filled-shell electron configuration $2p^6$) and an empty conduction band of metal $3d$ orbitals. Some of the problems encountered in quantifying the energies involved are discussed below. We shall also see that the ionic model gives valuable insight into the energy gap in magnetic insulators.[6] These simple ionic pictures need to be qualified in the light of more realistic theories which include orbital overlap. Nevertheless they are useful as 'zero-order' models for thinking about many types of transition metal oxides. This is especially true of another theory based primarily on ionic arguments, the *crystal field model*.

2.1.1 Crystal field theory

Crystal field (CF) theory is the name given to the model which describes how transition metal ions with d^n configurations are perturbed by their chemical environment.[2,7] The basic idea is shown in Fig. 2.1.

For a transition metal ion with octahedral coordination, the five d orbitals are seen to divide into two sets: two orbitals have lobes of maximum probability pointing directly at the near-neighbour oxygens, whereas the other three have nodal planes in these directions. These distinct sets of orbitals are conventionally denoted $\boldsymbol{e_g}$ and $\boldsymbol{t_{2g}}$ respectively, according their symmetry behaviour (or *irreducible representation*) in the symmetry point group O_h of an ideal octahedron.[8] Figure 2.1 also shows how the octahedral environment gives rise to a **crystal-field splitting** (Δ) between the lower t_{2g} and the higher energy e_g orbitals.

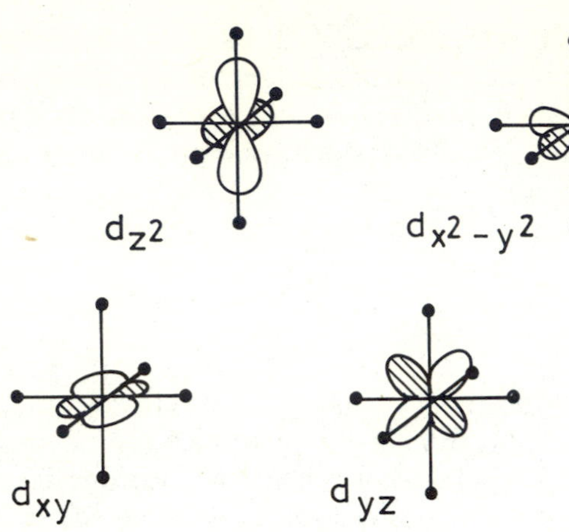

e_g

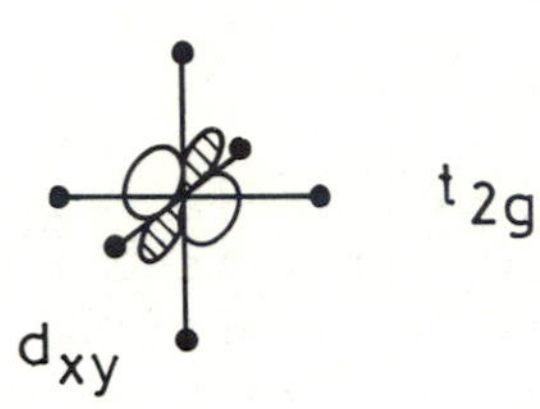

(a)

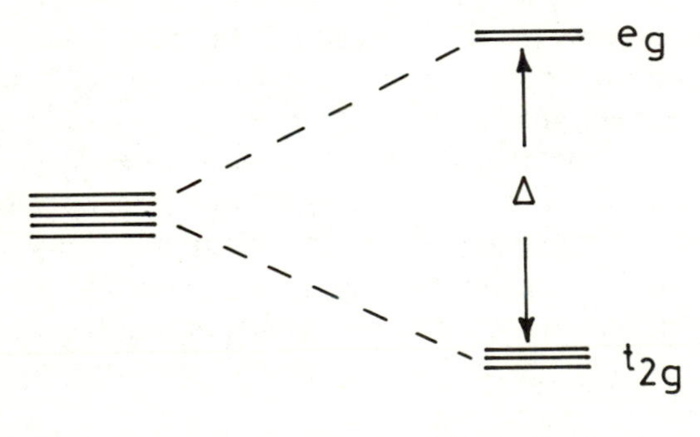

Spherical symmetry

Octahedral field

(b)

Fig. 2.1 The *d* orbitals of a transition metal ion in an octahedral site, showing (a) the different orientations of the e_g and t_{2g} sets, and (b) the resulting orbital energies with the *crystal field splitting*.

In the original version of the CF model, the energy splitting Δ was presumed to arise from a purely electrostatic perturbation of the transition metal *d* orbitals by the surrounding anions: i.e. the assumption of ionic character was rigorously applied.[9] It has long been recognized that this electrostatic approach is quite inadequate as a basis for the *ab initio* calculation of crystal-field splittings.[10] More sophisticated models, described later, show how the splittings can be interpreted in terms of orbital overlap and covalent interactions. Chemists tend to use the name **ligand field theory** for such 'non-ionic' versions of the model.[11,12] However, the term *crystal-field splitting* has been generally retained in the solid state literature,[7,13] and will be used in most of this book, without any implication of a purely ionic approach.

The electron configurations for ions in octahedral sites are obtained in the CF model by arranging the electrons in the t_{2g} and e_g orbitals in accordance with the Pauli exclusion principle. The ground states of the free ions satisfy **Hund's first rule**: i.e. the electrostatic repulsion between electrons is minimized by placing them, as far as possible, with parallel spins in different orbitals.[14] In crystal fields, different possibilities can often arise according to the relative magnitude of the splitting, Δ, and the exchange energy: if the former is large the lowest energy is obtained by complete filling of orbitals from the bottom up, whereas more favourable exchange energies are found with single filling of orbitals so that spins can be parallel. The resulting alternative **low-** and **high-spin** configurations are shown in Table 2.1. As well as the actual electron configurations, this table shows the **term symbol** giving the spin and overall symmetry behaviour of the ground state, and the **crystal field**

Table 2.1 *High- and low-spin states for* d^n *ions in octahedral sites*

	High spin			Low spin		
n	Configuration	State	CFSE/Δ	Configuration	State	CFSE/Δ
1	t_{2g}^1	$^2T_{2g}$	$\frac{2}{5}$	–	–	–
2	t_{2g}^2	$^3T_{1g}$	$\frac{4}{5}$	–	–	–
3	t_{2g}^3	$^4A_{2g}$	$\frac{6}{5}$	–	–	–
4	$t_{2g}^3 e_g^1$	^5E_g	$\frac{3}{5}$	t_{2g}^4	$^3T_{1g}$	$\frac{8}{5}$
5	$t_{2g}^3 e_g^2$	$^6A_{1g}$	0	t_{2g}^5	$^2T_{2g}$	2
6	$t_{2g}^4 e_g^2$	$^5T_{2g}$	$\frac{2}{5}$	t_{2g}^6	$^1A_{1g}$	$\frac{12}{5}$
7	$t_{2g}^5 e_g^2$	$^4T_{1g}$	$\frac{4}{5}$	$t_{2g}^6 e_g^1$	^2E_g	$\frac{9}{5}$
8	$t_{2g}^6 e_g^2$	$^3A_{2g}$	$\frac{6}{5}$	–	–	–
9	$t_{2g}^6 e_g^3$	^2E_g	$\frac{3}{5}$	–	–	–
10	$t_{2g}^6 e_g^4$	$^1A_{1g}$	0	–	–	–

Electron configurations, spectroscopic term symbols, and crystal field stabilization energies are given for $n = 1$ to 9. Low-spin states are specified only when they are different from high-spin. Note that d^4 (high spin) and d^9 are invariably found in sites distorted from regular octahedral geometry.

stabilization energy (CFSE) appropriate to the configuration. The term symbols are analogous to those for atoms, except that instead of specifying the total angular momentum appropriate to a free ion (*S*, *P*, *D*, etc.) they give the irreducible representation in the corresponding point group (O_h for full octahedral symmetry with an inversion centre). The CFSE values represent the stabilization expected for the particular configuration, relative to the hypothetical 'spherically averaged' case obtained by occupying each *d* orbital with an equal (and normally therefore fractional) number of electrons. The CFSE is given in units of the splitting, Δ, and is computed by assigning an energy $-\frac{2}{5}\Delta$ to the t_{2g} and $+\frac{3}{5}\Delta$ to the e_g orbitals; these values are designed to give an average (barycentre) *d* orbital energy of zero.[7]

The information in Table 2.1 can be used for rationalizing certain structural trends. Figure 1.5 showed that the contraction in ionic radius across each transition series is less regular than that expected for free ions. In the 3*d* series in particular, the 2+ and 3+ ions show an interruption of this contraction between the d^3 and d^5 configurations, and again between d^8 and d^{10}. These are high-spin ions, and the radii show the influence of the changing CFSE shown in the table. More rapid contraction occurs when the CFSE is increasing, and the interruption occurs when it decreases again as a result of filling the higher energy e_g orbitals.[15] These latter orbitals are, as we shall see later, quite strongly antibonding in nature, so that their occupation leads to a lengthening of the M–O bond. The influence of CFSE on bond length is shown most markedly by the comparison between high- and low-spin cases. In the 4*d* and 5*d* series, low-spin is the rule and the contraction is not interrupted in mid-series in the same way. Only Co^{3+} among 3*d* ions is commonly low-spin[2] (a fact at least partly attributable to the exceptional CFSE obtained with d^6) and is strikingly small in radius as a consequence.

As mentioned in Section 1.3.1, certain electron configurations have a notable tendency to occupy distorted environments. d^4 (high-spin) and d^9 ions in non-metallic solids invariably seem to give a tetragonal distortion of an otherwise octahedral site, with two lengthened M–O bonds, or occasionally a square planar coordination. The distortion is often regarded as a consequence of the **Jahn–Teller theorem**, according to which a non-linear molecule in an electronic state with orbital degeneracy will distort so as to lower its symmetry and remove the degeneracy.[11,16] All the ground states listed in Table 2.1 have orbital degeneracy except those with the term symbol *A*, which arise when both t_{2g} and e_g sub-shells are either empty, half filled, or completely full. In fact strictly regular octahedral sites in crystals are rare, and both careful crystallographic and spectroscopic investigations generally show low symmetries in most cases, even those not subject to Jahn–Teller effects. The reasons for such distortions must lie in a variety of ion-packing and chemical-bonding effects which have nothing to do with orbital degeneracy. Nevertheless, the strong tetragonal distortion seems to be confined to d^4 and d^9 configurations.[17] From Table 2.1, this behaviour can be traced to the occupation of e_g orbitals, which is uneven (neither half nor completely full) in these cases. Figure 2.2 shows how the initial octahedral splitting changes as a result of lengthening two M–O bonds to give a tetragonal field: the significant point is the splitting of the e_g orbitals, so that additional stabilization can be obtained when these are unevenly occupied. One might expect the same behaviour with low-spin d^7 ions, but this configuration is quite rare in oxides, being confined to Ni^{3+} compounds such as $LaNiO_3$. This compound is metallic, and the Jahn–Teller distortion is not present, showing that the effect is confined to *localized* electrons.[18]

The Jahn–Teller theorem should strictly apply also when orbital degeneracy comes from uneven occupancy of the t_{2g} orbitals. It appears, however, that the stabilization obtainable by distortion is much

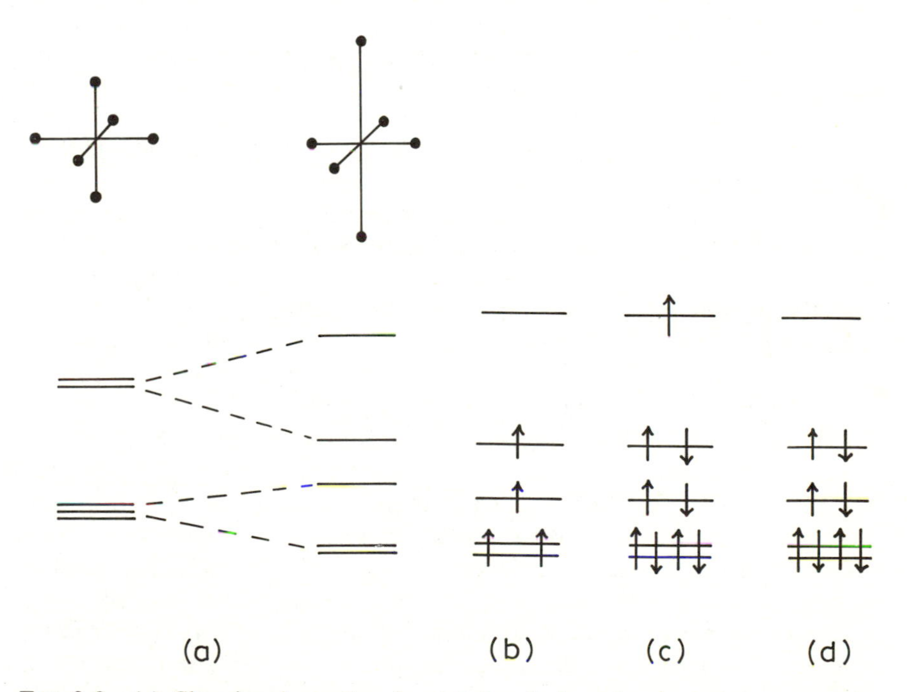

Fig. 2.2 (a) Showing how the d orbital splitting changes as two opposite metal–oxygen bond lengths are increased, lowering the symmetry to tetragonal; also the orbital occupancies in (b) high-spin d^4, (c) d^9, and (d) low-spin d^8 configurations.

smaller with these orbitals, and that any Jahn–Teller effect is generally swamped by other perturbations, including spin–orbit coupling and lattice vibrations. Jahn–Teller distortions involving t_{2g} orbitals may sometimes be *dynamic* in nature. They can have an influence on magnetic properties or in high-resolution spectroscopy,[19,20] but do not seem to be structurally significant. This is another illustration of the much stronger bonding interactions involving the e_g orbitals than is the case with t_{2g}; even structural Jahn–Teller distortions, however, may involve rather small changes in energy, perhaps an order of magnitude smaller than the overall crystal-field effect.

Although octahedral coordination is commonest for transition metal ions, other geometries do arise. We have seen above the effect of lowering the ideal octahedral to tetragonal symmetry. *Trigonal* distortion is also common, arising not as a result of Jahn–Teller-type distortion effects but as the normal site symmetry in some crystal structures, for instance corundum. Such a lowering of symmetry splits the otherwise degenerate sets of t_{2g} and e_g levels, and can be significant in controlling the electronic properties of some oxides;[18] we shall look at examples of this, e.g. in Ti_2O_3, in Chapter 5. *Tetrahedral* coordination is also found with some of the smaller ions such as Fe^{3+} (see Chapter 1, Fig. 1.5). Here the three- and two-fold degeneracy is preserved, but the splitting is reversed, between the lower e and the upper t_2 (the g subscript should no longer be used, as there is no longer a centre of symmetry). Both electrostatic and overlap-based calculations suggest that the CF splitting Δ is only about half that expected for octahedral sites[7,21] (the actual ratio predicted is $\frac{4}{9}$) and this is confirmed by experiment. A consequence is that crystal field stabilization energies for tetrahedral sites not only follow a slightly different pattern, but are also considerably smaller than the octahedral ones given in Table 2.1. Configurations with large octahedral CFSE values, such as d^3 and d^8, are more likely to be found in octahedral sites, and the common tetrahedral ion Fe^{3+} has d^5 and no CFSE whatsoever. These facts may be important in influencing the cation distribution in oxides, such as the spinels discussed in Section 1.3.2 (see Table 1.1), and are also thought to play a role in determining the distribution of transition elements in minerals.[22,23] However, size is also important, and since this is itself influenced by the CFSE, it is hard to separate the two factors.

One consequence of the different splitting patterns found with lower symmetries is that low-spin states may be found with some configurations not shown in Table 2.1. Low-spin ground states never arise with tetrahedral coordination in oxides, as the CF splitting is too small. d^8 ions in tetragonal symmetry, however, can have all electrons paired in the lower four orbitals, if the splitting of the upper levels shown in

Fig. 2.2 is large enough. This seems to be the case with the $4d^8$ compound PdO, which has square planar coordination.[24]

The quantitative power of CF theory is shown most clearly in the interpretation of spectroscopic and magnetic properties of d^n ions. The existence of characteristic crystal field (or '*d–d*') transitions of transition metal ions is well known.[25,26] As explained in Chapter 3, magnetic resonance techniques (mostly ESR) are another important source of information about transition metal impurities in oxides, including for example identification of the oxidation states and electron configurations.[27] Small splittings produced by lattice sites of low symmetry are important in the interpretation of ESR and magnetic properties, and some information about local geometry can be obtained.[28] The analysis of such measurements is highly technical and will not be discussed in detail here (but see Chapter 3 for a further discussion of magnetism). It is, however, worth discussing the interpretation of optical spectra, as these are an important source of quantitative data on CF splittings and other parameters.

The energy of a crystal-field state depends not only on the CF perturbation itself, but also on the various 'internal perturbations' that influence the energy of d^n states.[10–12] At a formal level, a solution to the CF problem entails solving an eigenvalue problem, with the *d* orbitals as a basis, and the Hamiltonian

$$H = H_{\mathrm{CF}} + H_{\mathrm{es}} + H_{\mathrm{so}}. \tag{2.1}$$

H_{CF} represents the CF splitting, its eigenvalues being simply the CFSE values of ground- and excited-state configurations. H_{es} is the electrostatic repulsion between *d* electrons, and H_{so} the spin–orbit interaction. Spin–orbit effects are important in magnetic and ESR studies, but for the 3*d* series their magnitude is considerably smaller than the other terms, and they are often ignored as a first approximation in the interpretation of optical spectra. (For 4*d* and 5*d* ions, spin–orbit coupling is stronger, and can no longer be ignored.) The CF and electrostatic terms are *comparable in magnitude*, and so must be treated on an equal footing. Thus no simple 'one-electron theory', which simply looks at orbital energies and ignores electron repulsion, can be used to interpret CF spectra.

The general treatment of electron-repulsion effects in solids is very hard, and this difficulty will be referred to again in a different context later. In the CF model the problem is tackled by assuming that the electrons are really still in atomic-like *d* orbitals. Thus the repulsion between electrons in different orbitals is treated by the theories of atomic structure.[11,29] For *d* orbitals they can be expressed in terms of three radial integrals F_0, F_2, and F_4, although it is more common to use the **Racah**

parameters denoted *A*, *B*, and **C**, which are linear combinations of the F_n integrals. *A* does not appear explicitly in the CF problem; *B* and *C* are treated as free parameters, to be fitted against spectroscopically observed energy levels. Values for free gas-phase ions can be found from atomic spectra, and it is interesting that such empirically determined parameters do not agree well with ones computed theoretically from radial wavefunctions of *d* orbitals.[30] The reason for this discrepancy has been traced to electron-correlation effects. Electrons can reduce their mutual repulsion by moving so as to avoid each other in ways not taken account of in a single-configuration wavefunction. Correlation thus acts partly to reduce the effective repulsion integrals, although additional corrections are sometimes applied. In transition metal compounds, Racah parameters appear to be further reduced by chemical bonding effects, as discussed below.[31]

The number of independent parameters in the CF problem is sometimes reduced by fixing the ratio *C/B* to a value (around 4.5) typical of all transition metal ions. When spin-orbit coupling is omitted the octahedral CF states then depend only on Δ and *B*. Spectra can be interpreted by using **Tanabe–Sugano diagrams**, where the calculated

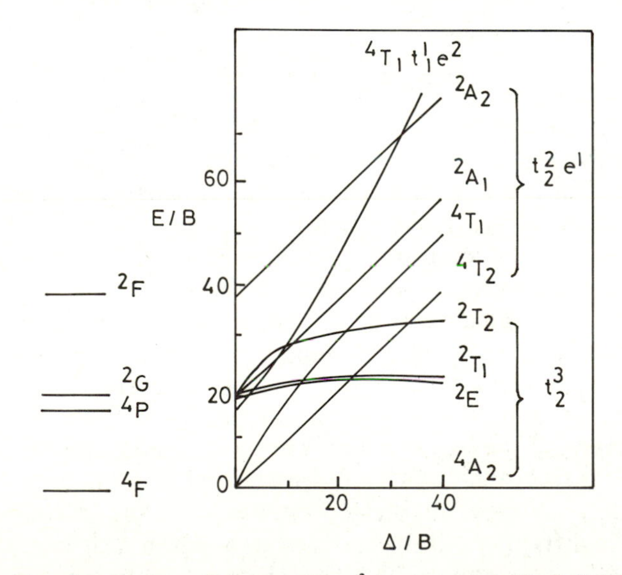

FIG. 2.3 Tanabe–Sugano diagram for d^3 in an octahedral field.[32] The lower energy free-ion terms are shown on the left, and the diagram shows how the states evolve as a function of the CF splitting Δ. All energies are expressed in terms of the Racah *B* parameter; a fixed ratio $C/B = 4.5$ is assumed, and spin-orbit coupling neglected.

state energies, in units of B, are plotted against the ratio Δ/B.[32] An example of such a diagram, for the d^3 configuration in an octahedral field, is shown in Fig. 2.3. The vertical axis gives the free-ion energies in terms of B; the various states such as ^4F and ^4P differ in the arrangement of electrons within the d orbitals, so that the inter-electron repulsions are different. The diagram shows the evolution of the states as a function of Δ/B. Some states may be unambiguously assigned to particular $t_{2g}^n e_g^m$ configurations: for example the ground state $^4A_{2g}$ comes uniquely from splitting the ^4F free-ion state, and corresponds to the configuration t_{2g}^3. In other cases, however, for example $^2T_{2g}$, more than one state of this spin and symmetry arises; not only the energies, but also the wavefunctions of these states are then determined by the competition between CF splitting and electron repulsion.

Figure 2.4 shows the optical absorption spectrum of ruby, a gemstone composed of Al_2O_3 with Cr^{3+} impurities. The $3d^3$ ion replaces Al^{3+} in its approximately octahedral site within the corundum lattice.[25] Although the real symmetry is low, the perturbation from a regular octahedron is small, and the diagram in Fig. 2.3 can be used. Starting from the $^4A_{2g}$ ground state, the **spin selection rule** ($\Delta S = 0$) shows that only excited quartet states will appear strongly in absorption. This allows the main features of the spectrum to be assigned as shown. The much weaker spin-forbidden transitions to ^2E_g and $^2T_{2g}$ states can also be seen; these are important in emission, that from the ^2E_g state being used in

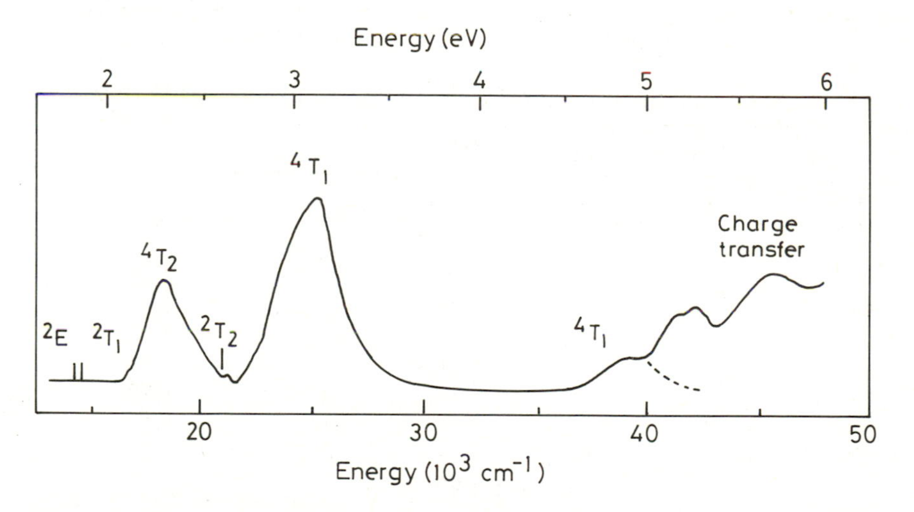

Fig. 2.4 Absorption spectrum of ruby, coming from Cr^{3+} impurities in corundum, Al_2O_3.[25] The bands are assigned to transitions from the $^4A_{2g}$ ground state, according the energy levels of Fig. 2.3.

the ruby laser (see Section 3.3.1). In octahedral or any centrosymmetric coordination, the CF transitions such as those shown in ruby are also forbidden by the **Laporte rule**: as all d^n states show g (*gerade* or even) behaviour under inversion, there should be no electric-dipole matrix elements between them. Although crystal sites are often not strictly centrosymmetric, CF transitions are always weak compared with the allowed charge-transfer and band-gap transitions that will be discussed later. In cases where a centre of symmetry is present, CF transitions become **vibronic** in nature: i.e. they are allowed by virtue of lattice vibrations which locally break the centre of symmetry.[11] In high-resolution spectra measured at very low temperatures, it is also possible to observe very weak magnetic-dipole allowed and electric-quadrupole transitions.[33]

Analysis of CF spectra is the normal way of determining the parameters of the model, although in principle other physical measurements, such as magnetic g values, can be used.[11,12] Typical Δ values for a selection of transition metals ions in oxide lattices are shown in Table 2.2. The values are for octahedral sites, and are compared with CF splittings found for ions in aqueous solution, where they are also octahedrally coordinated, as $M(H_2O)_6^{n+}$. Small variations are found from compound to compound. In particular, there is a strong dependence on the metal–oxygen distance: for example Cr^{3+} has a higher Δ value in ruby, where the average Cr–O distance is 191 pm, than in the pure oxide Cr_2O_3, where it is 199 pm.[34] (This accounts for the difference in colour between the two solids, ruby being red, and Cr_2O_3 green.)

Table 2.2 *Octahedral crystal field splitting parameters in oxides*

	Environment			
Ion	Binary oxide	MgO	Al_2O_3	$M(H_2O)_6^{n+}$
Ti^{3+}	–	–	2.36	2.52
V^{3+}	–	–	2.17	2.19
Cr^{3+}	2.06	2.01	2.25	2.16
Mn^{3+}	–	–	2.41	2.60
Fe^{2+}	1.2	1.31	–	1.29
Fe^{3+}	–	–	2.05	1.70
Co^{2+}	1.1	1.05	–	1.15
Co^{3+}	–	–	2.27	2.31
Ni^{2+}	1.1	1.06	–	1.05
Ni^{3+}	–	–	2.08	–

Values of Δ (in eV) are given for various octahedral environments. (Based on data froms refs 7, 34, 104.)

Certain other trends can be observed. Δ is larger in higher oxidation states. It also shows some, not entirely regular, decrease, along the $3d$ series from Ti to Ni. Although $4d$ and $5d$ elements are harder to study in the same way (most of the d^n oxides are metallic and less amenable to localized treatment) a comparison of molecular compounds and complexes suggests that we should expect considerably larger Δ values than in the $3d$ series.[2] Low-spin states are normally found in the lower rows, partly because CF splittings are larger, but also because electron repulsion effects are smaller for larger orbitals, so that the exchange terms favouring parallel-spin arrangements are less important.

The trends in CF splitting are consistent with the view, explained later in the context of molecular orbital cluster models, that the perturbation comes largely from bonding interactions between metal and ligand atoms, and is not pure electrostatics.[10] In particular, we expect stronger overlap with $4d$ and $5d$ orbitals, because they are not so contracted and 'core-like' as $3d$. In the $3d$ series itself, orbital overlap is smaller with the later elements, as the orbitals become contracted with increasing nuclear charge.[2] The same factors have other important consequences for the properties of oxides: for example, in the tendency to metal–metal bonding with some elements (see Section 1.3.1), and the trends in d band widths which influence the degree of metallic character (Section 5.2.4).

The other parameters emerging from fits to CF spectra are the electrostatic repulsion parameters B and C. As noted above, values for these are also available from the free ions, and one important observation is that B and C are invariably reduced in compounds, relative to their free-ion values. This has been called the **nephelauxetic effect**[31] (from the Greek, meaning 'cloud-expanding'), suggesting that it results from an apparent expansion of the d orbitals in compounds, leading to reduced repulsion between electrons. Such an expansion could be a natural result of lowering the effective charge on an ion through partial covalent bonding; but in the molecular orbital approach it is more natural to think rather of a partial *delocalization* of the 'd' electrons over the surrounding atoms. Typical values of B and C in oxides are in the region of 80 per cent of the free-ion values.[11,12]

2.1.2 Charge transfer and band gaps

Crystal field excitations are simple in the sense that they involve only rearrangement of electrons within orbitals on a single atom. They are therefore sensitive only to *local* bonding interactions. The contrast appears when we look at excitations of electrons between atoms, i.e. **charge transfer**. They are important not only because they give rise to strong spectroscopic transitions, but also because the band gaps in many

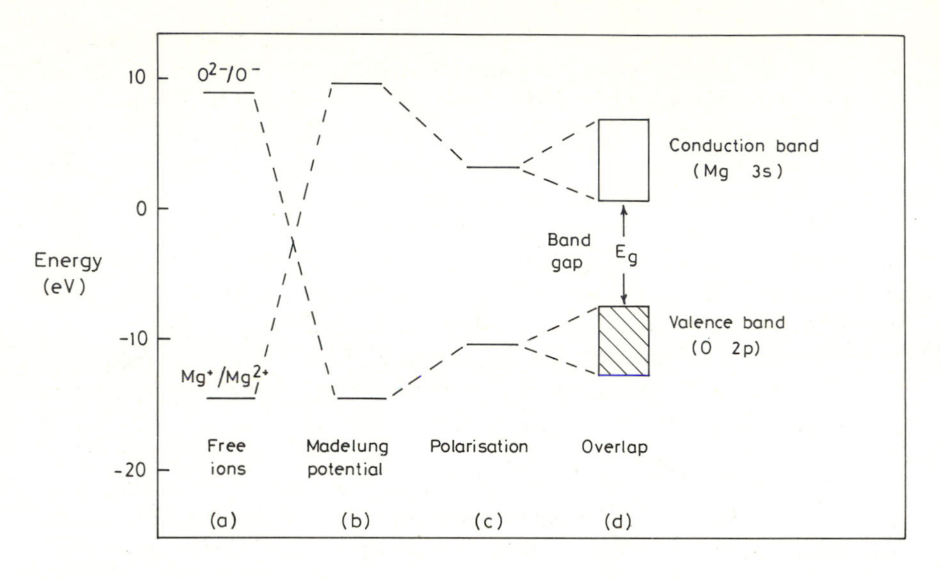

FIG. 2.5 Ionic-model interpretation of the band gap in MgO. The orbital energies of the filled oxygen $2p$ and the empty magnesium $3s$ levels are shown; the steps are (a) the free-ion energies; (b) ions in the Madelung potential of the lattice; (c) the effect of polarization; and (d) the band width coming from orbital overlap.

non-metallic oxides appear to be of this type. The ionic model can also provide a useful framework for interpreting charge transfer excitations, although quantitative predictions are much less certain.[1,4]

Consider first of all a simple non-transition-metal oxide such as MgO. The top-filled levels in the ionic picture are oxygen $2p$ orbitals, and the band-gap excitation, which experimentally is at 7.5 eV, should be from these orbitals into the empty magnesium $3s$. Figure 2.5 shows the steps required to estimate this band gap using ionic model arguments.[1] For isolated atoms in the gas phase, the ionic charge distribution Mg^{2+}–O^{2-} is not stable, and indeed the O^{2-} ion does not exist, as the reaction

$$O^- + e^- = O^{2-} \tag{2.2}$$

is highly *endothermic*. To stabilize the ions it is necessary to put them in a lattice, so that they experience an electrostatic **Madelung potential** of appropriate sign. This depends on summation of long-range Coulomb potentials from the ions in the lattice, and is expressed in terms of the same Madelung constant that arises in the estimations of ionic lattice energies. For example, the potential at the oxide site in MgO coming from 'point' ions Mg^{2+} and O^{2-} arranged in the rock-salt lattice is

$$V_M = +2e[6 - 12/\sqrt{2} + 8/\sqrt{3} - \ldots]/(4\pi\varepsilon_0 R). \tag{2.3}$$

The series in square brackets is the Madelung constant, and R is the interatomic Mg–O distance; e is the electronic charge and ε_0 is the dielectric permittivity of free space. At oxide sites, the Madelung potential is positive, and thus stabilizes electron energies as shown in Fig. 2.5; at cation sites it is negative and correspondingly destabilizing.

Column (b) in the diagram shows what might be called the 'simple ionic' estimate of the band gap. It is at least of the correct sign but is much too large numerically, as various effects have still been neglected. The first is the important **polarization** that occurs whenever charges are moved around in a lattice. Polarization energies may be estimated at different levels of sophistication. Simple electrostatics shows that a charge q, spread out over a sphere of radius R, is stabilized when placed in a continuum with relative dielectric constant ε_r, by an amount[35]

$$\Delta E_{\text{pol}} = \frac{-q^2}{8\pi\varepsilon_0 R}(1 - 1/\varepsilon_r). \tag{2.4}$$

A typical value calculated for one unit of electronic charge in an oxide is 2 eV, although this must be regarded as very approximate. A better method of calculating such energies is to represent the nearby ions explicitly as polarizable entities, and use the continuum electrostatic approximation only for more distant parts of the lattice.[4,36] More recently this idea has been elaborated into the **shell model** where each ion is treated as a light 'shell' of electrons connected to the heavy core by an imaginary 'spring'. The charges on the shell and the core, the strength of the 'spring', and various other parameters which model the short-range interactions between ions, are fitted against experimental quantities such as dielectric and elastic constants and vibrational data.[37] The resulting calculations are widely used for predictions of defect structure and properties of solids, and we shall refer to them later in this context. They may also be used to provide more elaborate estimates of polarization energies appropriate for band-gap calculations.

In Fig. 2.5 the effect of polarization has been shown by lowering the energy of empty (metal) levels and *raising* that of filled (oxygen) levels; this is consistent with the idea that *binding* energies for electrons in filled levels will be reduced.

The final effect shown in Fig. 2.5 is the broadening of electronic levels due to overlap between ions, giving bands, rather than sharp atomic levels. It is impossible to estimate such broadening from pure ionic model arguments; instead one must use values either derived from experiment or from other theories such as band-structure calculations. Photoelectron and X-ray spectroscopy results show typical oxygen $2p$

band widths of about 5 eV, and the conduction band formed from overlap of the quite diffuse Mg 3*s* orbitals is probably somewhat broader.[38,39] Incorporating reasonable values in Fig. 2.5 gives the experimental band gap.

Even in the simplest example like MgO the band gap can be seen to depend on many factors, some of which can only be estimated by semi-empirical arguments. This may seem a reason to doubt the utility of the model; it is worth pointing out, however, that a reliable estimate of the band gap of ionic compounds is hard to obtain from *any* theoretical model. The polarization energy is essentially a manifestation of electron correlation, and cannot easily be included in any *ab initio* calculation.[40]

The same procedure could in principle be used for d^0 transition metal compounds, for example to estimate the band gap in TiO_2 as the charge transfer energy from O^{2-} to Ti^{4+}. However, because of the higher ionic charges, all the terms in the calculation become correspondingly larger, and one could only hope to obtain the final band gap (3.0 eV) as a difference between terms of 50 eV or more. The outcome of the calculation would thus depend critically on values calculated, for example, for the polarization energy. With highly charged ions one might not in any case expect the ionic model to be reliable, and it is best regarded in these circumstances more as a qualitative framework than as a theory capable of providing quantitative estimates.

It is revealing, however, to apply the same type of calculation to a magnetic insulator. Figure 2.6 shows the result for NiO.[6] The crucial difference is that with magnesium the only energetically accessible electronic transition is

$$Mg^{2+} + e^- = Mg^+. \tag{2.5}$$

No others were plotted on Fig. 2.5 as they would have been off scale. However, with nickel the third ionization energy is much lower, and the process

$$Ni^{2+} = Ni^{3+} + e^- \tag{2.6}$$

must also be considered. Once again, the Madelung and polarization energies must be estimated, and also another effect specific to *d* orbitals: that is the crystal field terms which may give different CFSE values for the states involved. The band widths relate to states derived from 3*d* orbitals, and all the evidence for magnetic insulators (see Section 3.4) suggests that these widths are small, perhaps 1 eV. The final result shown in Fig. 2.6 agrees reasonably with the experimental band gap for NiO (3.8 eV) and suggests that the transition involved is quite different from that in MgO. In many magnetic insulators, ionic model calculations give a band gap deriving from the difference in successive ionization energies

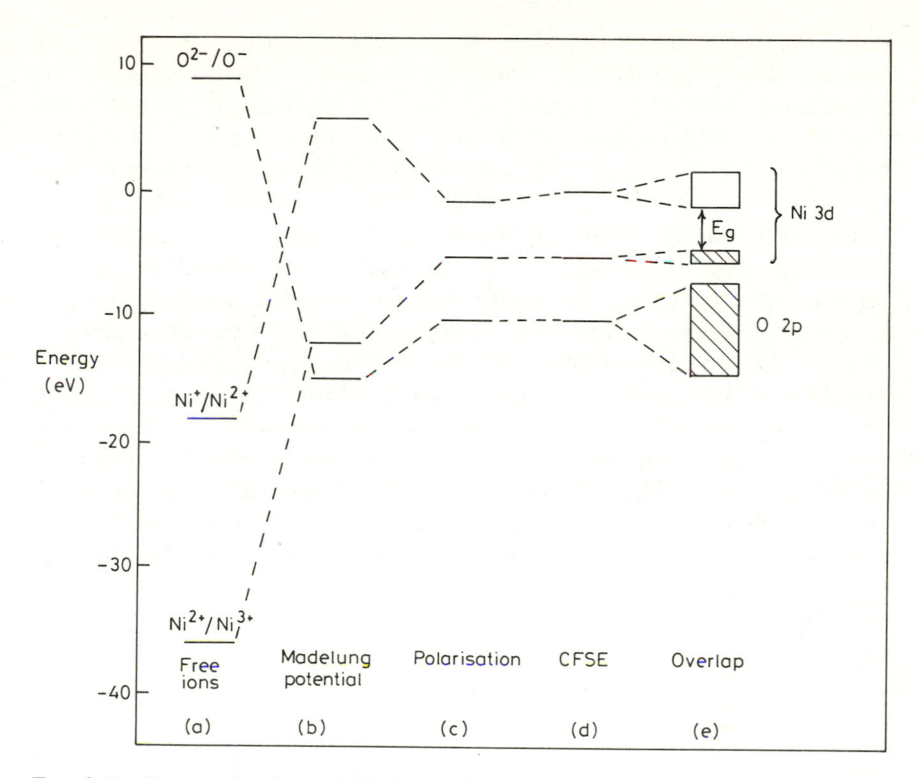

FIG. 2.6 Energy levels of NiO in the ionic model. Two different transitions involving the d levels of Ni^{2+} (corresponding to eqns (2.6) and (2.7) in the text) are shown. The steps in the calculation are the same as in Fig. 2.5, except that (d) is the change in crystal field stabilization of the different d configurations, and (e) the band width. (Based on ref. 6.)

of the *same* atom. Thus the energy of the 'top-filled level' corresponds to the ionization process (2.6): and the 'lowest empty level' to the transition

$$Ni^{2+} + e^- = Ni^+. \qquad (2.7)$$

For NiO itself there is some doubt as to the accuracy of this scheme, and many experiments suggest that the energy of the oxygen $2p$ levels may be higher than shown in Fig. 2.6, and so overlap the nickel $3d$ ionization energy (see Sections 2.2.2 and 3.4.2). Even if this is correct, however, the 'splitting' of the metal d levels, corresponding to the different energies of the two electronic transitions (2.6) and (2.7), is a crucial feature of the electronic structure in all magnetic insulators. The energy of the band-gap transition shown is essentially an inter-electron repulsion term. In the gas phase this is responsible for the difference between successive ionization energies, which is 19 eV for Ni^{2+}. In NiO the value is reduced to around 5 eV by solid-state polarization effects. Such a 'screened repulsion' term is an essential ingredient of the **Hubbard model** of magnetic insulators, discussed in detail in Sections 2.4.1 and 3.4.2.

The ionic model has been widely applied in recent years to the theoretical modelling of defects in oxides and other solids.[41] It is essential in these calculations to estimate the polarization terms which occur—as in charge transfer transitions—as soon as new charges are created in a lattice. The energies of point defects, and the different kinds of defect associations discussed in Chapter 1, all depend critically on relaxation terms; if it were not for these, defect energies would be very large, and non-stoichiometry in oxides much rarer than it is. The same type of modelling can also be applied to the calculation of electronic transitions involving defects. For example, Fig. 2.7 shows the result of calculations on charge-transfer processes for transition metal impurities in alumina.[42] Each level represents the energy associated with a transition, such as

$$Ti^{3+} = Ti^{4+} + e^- \tag{2.8}$$

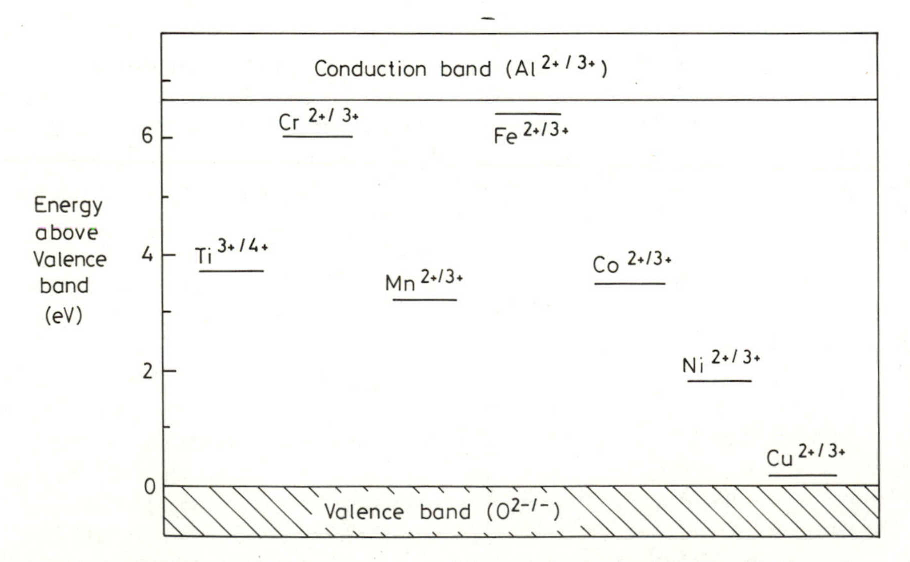

FIG. 2.7 Defect levels for some transition metal impurities in Al_2O_3, calculated using the ionic (shell) model.[42] Each level represents the energy of a defect ionization process, relative to the valence and conduction bands of the host.

relative to the valence and conduction band energies of the Al_2O_3 host. The trend in defect energies shown is characteristic of the transition elements, and reflects the increase in ionization energies with increasing atomic number along each series. For ionizations involving high spin d^n configurations this increase is interrupted after d^5; thus the ionization energy drops between Mn^{2+} and Fe^{2+}, as reflected by the defect levels in the diagram. This drop is a consequence of the loss in exchange energy resulting from pairing electron spins beyond the d^5 point.

In Section 3.3.2 the energies of defect levels associated with transition metal and other impurities will be discussed in more detail, and we shall see that there are still many experimental and theoretical uncertainties in this area. An especially difficult problem for the theoretical modelling is to know the true values of the band energies on an absolute scale, or in other words, the electron affinity of the solid relative to the vacuum.[43] As a result of this uncertainty it may be possible to calculate the relative energies of the different defect levels, without knowing exactly how they should be placed within the band gap of the host crystal.

There is another feature of the ionic model calculations which is very important. The polarization terms which make a significant contribution to the energies can be divided into two parts: that which comes from purely electronic motion, and that which involves the motion of atoms as a whole. Because of the difference in masses, the electronic contribution is rapid, and it is this that determines the high-frequency, or optical, dielectric constant of a solid. Atomic relaxations are much slower, and contribute only to the static dielectric constant.[13] For equilibrium processes, both types of relaxation are involved, but in spectroscopic transitions between electronic levels, the *vertical* excitation energies given by the Franck–Condon principle depend only on the rapid electronic component. Parametrizations of the ionic model allow a separate calculation of the two contributions, so that different energies can be calculated for fully relaxed equilibrium processes on the one hand, and spectroscopic transitions on the other. The results in Fig. 2.7 are of the former kind, and were aimed at predicting the thermodynamic stability of various transition metal ions in alumina. They are analogous, therefore, to the *redox* potentials used in electrochemistry,[44] and should not be used to predict the energies of spectroscopic transitions.

2.2 Cluster models

The next step beyond the ionic model is to include the overlap of orbitals within a limited number of atoms. The simplest cluster models treat a single metal atom interacting with its surrounding oxygens, although larger clusters can be considered. Often calculated within a molecular orbital (MO) framework, such models still give a local view of electronic structure, and are more applicable to non-metallic compounds, especially in connection with properties where band-structure effects are unimportant.

2.2.1 *The molecular orbital (MO) method*

Molecular orbital theory, like orbital theories of atoms and the band theory of solids, is an *independent electron* model, treating the behaviour of electrons moving in a potential field in which inter-electron repulsion is included in some average way.[45] The most common approach to the MO method is via the **linear combination of atomic orbital** (LCAO) approximation, which is analogous to the tight-binding method of solid-state physics.[46–48] Using an appropriate basis of valence atomic orbitals $\{\phi\}$, each molecular orbital is represented in the form

$$\Psi = \sum_i c_i \phi_i \tag{2.9}$$

where the coefficients $\{c\}$ and the **orbital energy** ε are found by a solution of the **secular equations**

$$\mathbf{H}c = \varepsilon \mathbf{S}c. \tag{2.10}$$

Here $\mathbf{H}$ and $\mathbf{S}$ are the Hamiltonian and overlap matrices expressed in the atomic orbital (AO) basis. With N basis AOs, eqn (2.10) has N independent solutions, thus giving N MOs; these may not all have different energies, however, as degeneracies can arise when the cluster has high symmetry.

In partially ionic systems, the MOs may be viewed most simply as perturbations of atomic orbitals, resulting from their overlap interaction. Figure 2.8 shows a qualitative MO energy scheme for an octahedral transition metal oxide cluster, that is an MO_6^{n-} unit. On the left- and right-hand sides respectively are represented the atomic orbital energies of the metal and oxygen atoms; in the centre are the resulting MO energies. A qualitative scheme of this kind is greatly aided by symmetry considerations.[8,46] One first constructs the different possible *symmetry adapted linear combinations* of oxygen orbitals. Combinations with the appropriate symmetry behaviour are then able to mix with metal orbitals. Of particular interest are the e_g and t_{2g} combinations of oxygen $2p$ illustrated in Fig. 2.9, which can combine with the corresponding metal d orbitals. The e_g set is composed of so-called **σ type** orbitals with lobes pointing along the metal–oxygen bond direction, whereas the t_{2g} set is **π type**, pointing in a perpendicular direction. Mixing between these

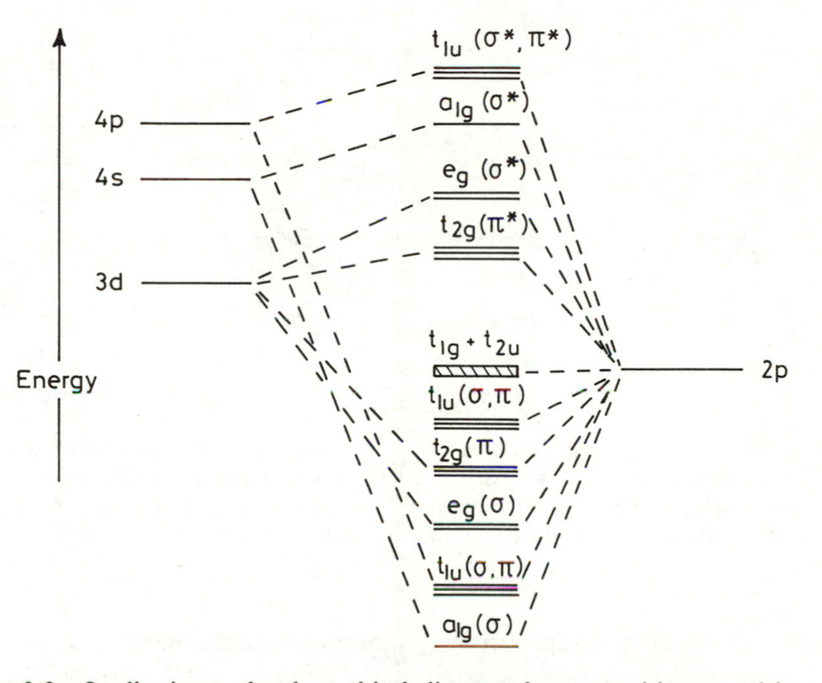

FIG. 2.8 Qualitative molecular orbital diagram for a transition metal in an octahedron of six oxygens. The atomic orbital energies of the metal and the oxygens are shown on the left and right respectively; in the centre are the energies of molecular orbitals resulting from their mixing. Symmetry labels corresponding to irreducible representations of the O_h point group are used.

oxygen $2p$ and the metal $3d$ orbitals produces bonding MOs, localized more on oxygen, and antibonding MOs with more metal character. For example,

$$\Psi_{\text{bonding}} = \alpha\phi(\text{M } 3d{:}e_g) + \beta\phi(\text{O } 2p{:}e_g) \tag{2.11}$$

and

$$\Psi_{\text{antibonding}} = \gamma\phi(\text{M } 3d{:}e_g) - \delta\phi(\text{O } 2p{:}e_g) \tag{2.12}$$

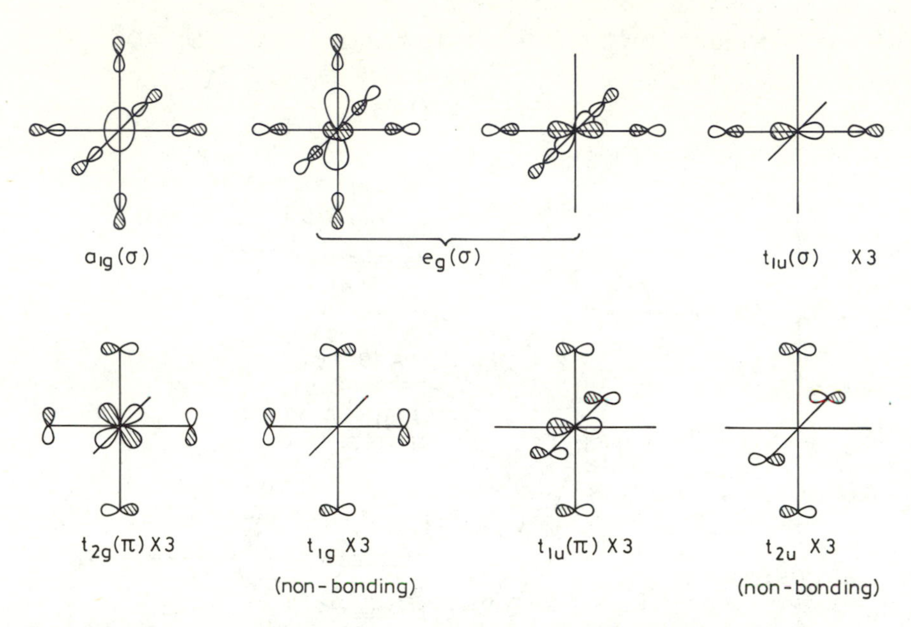

FIG. 2.9 Symmetry-adapted linear combinations of oxygen 2*p* orbitals arising in an octahedron, shown with a bonding combination of an appropriate metal orbital. The type of bonding (σ or π) is indicated. In the case of the triply degenerate *t* orbitals only one of each is drawn.

where the ϕs are the appropriate symmetry adapted atomic orbitals, and α, β, γ, and δ are positive coefficients. Similar combinations can be made, as shown in Fig. 2.9, to represent the interaction of the t_{2g} set of *d* orbitals and the higher-energy valence *s* and *p* with appropriate oxygen 2*p* combinations.

Although Fig. 2.8 shows a large number of bonding and non-bonding MOs derived from the oxygen 2*p* levels, the total number of these is just the number of original atomic orbitals. An electron count appropriate to a d^0 compound is therefore just enough to fill these MOs. Any additional electrons go into the antibonding orbitals, and specifically into the t_{2g} and e_g derived largely from the metal *d*. Thus the ionic model and *d* electron configuration arguments derive a rather more sophisticated justification from the MO picture. The 'oxygen 2*p*' orbitals are essentially metal–oxygen bonding, and the 'metal *d*' electrons somewhat antibonding in nature. The degree of mixing between the AOs depends inversely on the energy separation between them, but also directly on the amount of overlap. The valence *s* and *p* orbitals of the transition elements are much more spatially extended than the *d*, and overlap much more with their surroundings. Thus in spite of the greater energy gap to the oxygen orbitals, more covalent mixing takes place. This is especially true in the first transition series, where the 3*d* are highly contracted, and in the later elements almost core-like in spatial distribution. Thus mixing between 3*d* and oxygen 2*p* can be quite small, which partly explains the success of the crystal field model, in spite of the strong covalent mixing that can arise with the 4*s* and 4*p* orbitals.

One important feature of this model is the interpretation it gives to the crystal field splitting. As shown in Fig. 2.8 the perturbation resulting from orbital mixing is larger for the e_g orbitals than for t_{2g}. This is because overlaps of σ type are stronger than those of π type.[46,47] Of the 'metal *d*' orbitals, therefore, the e_g are more strongly antibonding than t_{2g}. The greater overlap interaction of the e_g orbitals gives rise to the octahedral CF splitting Δ, and also means that electrons in these orbitals have more influence on the metal–oxygen bonding. Two consequences, in the bond lengths trends for high-spin ions, and in the Jahn–Teller distortions caused by uneven occupancy of the e_g shell, were pointed out in Section 2.1.1.

The energy level scheme in Fig. 2.8 has been supported by a wide variety of approximate calculations, largely of the semi-empirical variety in which many of the necessary quantities are derived from experiment.[49] An example of this type of theory is the angular overlap model, which is useful for calculating CF splitting patterns in low-symmetry environments.[21,50] With modern computational facilities, however, highly sophisticated *ab initio* calculations are also possible.[51,52] Two examples are discussed below, to show the kind of information that can be provided by different techniques of calculation.

Figure 2.10 shows a small cluster taken from the structure of lithium niobate, $LiNbO_3$, and the orbital energies calculated for a possible defect configuration, in which lithium is replaced by iron.[53] It is known from experiment that iron impurities have an important effect on the optical properties of lithium niobate,[54] and the calculation was part of an attempt to calculate the electronic transitions of iron in different oxidation states, and at different substitutional positions in the $LiNbO_3$ lattice. This calculation employs the **Xα method**, which is very similar to a technique widely used for band-structure calculations, described below.[55] One important feature is that the electron repulsion terms are approximated; in particular the exchange part of the electron–electron potential is replaced by a local function of the electron density.[56] This makes the calculation easier than the alternative Hartree–Fock method; but a disadvantage of the local-density formalism is that the electron repulsion terms which determine the energies of different crystal field

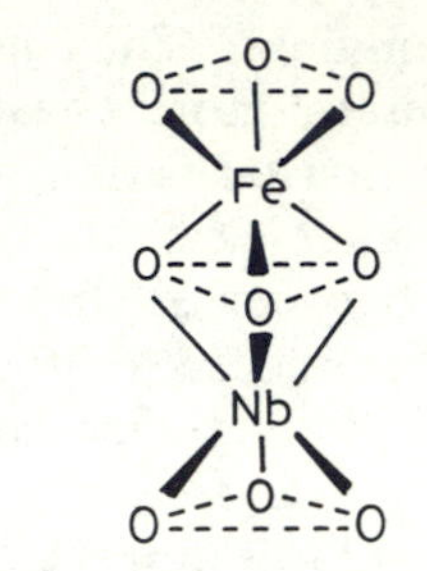

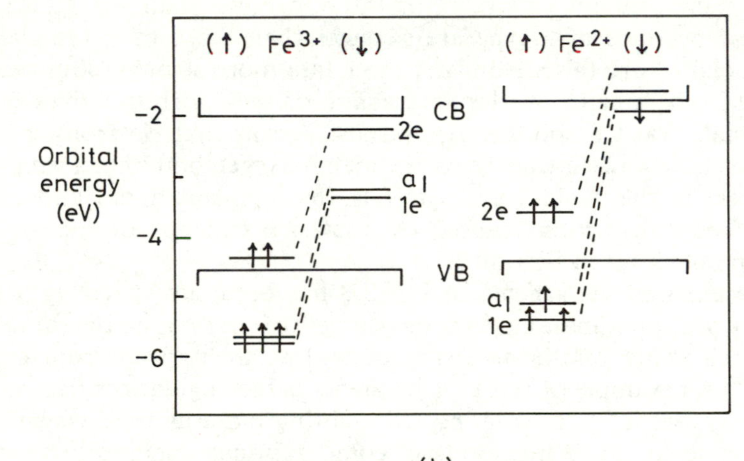

FIG. 2.10 Energy levels calculated by the Xα method for a cluster based on the $LiNbO_3$ structure, with Fe replacing Li.[53] The cluster geometry is shown in (a) and (b) shows the orbital energies for Fe^{2+} and Fe^{3+}. Only the Fe 3*d*-derived levels are shown explicitly; VB and CB denote calculated band edges. Note that different orbital energies are given for spin-up and spin-down electrons.

states are not correctly calculated. In a high-spin system, this difficulty is partly overcome by calculating *different orbitals for different spins*: that is, different exchange potentials are used for spin-up and spin-down electrons. Figure 2.10 shows that as a consequence, the orbital energies for the majority (up) spin electrons are lower than for minority (down) spin. The CF splittings (which also show the effect of the trigonal symmetry, splitting the octahedral t_{2g} levels into two components) are calculated to be smaller than the spin-up/spin-down exchange splitting, so that the ground states of Fe^{2+} and Fe^{3+} are correctly estimated to be high spin.

The calculated orbital energies shown in Fig. 2.10 are different for the two oxidation states Fe^{2+} and Fe^{3+}. This is a feature of all fully theoretical calculations, and highlights one limitation of qualitative schemes where the same orbital energies are generally assumed irrespective of the orbital occupancy. In any accurate calculations it is never correct to estimate transition energies simply by taking orbital-energy differences. In the $X\alpha$ method the corrections to this 'naïve' estimate are generally rather small, however,[56] and it is a reasonable approximation to use the orbital energies to look at possible electronic excitations. Only the Fe 3*d* levels are shown explicitly in Fig 2.10, but the top of the valence band (VB) and the bottom of the conduction band (CB) are also indicated. The experimental band gap of $LiNbO_3$ is 3.8 eV, and therefore slightly underestimated in the calculation. For the Fe^{3+} impurity, a transition at about 1.5 eV is predicted for a down-spin electron going from VB (the top of the largely oxygen 2*p*-derived valence band) into the iron 3*d* level 1*e*. In the case of the Fe^{2+} impurity the lowest-energy transition (0.2 eV) comes from a down-spin transition from iron 2*e* orbital to CB, the conduction-band minimum. These defect transitions are of the charge-transfer type referred to in Section 2.1.2 and discussed in more detail in Section 3.3.2. There are considerable experimental problems in the assignment of defect transitions in oxides, and it is hard to check the predictions of this type of calculation.

For a contrasting calculation, we may cite one performed on a cluster of type $(NiO_6)^{n-}$, intended to give a local representation of the energy levels in NiO.[52] The Hartree–Fock method employed here represents the exchange terms correctly, without a local-density approximation.[51] Unfortunately, in HF calculations the naïve estimation of excitation energies by orbital energy differences is much worse than in local-density theory. Part of the reason has to do with the different treatment of empty orbitals in the two methods. In local-density calculations both filled and empty levels are determined by the *same* potential.[56] However, the calculation of exchange terms in HF theory has the consequence that in an N-electron system, repulsions from only $N-1$ electrons contribute to filled orbital energies, but all N electrons contribute to empty orbitals. This is physically reasonable in the calculation of a true band gap, but even in band theory the uncorrected HF method seriously *overestimates* excitation energies whereas local-density theory generally *underestimates* them.[40] In the case of a cluster calculation, however, (as in the exciton problem discussed in Section 2.4.2) there is a *missing* electron in the excited state, the hole left behind in a previously filled level. The first correction necessary is to subtract the spurious repulsion from this hole. Unfortunately, for an accurate calculation of excited states, it is generally also necessary to supplement the HF method with a **configuration interaction** calculation, in which configurations with different orbital occupancies are mixed together. This overcomes some of the deficiencies of the single-configuration approach, and recovers part of the **correlation energy** neglected in the HF method.[57,58]

We can see from the above discussion that the penalty for using more accurate methods is that the simplicity of qualitative pictures disappears. Both localized CF transitions and band gaps must be calculated as the energy difference between different states, and the orbital energies themselves lose any direct experimental significance. Such calculations can, however, be very valuable if properly carried out and interpreted. In the case of NiO the cluster calculation seems to confirm many of the features of the ionic-model picture of Fig. 2.6.[52] Thus the band gap—estimated by looking at the energy difference between states with one electron removed and one added to the cluster—is found to depend crucially on the repulsion between electrons in the 3*d* orbitals. Ionized states with electrons removed from either the nickel 3*d* orbitals or the oxygen 2*p* are predicted to be very similar in energy, although the 3*d* ionization is slightly lower in energy as in the ionic calculation. The effect of polarization is also significant. In all accurate cluster calculations the potential from the remainder of the lattice must be included. Often this is done using a set of point ions, without any polarizability. Such a method gives a band gap of 10 eV in the NiO cluster calculation, but a value much nearer the correct one (4.6 eV against the experimental 3.8 eV) is obtained by including the polarization of the surrounding crystal in a way similar to that discussed above. As we have mentioned already, some features of the electronic levels in nickel oxide are still controversial, and in many ways this compound represents a 'test case' for different models.

2.2.2 Configuration interaction models

Configuration interaction (CI) is often used as a way of improving some of the deficiencies of the Hartree–Fock method, by taking mixtures of different MO configurations, chosen to give a better representation of the wavefunction than a single orbital configuration. An alternative approach is to bypass the calculation of molecular orbitals altogether, and to construct a wavefunction by mixing different configurations of *atomic orbitals*.[59] Earlier attempts to use this approach in a non-empirical way for transition metal compounds do not seem to have been followed up.[60] The idea has been revived, however, as an *empirical* scheme, with the quantities required treated as parameters to be fitted against experiment.[61-63] To illustrate the method, we once again refer to nickel oxide (see Fig. 2.11).

Using the ionic model as a first approximation, the wavefunction can be described as $\Psi(d^8)$; the filled shells of oxygen $2p$ orbitals are understood to be included implicitly here. Interaction between nickel $3d$ orbitals and the $2p$ of surrounding oxygens can then be included by mixing in different configurations, such as $\Psi(d^9\underline{L})$ and $\Psi(d^{10}\underline{L}^2)$, where $\underline{L}$ is used to represent a hole in the ligand (oxygen) orbitals. Thus overlap interactions between metal and oxygen are included by making linear combinations of these different many-electron *states*, rather than *orbitals* as in MO theory. The effect of this mixing is represented in Fig. 2.11 (a). The ground state, for example, becomes

$$\Psi_0 = \alpha\Psi(d^8) + \beta\Psi(d^9\underline{L}) + \gamma\Psi(d^{10}\underline{L}) \tag{2.13}$$

and different excited states can also be constructed. The energies and mixing coefficients for these states are obtained from secular equations like (2.10), and three kinds of parameters are required, as in semi-empirical MO calculations [49]: the unperturbed energies H_{nn} of each configuration, the off-diagonal Hamiltonian elements H_{mn}, and the (many-electron) overlap integrals S_{mn} between configurations. In the semi-empirical method the overlaps are neglected; the off-diagonal H_{nm} terms can be related to matrix elements involving atomic orbitals, and estimated from band-structure calculations. The diagonal energies could, in principle, be obtained from the ionic model, but are more often treated as adjustable parameters, to be fitted against experiment.

Figure 2.11 (b) shows the estimated energies for CF transitions, compared with a conventional CF calculation and the optical spectrum of NiO. The crystal field splitting arises in the configuration interaction (CI) model in the same kind of way as in MO theory: that is because the metal e_g and t_{2g} orbitals interact differently with the ligands. d^8 states with two vacant e_g orbitals are depressed in energy more than those with only one or no such vacancies, because stronger mixing takes place with the excited charge-transfer configurations. Simply as a model for calculating CF transitions, there is rather little advantage over conventional theory, although charge-transfer excited states are also calculated. The real power of the method becomes apparent, however, when other types of excitation are considered, for example the ionized states observed in photoelectron spectroscopy.[63] This is done by calculating configurations with different numbers of electrons: for ionized states in NiO, the most important low-energy configurations are d^7 ('nickel $3d$ ionization') and $d^8\underline{L}$ ('oxygen $2p$ ionization'). Figure 2.11 (c) shows a fit to the observed photoelectron spectrum. It is interesting that this model is able to account nicely for the peak at a binding energy of 8 eV which

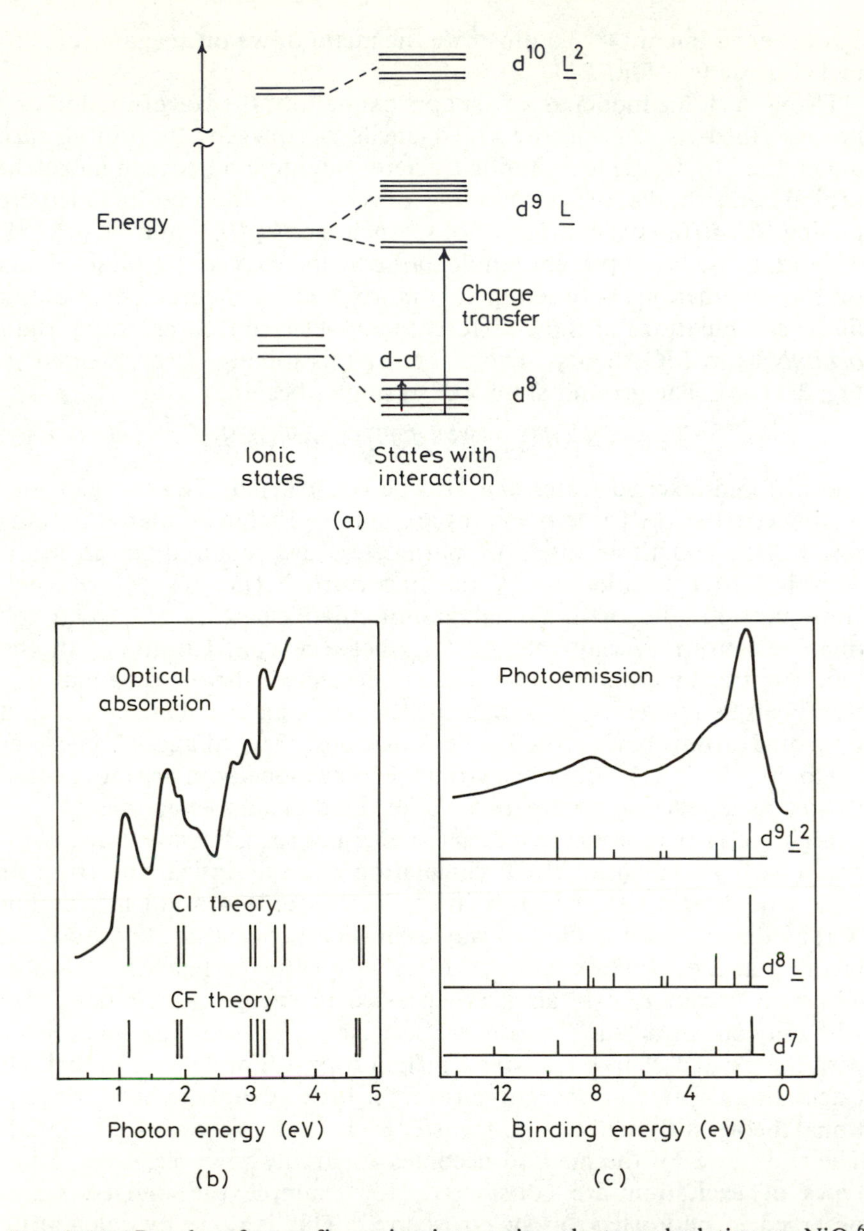

FIG. 2.11 Results of a configuration interaction cluster calculation on NiO.[62] (a) The 'ionic' basis states, and the result of mixing between them; $\underline{L}$ represents a 'ligand' (oxygen) hole. (b) Calculated excitation energies, compared with a conventional crystal field fit and with the experimental absorption spectrum. (c) The ionized states, expressed in terms of contributions from different ionic configurations, compared with a measured photoelectron spectrum.

has no simple interpretation in one-electron models: that is to say it does not correspond to the energy of any occupied orbital in NiO, but is a many-electron 'satellite' peak. The figure shows the breakdown of the ionized states in terms of contributions from the different configurations. The surprising prediction is that the lowest-energy ionized state is mostly $d^8\underline{L}$, in character: i.e. it corresponds to removal of an electron predominantly from an oxygen rather than a nickel orbital. This result is contrary to the predictions of both the ionic and molecular orbital calculations discussed above, and to most previous interpretations of the electronic structure of NiO. If correct it has important implications generally to the oxides of later $3d$ series elements such as copper, as discussed later (Sections 3.4.2 and 5.4.2).

In spite of some superficial similarities between the MO and CI cluster methods, the underlying philosophy is profoundly different. By concentrating on directly observable many-electron states, rather than on theoretical orbitals, the CI approach is more closely adapted to the interpretation of spectroscopic results. At the same time, some of the electron correlation effects neglected in orbital wavefunctions can, in principle, be included. For example, by using empirical values for the unperturbed configuration energies, the intra-atomic correlations and the longer-range polarization terms that are difficult to treat in MO methods are implicitly included. Correlation within the metal–oxygen bond is also better taken account of, for a CI wavefunction such as that in eqn (2.13) is *not* the same as a single-configuration wavefunction with covalent mixing included in the molecular orbitals. This difference may be particularly important in transition metal compounds, especially of the $3d$ series, where the very contracted metal orbitals lead to large intra-atomic repulsion terms, and thus rather important correlation corrections to MO wavefunctions.[64] At the same time, however, the empirical CI model described above is too crude to give a serious cluster wavefunction. The simple parametrization and the neglect of overlap lead to a very tractable model by neglecting many terms that should be included in an accurate calculation. In particular, covalent bonding with metal *s* and *p* orbitals which is important in the overall bonding in transition metal compounds, is totally neglected. Presumably these effects can be regarded as 'hidden' in the parametrization, in the same way that various results of covalency are probably hidden in the parametrization of the ionic model. Although models such as these seem to be valuable for the interpretation of experimental results, the parameters which emerge from the empirical fits should not be regarded as 'absolute quantities', and care should be taken in applying them outside the context of a particular model.

2.3 Band theory

The band model is the theory of electrons moving in a periodic lattice.[1,13] As with the molecular orbital approach, our discussion here will pass over many of the theoretical foundations of the model, which can be found in other books.[1,13,48] We shall concentrate rather on its particular application to transition metal oxides. Before describing the kind of output that can be expected from a band calculation, we shall look at how the band structure of an oxide can be interpreted in terms of chemical bonding, and in particular the relationship between band theory and the more chemically based models described earlier in this chapter.

2.3.1 Band structure and chemical bonding

The traditional approach to band theory starts with a basis of free-electron states, and looks at how these are perturbed by the periodic potential of a crystal lattice.[13] Such a method works well for 'simple' metallic elements such as alkalis, and can even be applied to a wide range of compounds containing non-transition elements.[65] It is not applicable to transition metals or their compounds, however, because the *d* states cannot be represented satisfactorily by a free-electron basis.[48] The simplest approach is based on the same linear combination of atomic orbital (LCAO) idea that is used for molecular orbitals:[1] in the context of band calculations this is known as the **tight binding** method.

To illustrate the application of the method to band theory, it is simplest to imagine a one-dimensional model which might represent part of the structure of a metal oxide. Consider the linear chain of alternating metal and oxygen atoms shown in Fig. 2.12. As in MO theory we form orbitals for the electrons in the chain by making linear combinations of the AO basis functions. This problem is soluble only because of the periodic nature of the chain, which imposes a symmetry constraint requiring that the crystal orbitals must be of the form

$$\Psi_k = \sum_n \exp(ikna)\,\phi_n. \tag{2.14}$$

In this expression ϕ_n is an atomic orbital located in unit cell number n in the lattice, and a is the spacing between unit cells. k is a quantum number for crystal orbitals, known as the **wavevector,** as the orbitals given by eqn (2.14) have a wave-like form, with wavelength equal to

$$\lambda = 2\pi/k. \tag{2.15}$$

Equation (2.14) is effectively a statement of **Bloch's theorem** for states

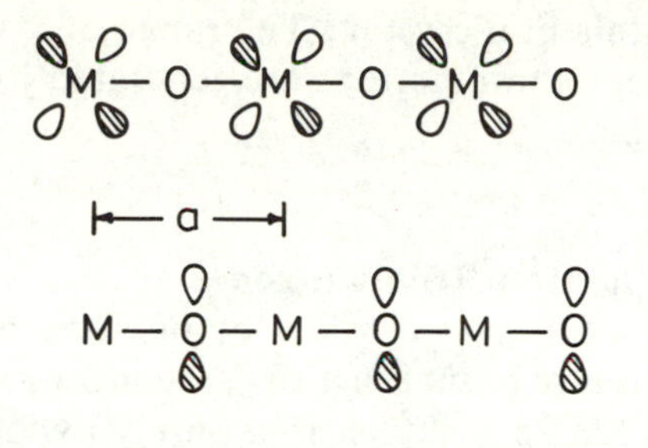

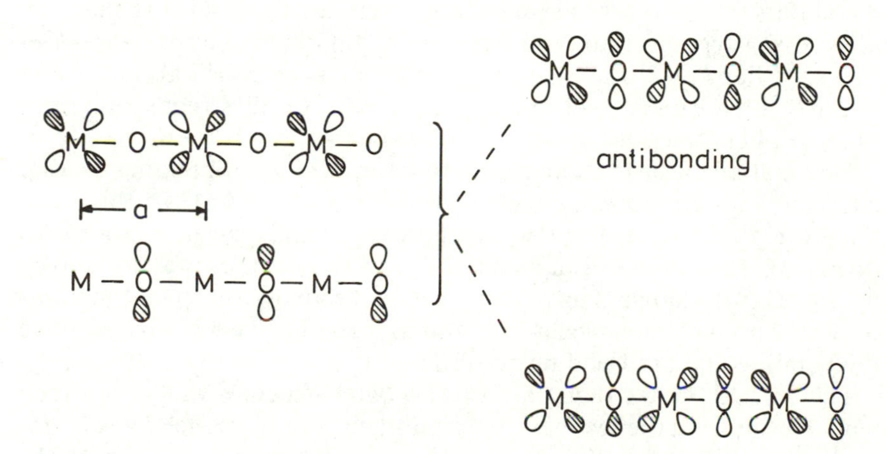

FIG. 2.12 One-dimensional band states for a metal oxide. A set of valence orbitals is shown for a chain of alternating metal and oxygen atoms. Bloch functions with $k = 0$, shown in (a), cannot mix because of their different symmetry. Atomic orbital combinations with $k = \pi/a$, shown in (b), can, however, interact to form the bonding and antibonding combinations illustrated.

constructed from atomic orbitals in a crystal.[1] The range of k values is chosen so that all possible linear combinations are generated just once; in our one-dimensional example

$$-\pi/a < k \leq \pi/a. \tag{2.16}$$

This range of k is known as the **first Brillouin zone**.

The form given for Ψ_k would be appropriate for just one basis AO per unit cell; in fact there are many basis functions, the most important being the metal d and the oxygen $2p$ orbitals. In a purely ionic picture, each such atomic orbital would give one band: that is one orbital of type Ψ_k for every k value. When the overlap between orbitals is included, these orbitals may mix together, giving bonding and antibonding combinations just as in MO theory. Such mixing is governed by symmetry constraints, and one of the consequences of this can be seen from the functions shown in Fig. 2.12. We have chosen to illustrate just one metal d and one oxygen p orbital in each unit cell, arranged so that they can achieve a π-type overlap (compare the t_{2g} MO combinations shown in Fig. 2.9). For $k = 0$, the phase factors of all atomic orbitals according to eqn (2.14) are unity: then as a consequence of the different symmetries of the p and d functions they have no net interaction. If we were to form a combination of the two, we would find that regions of bonding overlap alternate with antibonding regions which exactly cancel. In this case, therefore, the bands at $k = 0$ will be pure metal and oxygen orbitals with no mixing. But at the Brillouin zone edge, with $k = \pi/a$, functions given by eqn (2.14) change sign in alternate cells. Now the two functions overlap without any cancellation, and as a result they will mix to make the bonding and antibonding combinations shown in Fig. 2.12.

Figure 2.13 shows that part of the **band-structure diagram** which would result from the two basis orbitals in our simple model. We did not explicitly consider functions with intermediate k values, and indeed the coefficients in eqn (2.14) are complex numbers; but we can guess that the mixing between oxygen p and metal d orbitals increases smoothly between $k = 0$ and $k = \pi/a$. Thus the band energies are forced apart progressively as k changes, by the formation of bonding and antibonding orbitals.

Although the model just discussed is highly simplified, it exemplifies many of the important features of the band structure of real compounds. Each atomic orbital in the unit cell gives one band; over an extended energy range these would include the lower-lying oxygen $2s$ functions, as well the metal s and p orbitals at higher energy. The crystal orbitals are made up of combinations of these AOs with differing bonding and antibonding character, and the variation of band energy with wavevector—known as the **dispersion**—reflects the way in which bon-

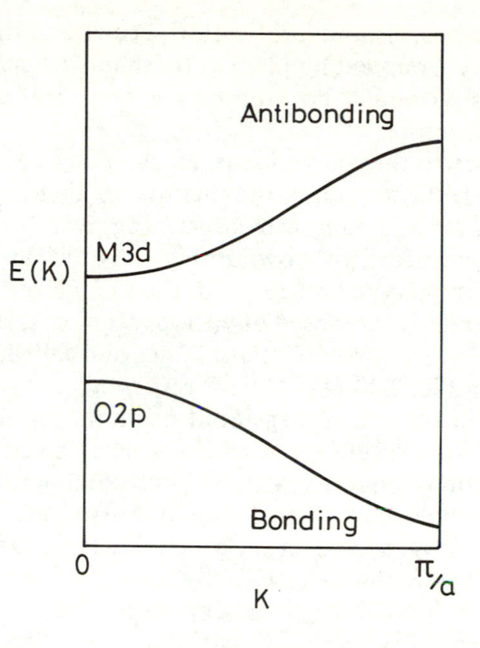

FIG. 2.13 Band structure resulting from interaction between the atomic orbitals of the one-dimensional picture of Fig. 2.12. The band energies at $k = 0$ are of non-bonding oxygen $2p$ (low energy) and metal $3d$ orbitals; the progressively stronger interaction between these as k increases leads to the formation of bonding and antibonding combinations, with the energy trends shown.

ding character changes with k. In compounds with considerable ionic character, the bonding orbitals will have a predominant oxygen contribution, and the antibonding ones will be largely metal d. This is analogous to the MO picture, and the interpretation which MO theory provides for the 'd^n' configuration also carries over: a d^0 compound has just enough electrons to fill the largely oxygen bonding bands, and otherwise n is the number of electrons per metal atom in the antibonding bands at higher energy.

Of course, band structures in three dimensions have certain complications not apparent from the one-dimensional model.[1] k becomes a vector quantity $\boldsymbol{k}$ which shows both the wavelength and direction of the 'electron waves' that make the crystal orbitals. This creates problems of representation, and the normal way of presenting a band structure is to show a selection of $\boldsymbol{k}$ values: a kind of 'guided tour' intended to illustrate the range of energies found. The first Brillouin zone, which as in one dimension defines the range of $\boldsymbol{k}$ necessary to generate all allowed orbital combinations without duplication, becomes a three-dimensional region of '**k**-space' bounded by planes. Its shape depends on the crystal lattice, and must generally be found by geometrical arguments based on the reciprocal lattice.[13,66]

We shall illustrate the above ideas by referring to a fairly accurate calculation on a real compound, the metallic oxide RuO_2.[67] This has the rutile structure, with a unit cell containing two formula units (see Fig. 1.7). The band structure shown in Fig. 2.14 has, as expected, a total of 10 bands of Ru 4*d* type and 12 from the O 2*p* levels. The horizontal **k** axis in Fig. 2.14 (b) is composed of various segments, showing different **k** directions within the first Brillouin zone, and labelled according the scheme shown in Fig. 2.14 (a). Thus Γ represents the point $\mathbf{k} = (0,0,0)$, and so on. As in the one-dimensional example the dispersion of the bands—the variation of energy with **k**—is determined by the different bonding interactions possible with orbital combinations of different wavevector. The only such interactions in our model system were between metal and oxygen, and in many oxides the *d* band dispersion is probably controlled in the same way, by covalent metal–oxygen interaction. However, in three-dimensional crystal structures, some direct O–O and M–M overlap also occurs. Occasionally, as discussed in Sections 1.3.3 and 5.3.2, metal–metal bonds have a structural effect, and this is the case with distorted variants of the rutile structure found with some d^1 and d^2 oxides. Metal–metal bonding is not structurally significant in RuO_2, but nevertheless some Ru–Ru 4*d* overlap probably occurs, so that the 4*d* band energy is determined by a competition between this and Ru–O bonding.

Although the band structures of oxides are most easily interpreted in terms of an atomic orbital basis, the LCAO method is not always used as a means of calculation. More common are variants of the **augmented plane wave** (APW) method, in which crystal orbitals are obtained from a numerical integration of Schrödinger's equation.[68] This is made possible by using suitable approximations to the actual three-dimensional potential experienced by the electrons. Exchange is treated by the local-density method referred to in Section 2.2.1. The **muffin-tin approximation** is then used to provide a spherical average around each atom, so that the potential is a function of distance from the centre only.[48] The orbitals within each atomic region are joined by suitable boundary conditions, often using a region of plane waves between them, where the potential is approximated by a constant value. Although explicit atomic orbitals are not therefore required as input to the calculation, they are generated during the course of it, and the results can be analysed in terms of an effective atomic-orbital basis.

One way of presenting the results from the RuO_2 calculation is shown

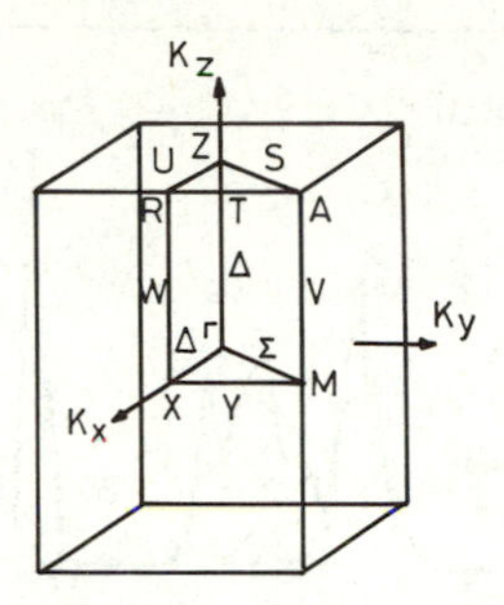

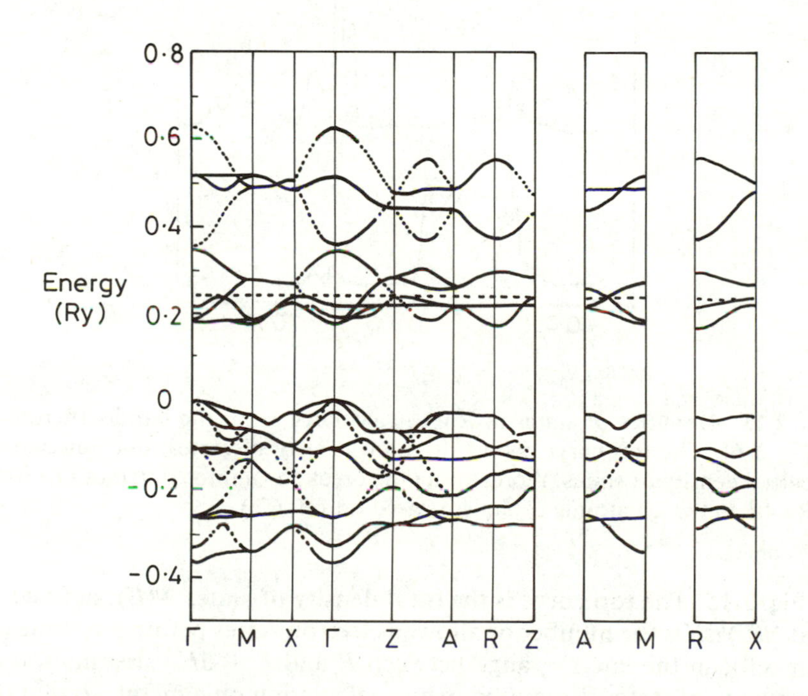

FIG. 2.14 Band-structure calculation on RuO_2, performed by the augmented plane wave (APW) method:[67] (a) shows the Brillouin zone for the tetragonal rutile lattice (see structure in Fig. 1.7); (b) shows the band energies, as a function of ***k*** vector according to the labelling scheme of (a). (Energy scale 1 Ry = 13.6 eV).

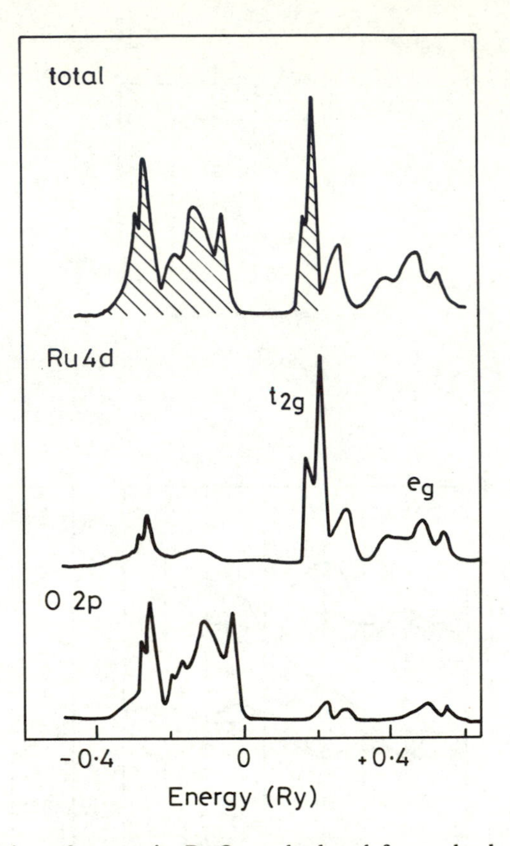

FIG. 2.15 Densities of states in RuO_2, calculated from the band structure in Fig. 2.14. The top curve shows the total density of states, and lower ones partial density of states, representing the breakdown into contributions from Ru 4*d* and O 2*p* atomic orbitals. (Based on ref. 67.)

in Fig. 2.15. The top curve is the total **density of states** $N(E)$, defined so that $N(E)\,dE$ is the number of allowed electron states per unit volume (or unit cell), in the energy range between E and $E + dE$.[13] Estimation of the density of states is required in any calculation on a metal, so that the **Fermi level**—i.e. the top-filled level in the ground state—can be located in the band structure. As expected for a Ru^{4+} compound with the electron configuration $4d^4$, the Fermi level in RuO_2 lies within the upper band, having majority Ru 4*d* character. The atomic character of these bands is shown explicitly in the lower plots, as the **partial densities of states,** or contributions made to $N(E)$ by the different atomic orbitals. Oxygen 2*p* orbitals make the major contribution to the lower energy bands, and ruthenium 4*d* to the upper ones, as in the MO picture. The similarity with localized models also carries over into the division of the Ru 4*d* band into two sets, labelled t_{2g} and e_g in Fig. 2.15. In fact the site symmetry of each metal atom in the rutile structure is quite low, and these labels are not strictly applicable: a detailed breakdown of the density of states of RuO_2 shows that each Ru 4*d* orbital makes a slightly different contribution.[67] Nevertheless, the splitting of the *d* band in a way similar to that predicted by the crystal field model is a general feature of compounds with a transition metal in octahedral coordination, and shows the importance of the local bonding interactions discussed earlier.

The main features of the density of states predicted by the band calculation are quite nicely confirmed by photoelectron spectroscopy (PES).[69] The two spectra shown in Fig. 2.16 cover the energy region of the Ru 4*d* and O 2*p* bands. In a simple one-electron interpretation of these experiments, valid to a good approximation for many metallic compounds, the experimental angular-integrated PES curves should

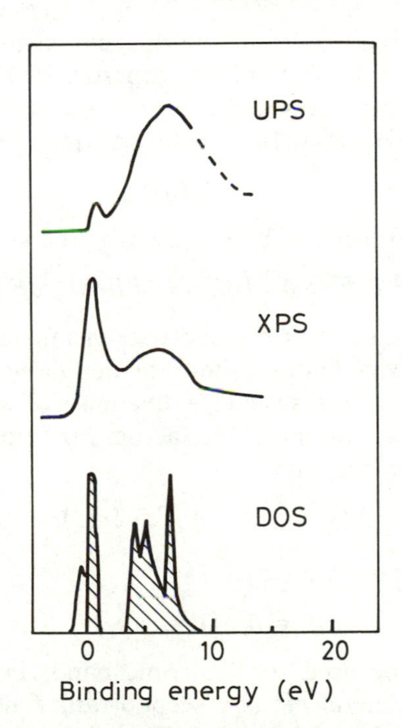

FIG. 2.16 Comparison of the calculated density of states (DOS) of RuO_2 with experimental photoelectron spectra.[69] In the UPS measured at 21 eV energy, ionization cross-sections for oxygen 2*p* and ruthenium 4*d* orbitals are comparable in magnitude. The XPS, however, is dominated by Ru 4*d* ionization.

reflect the density of states, weighted by the ionization cross-sections of the orbitals concerned.[1,39] For the upper spectrum in Fig. 2.16, obtained with UV photons of energy 21.2 eV, the cross-sections for O 2*p* and Ru 4*d* atomic orbitals have comparable values, so that the spectrum is similar to the total density of states. The sharp Fermi-level cut-off, characteristic of a metal, can be clearly seen in this spectrum, which is measured at quite high energy resolution. The other spectrum is obtained with X-ray photons, and the Ru 4*d* ionization cross-section dominates. The increase of relative intensity of the '4*d*' band is apparent.

It is possible to measure the $E(\boldsymbol{k})$ curves of the band structure directly, using angular-resolved photoemission techniques with single-crystal samples.[70] Unfortunately, this type of study does not seem to have been carried out on RuO_2 or the compounds discussed in Section 5.1, where the straightforward band model is expected to work.

2.3.2 The properties of conduction electrons

The direct physical significance of the $E(\boldsymbol{k})$ curves in the band structure is that they determine the dynamical properties of electrons in solids.[48] The slope of the curve gives directly the velocity of an electron (v) moving in a particular band. In one dimension,

$$v = \hbar^{-1}\mathrm{d}E/\mathrm{d}k. \qquad (2.17)$$

In three dimensions this must be replaced by the vector equation

$$v = \hbar^{-1}\nabla_k E \equiv \hbar^{-1}(\partial E/\partial k_x, \partial E/\partial k_y, \partial E/\partial k_z). \qquad (2.18)$$

It can be seen that electronic conductivity in a metal or semiconductor will be favoured by wide bands, where the derivative of $E(\boldsymbol{k})$ is large. A related theoretical quantity is the **effective mass** of an electron, defined as the value m^* required to match the actual $E(\boldsymbol{k})$ curve in a band to the adjusted free-electron formula

$$E(\boldsymbol{k}) = V_0 + \hbar^2 \boldsymbol{k}^2/(2m^*). \qquad (2.19)$$

In one dimension, therefore:

$$m^* = \hbar^2/(\mathrm{d}^2E/dk^2). \qquad (2.20)$$

This equation may be used for isotropic bands in three-dimensional solids, although in general m^* is a second-rank tensor. Clearly, 'light' highly mobile electrons require wide bands; 'heavy' electrons in narrow bands are more strongly scattered by impurities, and are also more easily trapped by defects and lattice polarizations, sometimes leading to a breakdown of metallic behaviour.

The effective mass is often used as a way of adapting predictions from the free-electron theory, replacing the real electron mass in order to correct for band structure effects.[13] One formula arising in this way gives the conductivity

$$\sigma = ne^2\tau/m^*. \tag{2.21}$$

Here n is the number of free carriers with charge e per unit volume, and τ is the average relaxation time, the interval between scattering events coming from vibrations or collisions with defects. Another simple prediction is the Drude theory which gives the dielectric function of a simple metal:

$$\varepsilon(\omega) = \varepsilon(\infty)\,[1 - \Omega_p^2/(\omega^2 + i\omega/\tau)] \tag{2.22}$$

where $\varepsilon(\infty)$ is the high-frequency dielectric constant arising from the polarizability of electrons other than those in the conduction band, and Ω_p is the plasma frequency, given by

$$\Omega_p^2 = ne^2/[\varepsilon(\infty)\varepsilon_0 m^*]. \tag{2.23}$$

In free-electron theory, Ω_p is the photon energy beyond which a metal becomes transparent. Sometimes in oxides the plasma frequency is below the energy of other electronic excitations, giving in principle a 'window' of transparency in the visible or near-UV region of the spectrum.[71] In fact in transition metal oxides the free carrier lifetime τ is often rather short; this gives a large imaginary contribution to the dielectric function (2.22), which causes strong absorption in the optical energy range. The optical properties of metallic oxides will be discussed in Section 5.1.3, where we shall also see how plasma frequencies of metallic oxides may also be studied conveniently by electron energy-loss spectroscopy.

The above formulae, and others obtained from free-electron theory by simply substituting m^* for m_0, are often used on a purely empirical basis for interpreting the properties of metallic compounds. They are strictly applicable if the bands are isotropic: i.e. when the conduction band dispersion is independent of the direction of $\boldsymbol{k}$. This is never true in transition metal oxides, and it is theoretically more satisfactory to work directly from the band structure, taking account of the specific dynamics predicted by the $E(\boldsymbol{k})$ curves.[48] Many physical properties of a metal are determined by electron states close to the **Fermi surface**: that is the locus of $\boldsymbol{k}$ values corresponding to the boundary between occupied and empty levels. Figure 2.17 shows the Fermi surface predicted by the band-structure calculation on RuO_2; $\boldsymbol{k}$ values within the shape shown correspond to occupied levels in the conduction band, and values outside to empty levels. In this case it has been possible to compare the calculation directly with Fermi surface 'mapping' experiments measuring the dynamics of electrons in a magnetic field.[72] A good agreement is found,

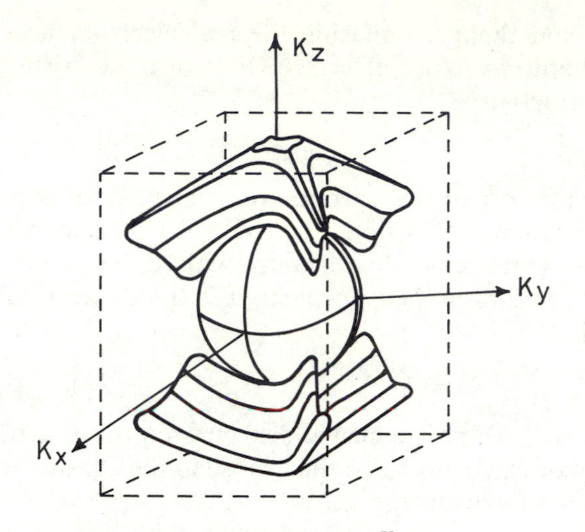

Fig. 2.17 Calculated Fermi surface for RuO_2.[72] Occupied states have k values lying within the shape illustrated.

but this type of experiment is rarely possible with oxides, as the strong carrier scattering often leads to short mean-free paths, so that the Fermi surface is not well defined.

Two important properties of a metal depend only on the density of states at the Fermi level, $N(E_F)$, and not on the detailed Fermi surface shape. The electronic contribution to the specific heat is

$$C_{el} = (\pi^2/3)k^2TN(E_F) \tag{2.24}$$

where T is the absolute temperature and k is Boltzmann's constant; and the temperature-independent Pauli paramagnetic susceptibility is

$$\chi_p = \mu_0\mu_B^2N(E_F) \tag{2.25}$$

where μ_B is the Bohr magneton. Both of these quantities can be used for an 'experimental' determination of $N(E_F)$. The specific-heat contribution must be obtained at low temperatures where it is not swamped by the normal lattice (Debye) contribution; and in the case of the Pauli paramagnetism a correction for the Landau diamagnetic term must be applied to the measured susceptibility. $N(E_F)$ can also be estimated from the height of the Fermi edge in photoelectron spectra. Chapter 5 discusses some examples where these different estimates give reasonable agreement, as well as cases where serious deviations are found from such simple 'independent-electron' models.

Other properties depend more explicitly on the behaviour of electrons at the Fermi surface, through their velocity ν given by eqn (2.18) above. Thus the conductivity is really a tensor with components given by

$$\sigma_{ij} = \tfrac{1}{3}e^2 N(E_F)\tau\langle \nu_i \nu_j \rangle_F \tag{2.26}$$

where $\langle\,\rangle_F$ means an average over the Fermi surface. The plasma frequency is similarly given by

$$\Omega_p = 2e^2\hbar^2 N(E_F)\langle \nu^2 \rangle_F/(3\varepsilon_0). \tag{2.27}$$

The 'simple' predictions of the band theory of metals depend on the assumption that electrons move independently through the crystal, unaffected by each other except for the constraints imposed by the exclusion principle. That this can ever be a good approximation is due to the screening effects of metallic electrons, which 'smear out' individual electron–electron repulsions into an effective potential.[73] The efficiency of such screening depends both on the concentration of metallic electrons and on the width of the band, and for many oxides one or both of these quantities is insufficient for the independent electron picture to be valid. In the limit, metallic character breaks down altogether, but there are many intermediate cases where metallic oxides show anomalous behaviour, frequently in their magnetic properties.[18] Some of the ways in which these can be treated by band theory are described in the following section.

2.3.3 Magnetic band structures

The normal band picture assumes that crystal orbitals are *spin independent*, in the sense that spin-up and spin-down electrons have the same spatial functions. The Pauli susceptibility comes from a calculation of the Zeeman splitting of spin-up versus spin-down energies in a magnetic field, and the small alteration of orbital occupancies which is a result of this.[13,48] But a change in the relative numbers of spin-up and spin-down electrons also has an indirect effect on their energies, through the exchange interaction which is effectively attractive between electrons with parallel spin. This in turn gives a correction to the paramagnetic susceptibility, which becomes

$$\chi' = \chi_p[1 - KN(E_F)/2]^{-1} \tag{2.28}$$

where χ_P is the uncorrected Pauli term and K is the exchange interaction between two parallel-spin electrons. (The latter quantity is often denoted by U;[48] we reserve this symbol here for the intra-atomic *Coulomb* term arising in the Hubbard model discussed later.) The **Stoner enhancement** term $[1 - KN(E_F)/2]^{-1}$ is often significant for transition metal compounds, as the d band may be narrow so that the density

of states $N(E_F)$ is large;[18] and at the same time the exchange integrals between electrons in the contracted d orbitals (especially in the $3d$ series) are much larger than for the s and p valence orbitals of the non-transition elements.
If

$$\tfrac{1}{2}KN(E_F) > 1 \qquad (2.29)$$

the normal paramagnetic state of the metal is unstable, and a splitting between spin-up and spin-down electrons is predicted even in the absence of a magnetic field. The most obvious result is **ferromagnetism** of the band electrons, which is found in the metallic elements Fe, Co, and Ni; for these later $3d$ elements the d orbitals are quite contracted and the band widths are small even in the elements[1]. Ferromagnetism is found in several oxides, for example CrO_2 (discussed in Section 5.2.1). It is not obvious, however, that the ferromagnetic arrangement of spins will necessarily give the lowest energy. Sometimes a full calculation shows that the direction of spin alignment will rotate from one atom to the next. In the simplest type of **antiferromagnetic** arrangement an excess of spin-up and spin-down density is found on alternate atoms; more complex cases of spiral-spin structures can also be found.[56] Such **spin-density waves** may even have a wavelength incommensurate with the lattice spacing. This is found for example in metallic Cr, where it seems to be associated with **Fermi-surface nesting** effects of the kind described in the following section.[74]

The idea of using *different orbitals for different spins* was described in Section 2.2.1 in the context of Xα cluster calculations. It has also been used in band calculations on magnetic solids, including even insulators.[56,75] To see how this can work, we return to the one-dimensional model discussed earlier (see Fig. 2.13). In a d^1 compound, the Fermi level would lie in the middle of the upper band shown (the 'd band'), and a metal would normally be expected. Suppose, however, that the ground state of this imaginary system was actually a magnetic insulator, with the antiferromagnetic spin ordering shown in Fig. 2.18 (a). The unit cell is doubled as a result of the magnetic ordering, and a band calculation which takes this into account would have to include two transition metal atoms, with their respective d orbitals, per unit cell. The result of this cell doubling may therefore be to give a *splitting* of the d band, as shown in Fig. 2.18 (b); the lower-energy bonding, or 'oxygen $2p$' bands, are omitted here for clarity. The lower d band of the magnetic structure can now hold two electrons per *pair* of metal atoms, and will therefore be filled in the d^1 case, giving an insulator.

The two d bands in Fig. 2.18 (b) could be labelled 'spin-up' and 'spin-down' but this would be rather inaccurate, as the occupied spin directions

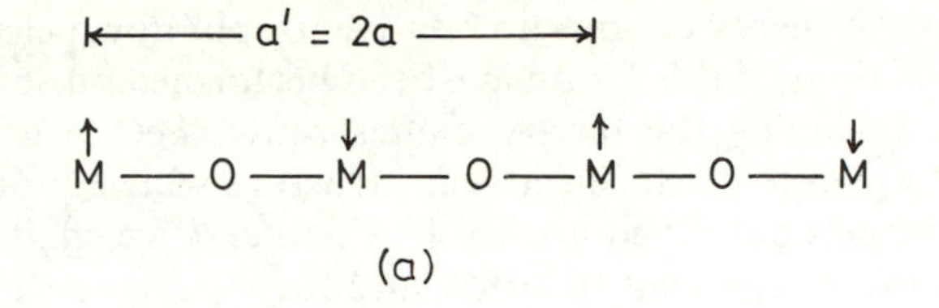

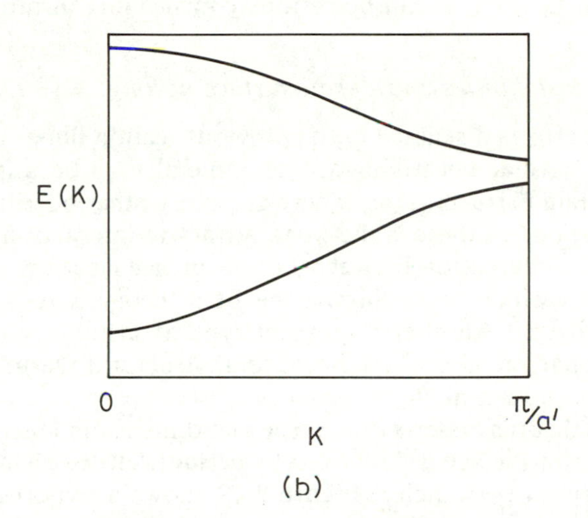

FIG. 2.18 (a) The metal–oxygen chain of Fig. 2.12 shown as a magnetic insulator, with an antiferromagnetic d^1 ground state. (b) The enlarged unit cell ($a' = 2a$) leads to a splitting of the bands. Only the 'metal' conduction band is shown; the lower-energy (occupied) band corresponds to wave-functions giving the spin distribution shown in (a), and the higher (empty) band to the reverse spin distribution.

alternate from site to site. The lower band really represents two functions: one for spin-up electrons with maximum density on the odd-numbered metal atoms in the structure of Fig. 2.18 (a), and one for spin-down electrons on the even-numbered atoms. The upper band, which is empty in the ground state shown, corresponds to functions with the opposite pattern of spin distribution.

This example shows how band theory can represent a magnetically ordered insulator in terms of a **spin-density wave** in the structure. The actual band gap predicted by the theory would depend on how the calculation were performed. In the commonly used local-density method, the bands are split by exchange effects: thus the predominance of spin-up density on the appropriate sites gives an exchange potential that lowers the energy of spin-up relative to spin-down electrons there.[56] Band calculations of this kind have been performed on several magnetic insulators, including the binary oxides MnO, FeO, CoO, and NiO.[75] They give a picture of the electronic structure which is rather different from the models described above. In Chapter 3 we shall give a critical discussion of the different theories, and suggest that although the band picture may sometimes give a reasonable description of the ground state of magnetic insulators, it can be seriously misleading in other ways.

2.3.4 Peierls instabilities and Fermi-surface nesting

The magnetic effects discussed in the previous section illustrate one way in which the normal Fermi surface of a metal may be unstable with respect to certain perturbations. There are many other possible perturbing influences; one of these is the weak attractive interaction which can result from the interaction between Fermi-surface electrons and lattice vibrations, and which according to the BCS theory is responsible for superconductivity.[48] Another important type of instability operates in 'low-dimensional' metals, where because of structural features the conduction bands show a highly anisotropic dispersion.[76]

The **Peierls theorem** asserts that a true one-dimensional metal does not exist, as its Fermi surface is destroyed by **periodic lattice distortions** that change the crystal periodicity. Figure 2.19 shows a hypothetical band structure for a one-dimensional system, and illustrates the effect of such a periodic distortion. Corresponding to the enlarged 'superlattice periodicity' a', new Brillouin-zone boundaries appear, and lead to additional band gaps at k values which are a multiple of π/a'. If the Fermi surface of the unperturbed crystal is at one of the new zone boundaries, the new band gap resulting from the distortion will destroy the metallic properties of the conduction electrons.

Opening a band gap at the Fermi level leads to some stabilization of occupied electron states. But it is not obvious that a periodic lattice distortion is energetically favourable, as it will generally be opposed by the elastic energy resulting from short-range forces, which normally favour a regular lattice spacing. The essence of the Peierls theorem is that in a *one-dimensional system* the electronic perturbation always wins, and that an appropriate **Peierls distortion** should destroy the metallic ground state. The perturbation of electronic levels also leads to a **charge-density wave** (CDW), with a periodic accumulation of electron density in phase with the lattice distortion. A simple way of imagining the CDW is to think of electrons concentrating in regions where there is extra bonding character as a result of the periodic distortion.

The periodicity, a', of the lattice distortion and the charge-density

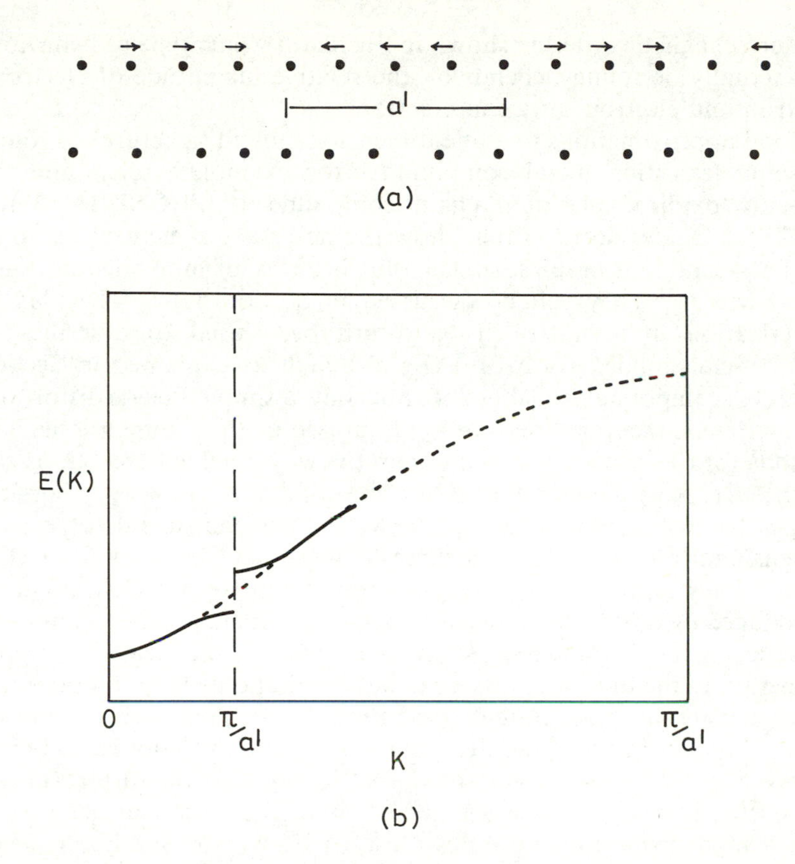

FIG. 2.19 A periodic lattice distortion of a chain of atoms (a), and its effect on the band structure (b). New band gaps appear at $k = \pi/a'$, where a' is the periodicity of the distortion (and also at multiples of this value of k, which are not, however, shown).

wave resulting from the Peierls instability is related to the wavevector k_F at the Fermi surface by

$$a' = \pi/k_F. \tag{2.30}$$

Depending on the value of k_F, a' may or may not be commensurate with the underlying lattice spacing a. A particularly simple example is that of a half-filled band in one dimension. Then $k_F = \pi/2a$ and a' is just twice a. The periodic distortion can be simply described as a 'dimerization' of the original lattice, with electrons in the charge-density wave concentrating in the 'bonds' between the more closely spaced pairs of atoms.[1] This is clearly an alternative form of instability to the antiferromagnetic ordering shown in Fig. 2.18; which type of behaviour will actually be found depends on the relative magnitude of electron–electron and electron–lattice interactions.

Good approximations to a one-dimensional band structure are found in some transition metal compounds, for example $NbSe_3$, and the partially oxidized platinum chain compound $K_2[Pt(CN)_4]Br_{0.3}3H_2O$ ('KCP').[76] Oxides do not often show the necessary structural features, but an exception is in the so-called 'blue bronzes' of molybdenum, such as $K_{0.3}MoO_3$.[77] They will be discussed in Section 5.3.1. The idea of dimerization in a half-filled band has been used to describe the metal–metal bonded form of VO_2, although as explained in Section 5.3.2 this compound probably does not show a simple Peierls distortion.

The Peierls theorem does not apply in two or three dimensions. The reasons for this general statement, and the way in which exceptions can occur, are illustrated in Fig. 2.20. The diagrams show two possible shapes for the Fermi surface of a two-dimensional metal (or a two-dimensional cross-section of a three-dimensional Fermi surface). The dashed lines represent new Brillouin-zone boundaries which might be introduced by a periodic lattice distortion. An energy discontinuity will occur along these boundaries, so that states just inside them are lowered in energy. In the first case, (a), it can be seen that only very few occupied electron states are close enough to be strongly affected; in this situation, which is the normal one, the electronic stabilization will not be sufficient to overcome the elastic forces of the lattice, so that the distortion will not occur. In case (b), however, the shape of the Fermi surface is such that a high proportion of states close to E_F will be stabilized. Then the electronic effect is much larger and a distortion may now take place. As in one dimension, charge-density waves will result, but there is an important difference in the two-dimensional case. In the example shown, not all the Fermi surface will be destroyed by the perturbation, and the compound may remain metallic.

The hypothetical case just described shows an example of **Fermi-surface nesting**: the existence of nearly parallel planes of Fermi surface which can be strongly perturbed by a periodic lattice distortion. A one-dimensional metal can also be thought of in this way. Real solids are of course three dimensional. A 'one-dimensional' band structure simply means that dispersion of the $E(\boldsymbol{k})$ curves is only significant in one direction of the wavevector $\boldsymbol{k}$. If such a band structure were plotted in three dimensions it would be completely flat in the two perpendicular directions. Thus the Fermi surface is perfectly nested, consisting of flat parallel planes. Good nesting is not so common in materials with two- or three-dimensional interactions, although some examples are known. Some layer chalcogenides such as TaS_2 show strong periodic lattice

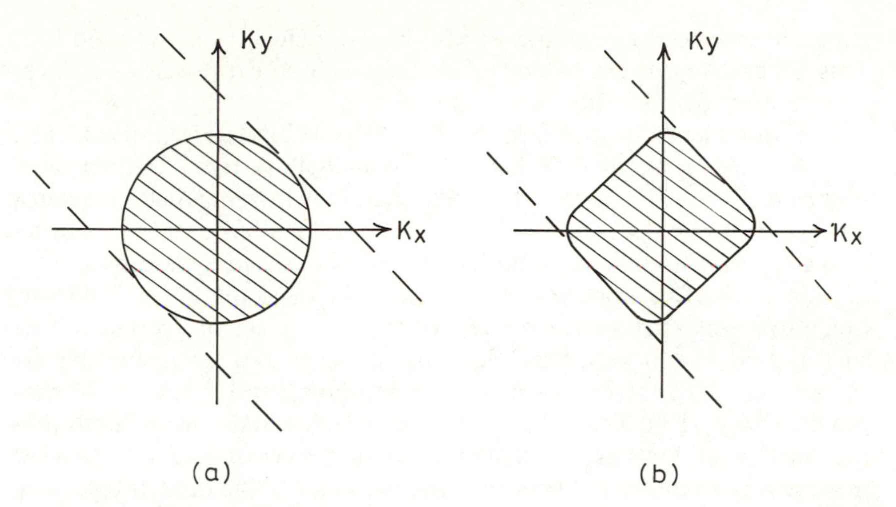

FIG. 2.20 Fermi-surface nesting in two dimensions. The broken lines represent new zone boundaries that might be introduced by a periodic lattice distortion. In (a), only a few states close to the Fermi surface are strongly perturbed, whereas in the *nested* Fermi surface of (b), many states are perturbed and the distortion may be energetically favourable.

distortion and charge-density wave anomalies.[78] Once again, transition metal oxide examples seem to be rare, but some molybdenum-bronze compounds of slightly different composition and structure to the one-dimensional 'blue bronze' show conductivity and structural anomalies of this kind.[77]

2.4 Intermediate models

The two kinds of model discussed above – localized (ionic or cluster type) and delocalized (band theory) – give very different pictures of the electronic structure of a solid. They are not always incompatible. The orbital picture is extremely useful for thinking about a complex many-electron wavefunction, but it has some arbitrary features that are not always realized. An important theorem shows that we can perform a linear transformation among the occupied orbitals of a system without in any way changing the overall wavefunction.[9] This arbitrariness is important in the bonding theory of molecules, where localized bonds and delocalized MO models may simply represent different, but equivalent, ways of specifying the orbitals. In a d^0 or other closed-shell insulator a similar equivalence applies, and we are at liberty to regard the occupied states either as localized (ionic- or MO-like) or delocalized and band-like. This is because one set of occupied states can be expressed as a linear combination of the other type.

The equivalence of localized and band-like orbitals often disappears, however, when we do not have a closed-shell system. The partially occupied *d* states have very different character in a magnetic insulator and in a metal: in the former they appear to be localized, and in the latter delocalized (although the magnetic band-structure models discussed in Section 2.3.3 offer a possible way to reconcile these pictures). A distinction also arises as soon as we excite or ionize an electron from a closed-shell system; the localized and delocalized pictures of the resulting state correspond to different electron distributions, and predict different physical properties. The problem now is to decide under what conditions the different models are valid. We might also ask whether it is possible in some way to interpolate between the two limits. The models discussed below attempt to answer these questions, by considering the competition between the different interactions which favour either localization or delocalization of electrons. Band-like properties, as we have seen above, result from the bonding overlap of atomic orbitals in a solid. Localization, on the other hand, can be produced by the repulsion between electrons, or by their interaction either with static defects or disorder in a lattice, or with vibrational modes of the lattice.[1] The proper treatment of all these interactions represents one of the most important and difficult problems in theoretical solid-state physics. Highly mathematical theories have been developed, but none of them can claim to be in any way 'exact', as it is necessary to isolate and idealize the various effects in a more-or-less unrealistic way. The discussion which follows will be be largely qualitative and descriptive; as with the theories discussed already in this chapter, the aim is to describe the physical content of the deferent models, and to avoid the technical details.

2.4.1 The Hubbard model

In band theory, as in the simple orbital theories of atoms and molecules, it is assumed that repulsion between electrons can be represented by using an average *effective* potential. In metals, and especially for states near the Fermi surface, this often works very well, because of the good screening properties of the 'mobile' electrons[73]. It breaks down, however, in very narrow bands, and when the concentration of conduction electrons is low. In these circumstances **electron correlation** becomes important: this term describes the way in which electrons alter their motion so as to avoid each other more effectively than an 'independent-electron' orbital wavefunction can describe. The treatment of electron correlation, even in idealized models of simple metals, has a long and difficult history[80]. An approximation especially useful in transition metal compounds is based on the ideas of Mott and Hubbard,[81,82] and consists of neglecting the repulsion between electrons except when they are on the *same* atom.

Some of the ideas of this so-called **Hubbard model** have been introduced already, in the context of the ionic picture of a magnetic insulator such as NiO (see Fig. 2.6). There we saw that the band gap is predicted to come from the repulsion between electrons in the partially occupied d shell. In this specific case, the band-gap excitation corresponds to

$$Ni^{2+} + Ni^{2+} = Ni^{+} + Ni^{3+} \tag{2.31}$$

or equivalently;

$$d^8 + d^8 = d^9 + d^7. \tag{2.32}$$

The energy input required, coming from the extra electron repulsion term in the d^9 configuration, is the most important parameter of the model, known as the **Hubbard *U***.

In Fig. 2.6 the effect of some interatomic overlap was represented as giving a small band width to the states involved in the transition. Fig. 2.21 illustrates what happens as this band width W is increased. At a critical value around $W \sim U$, the upper and lower states overlap and

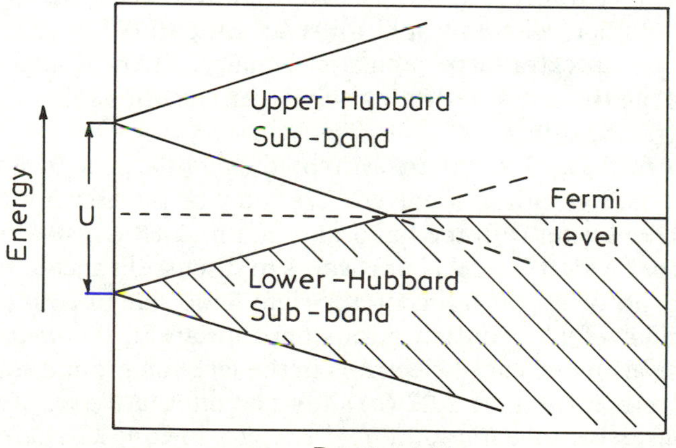

Fig. 2.21 The Hubbard sub-bands, and their overlap to give a metallic state when the band-width W exceeds the Hubbard U. For smaller values of W, the lower sub-band corresponds to the binding energy of occupied levels, and the upper sub-band to the energy of empty levels.

the gap vanishes. We now have a metallic system, with the Fermi level in the middle of a partially occupied band. For $W < U$, as in NiO, there is a splitting between a lower and an upper **Hubbard sub-band**.

The important conclusion from the Hubbard model is, therefore, that the transition between the localized behaviour of a magnetic insulator and the band-like characteristics of a metal is determined by the relative magnitudes of the band width W and the Hubbard U. Not surprisingly a great deal of work has concentrated on attempting to estimate these parameters for real systems.[83,84] It is often assumed that W can be obtained from a band-structure calculation, but it is not clear how reliable this is. Most calculations cannot incorporate the effects of the Hubbard term, and by altering the relative energies of metal d and oxygen p states, this could introduce a significant error. In principle the d band widths might be found from photoelectron spectroscopy, but as we shall see in Chapter 3 the interpretation of these spectra is beset with difficulties.

Even greater uncertainties surround the Hubbard U itself. In the gas phase it can be estimated as the difference between successive ionization energies I_n: for example for a ground state M^{2+} as in NiO we would have

$$U_{\text{gas phase}} = I_3 - I_2. \tag{2.33}$$

Such gas-phase atomic values for the $3d$ elements in the 2+ and 3+ states are shown in Fig. 2.22 (a). Quite notable is the peak associated with a d^5 ground state. This is a consequence of the exchange stabilization of electrons with parallel spin: an electron removed from one d^5 ion and placed in another, where its spin must be antiparallel to those already present, pays an extra large repulsion penalty. There is also a general increase along the series, reflecting the larger repulsions that come from more contracted orbitals.

The values in Fig. 2.22 (a) are estimated using the ground states of all the atomic configurations involved. This may be the appropriate model to use in connection with the band gap in a magnetic insulator such as NiO (but see Section 3.4.2). If one wants to discuss the metal/non-metal borderline, however, the terms on the right hand side of eqns (2.31) and (2.32) should refer to configurations appropriate to the *metallic* state. The precise arrangement of electrons in the free-ion ground state is now no longer relevant. Figure 2.22 (b) shows an alternative set of estimates for 2+ ions, where the d^{n-1} and d^{n+1} energies in eqn (2.32) are taken as *average of configurations*, with essentially randomized spin and orbital arrangements of the d electrons.[85] In these estimates the exchange stabilization is found only in the initial d^n state. Thus there is a progressive increase towards the half-filled d^5 configuration, and a decrease away from it. Other factors being equal, we would therefore

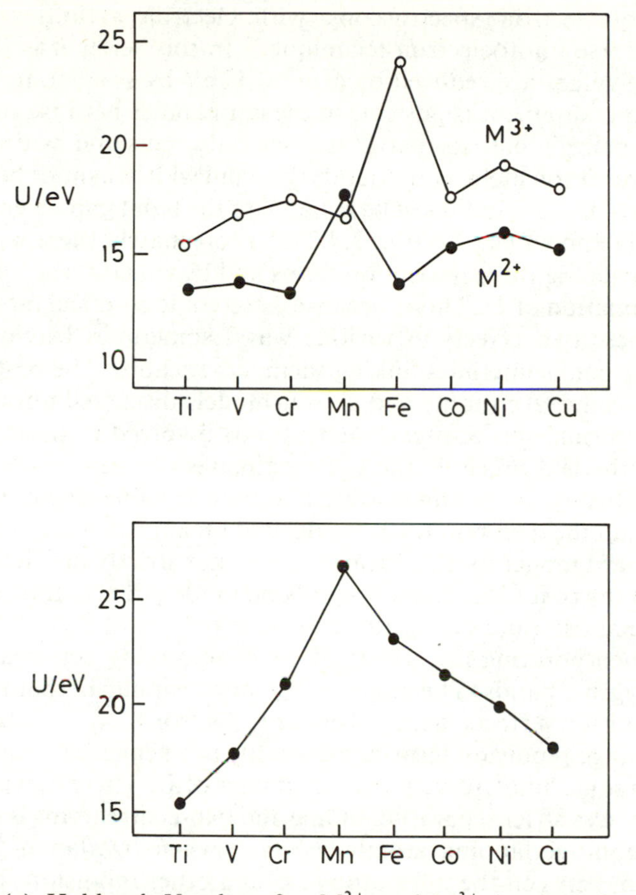

FIG. 2.22 (a) Hubbard *U* values for M^{2+} and M^{3+} ions of the 3*d* series, estimated from atomic ionization potentials. (b) Values for M^{2+} ions taking the average of configuration final states as described in the text.

predict that magnetic insulators in the 3*d* series are more likely to be found not just with d^5 ions, but also with neighbouring ones. This expectation is borne out in practice.[18,86]

Although the trends suggested by gas-phase data are undoubtedly significant in solid compounds, it is generally agreed that the actual *U* values are far too large. All charge-transfer energies in solids are screened by polarization effects; in the ionic-model calculation for NiO (Fig. 2.6), these lead to a reduction from the gas phase *U* value of 19 eV to an effective solid-state value around 5 eV. Similar effects are seen in the lanthanides, where *U* for the 4*f* orbitals can be estimated directly from spectroscopic measurements, combining ionization energy data from photoelectron spectroscopy with electron affinities measured by the inverse photoelectron technique.[87] In this series it is found that gas-phase values are reduced by around 12 eV by solid-state screening. A direct measurement is possible in the lanthanides because the $4f$ band widths are negligible, so that U is essentially the same as the '$4f$ band gap'. In transition metal compounds the band widths cannot be neglected in this way, and U should *not* be equated to the band gap, even when this is properly known (see Section 3.4.2). Unfortunately there is no unique way of including the polarization terms and leaving out the band widths in the estimation of U. This is because different theoretical models incorporate these two effects in various ways, sometimes leaving one out altogether, and sometimes linking them inextricably. The best estimates probably come from ionic and cluster models discussed above, but the different formalisms and parametrizations involved in these cannot be expected to yield precisely the same estimates.

The difficulty in finding reliable Hubbard U values is not surprising. Not only do the different methods depend on approximate theories, but the Hubbard model itself is idealized, and not strictly applicable to real solids. In the first place it is a *single-band* model, concentrating (in transition metal compounds) on the d orbitals only, and there is no unique way of incorporating the effects of covalent mixing between the metal d and oxygen p bands. The model is also approximate in that it considers only repulsion effects acting between electrons on a single centre. Longer-range repulsion terms are certainly not negligible, and may contribute in some 'hidden' way to the estimates of U. There is even a *formal* difficulty, as a strict separation of one and two-centre terms is only possible for non-overlapping atomic orbitals; when overlap is included a distinction between the intra-atomic U and other repulsion terms is not unique, but depends on the orbital basis employed.

In spite of these difficulties, the Hubbard model is widely used in a qualitative and a semi-quantitative way to discuss the properties of magnetic insulators.[1] We have already noted that it gives quite a different picture of the band gap in these compounds from that provided by spin-dependent band-structure calculations, and in Chapter 3 we shall compare the merits of the two pictures and argue that in the present state of theory the Hubbard model is the better approach. Much theoretical work is currently in progress to extend the model, both to a many-band system where covalent interactions are included, and to include other repulsion terms, such as those between near-neighbour atoms.[88,89] This activity has been largely stimulated by a desire to understand high T_c superconductivity in copper oxides, discussed in Section 5.4.

It has been implicitly assumed so far that all transition metal ions have the same electron configuration. Suppose, however, that we consider a mixed-valency compound, in particular a magnetic insulator doped so as to add or subtract some electrons. The extra carriers or holes can move without incurring the Hubbard repulsion. For example, in an oxide containing all Cu^{2+}, the transfer of an electron involves the same type of Hubbard U as in NiO. But if doping produces some Cu^{3+}, electron transfer can take place via

$$Cu^{2+} + Cu^{3+} = Cu^{3+} + Cu^{2+}. \tag{2.34}$$

At the simplest level, therefore, we expect mixed-valency compounds such as doped magnetic insulators to become metallic. This is not always the case, as the carriers are often trapped by the effects considered in the sections below. When metallic conduction does take place, it will also interfere with the antiferromagnetic order of the spins (see Sections 5.2.2 and 5.4.3).

2.4.2 Excitons

Electronic excitation of an insulator yields an electron in the conduction band and a hole in the valence band. Although these sometimes behave as free carriers, the electrostatic interaction between them can lead to a bound state known as an **exciton**.[1,13] The formation of excitons can be recognized from features which appear in the absorption spectra of many solids, at lower energies than the genuine band-to-band continuum of excited states. Exciton binding energies depend both on details of the band structure—in particular the widths of the valence and conduction bands and hence the effective mass of the carriers—and on the dielectric properties of the solid which control the magnitude of the electrostatic attraction. It is normal to consider two extreme cases. If the carriers are light (wide bands) and the attraction weak (high dielectric constant) the weak binding leads to a spatially extended **Wannier exciton**. An approximate but useful energy expression can be obtained by using a quasi-hydrogenic approximation:

$$E_n = -R\mu/(\varepsilon_r^2 n^2). \tag{2.35}$$

In this formula R is the Rydberg constant (13.6 eV) which gives the energy levels of a hydrogen atom as $-R/n^2$ with $n = 1,2,3, \ldots$. The exciton is imagined as a kind of 'positronium' atom, embedded in a lattice which modifies its energy through the screening effect of the dielectric constant ε_r, and by band-structure effects which determine electron and hole effective masses (m_e and m_h), and thus the *reduced mass* μ:

$$1/\mu = 1/m_e + 1/m_h. \tag{2.36}$$

The formula gives a succession of hydrogenic levels. E_1 is the energy of the lowest exciton state relative to the free electron and hole; states with higher values of n can sometimes be seen in an absorption spectrum.[90]

The Wannier model is valid under the same conditions as the hydrogenic impurity-state model discussed below. It is necessary that either the electron or the hole 'orbitals' (or both of them) are sufficiently extended that the details of the crystal lattice are unimportant, so that the bound carriers experience an average crystal environment. The opposite limit arises when the attraction is so strong that both electron and hole remain confined essentially to the same atom or molecular group. This is known as a **Frenkel exciton**, and will be formed when the electron–hole binding is large compared with the conduction and valence band widths. In this case, the solid will have excitation energies substantially less than its band gap.[91]

Both Wannier and Frenkel excitons can be seen in transition metal oxides. One of the classic examples of the former is Cu_2O, where the absorption spectrum measured at low temperatures shows a progression of exciton levels just below the band gap at 2.1 eV[90] In this d^{10} insulator, the gap is between the narrow filled $3d$ band and a conduction band of largely copper $4s$ character. We might also expect to see Wannier excitons in d^0 insulators, and the example of TiO_2 is described in Section 3.1.2. But this does not seem to be a general rule, for reasons that are not entirely clear. Weakly bound excitons are strongly perturbed by lattice vibrations, and can only be observed at low temperatures. It may be that in most transition metal oxides the exciton structure is lost because strong interaction of electrons with the lattice gives rise to a considerable vibrational broadening of electronic transitions.

The crystal field, or *d–d* tansitions of magnetic insulators, are good examples of Frenkel excitons. They occur at energies well below the band gap, and show very little extra broadening compared with similar transitions in isolated impurity ions. The localized nature of a Frenkel exciton means that any dispersion shown individually by the occupied and empty bands involved will be lost. Thus the apparent narrowness of *d–d* transitions in magnetic insulators *cannot* be used to infer that the *d* bands are similarly narrow, although the very existence of such an exciton may set an upper limit to the band width. There is an important connection between Frenkel excitons and the Hubbard model described above. It is normal to use the language of electrons and holes when talking about excited states. But of course the 'electron–hole attraction' is nothing other than an electron repulsion term which is *missing* in the excited state. Thus the electron–hole attraction which forms Frenkel excitons in the excited states of magnetics insulators is essentially the same as the Hubbard U parameter important for the ground state.

2.4.3 *Impurity states*

Defects and impurities are sites where the otherwise periodic potential of a perfect lattice is perturbed. One consequence is that they act as scattering centres, and thus lower the mobility of electronic carriers in a metal or a semiconductor. In non-metalic solids defects may have a more serious effect on the electronic properties, by introducing localized **impurity states** within the band gap. As with excitons, the appropriate model depends on the strength of the perturbation provided by the defect.[92]

At one extreme lies the hydrogenic model, widely used as a first approximation for non-transition metal impurities in semiconductors such as silicon.[13] The predicted binding energy of a conduction band electron to a positively charged impurity is

$$E_b = -Rm_e/\varepsilon_r^2. \tag{2.37}$$

In a similar way, a hole will be bound to a negative charge, giving an acceptor level just above the valence band. To give an illustration of this formula, suppose we have a positively charged impurity, such as an interstitial alkali ion M^+ in an oxide bronze. For a 4*d* or 5*d* oxide we might expect an effective mass $m_e \sim 1$, and a dielectric constant ε_r between 5 and 10. Then the predicted binding energy is in the range 0.1–0.5 eV. This is a very useful rough estimate, but there are various reasons why the hydrogenic model cannot be expected to be quantitatively useful for oxides. In the first place, even in the relatively favourable example we have chosen, the ground-state Bohr radius of the impurity orbital is only 300–600 pm. This is of the order of a crystal lattice spacing, and so the basic assumption of the hydrogenic model, that the bound electron experiences only an average lattice environment, is not valid. There are other possible complications. In the case of transition metal impurities, they will have their own complicated energy-level structure, for example crystal field levels. In addition, most oxides have very different values for the static and high-frequency dielectric constants. Not only is it not obvious which one should be used, but the difference between these two quantities implies that impurity electrons are likely to be strongly perturbed by polaron formation, as discussed below.

A very different picture is that provided by the ionic model, although interestingly the magnitudes predicted are similar. We now imagine the impurity electron or hole to be entirely localized on one ion. Binding to the defect is provided by electrostatic effects, as can be seen in the example of lithium-doped NiO. Each Li^+ replaces one Ni^{2+}, and charge conservation requires a compensating hole, in the form of Ni^{3+}.

It is electrostatically favourable for the hole to remain next to Li^+, as this impurity has an effective negative charge compared with the Ni^{2+} found on regular sites. Estimates of this binding energy must include polarization terms. A value of about 0.5 eV has been calculated using the polarizable-ion methods discussed in Section 2.1.2.[93]

In between these two extreme models is a gap which can, in principle, be filled by theories similar in philosophy to the cluster calculations described in Section 2.2. We saw indeed that cluster calculations have been used to treat the electronic structure of defects. It is, however, possible to calculate the levels arising from the perturbation of the band structure by the defect site, without resorting either to a finite cluster, or to a hydrogneic type of approximation. Such models are quite complicated, as they involve the interaction of discrete levels (the impurity orbitals) with a continuum of states (the bands of the host). This resembles a scattering problem, and is often formulated in terms of Green's functions.[94] But the physical input required is quite similar to that in a cluster calculation, and involves the unperturbed orbital energies (including the density of states for the valence and conduction bands of the host crystal), and the mixing interaction between them.

The results of such an impurity-state calculation, on chromium as an impurity atom replacing titanium in $SrTiO_3$, are shown in Fig. 2.23.[95] The host density of states, and the off-diagonal matrix elements required, were estimated from a tight-binding parametrization of a band structure calculation, and the unperturbed Cr 3*d* energies from a self-consistency argument. As in the cluster calculation shown in Fig. 2.10,

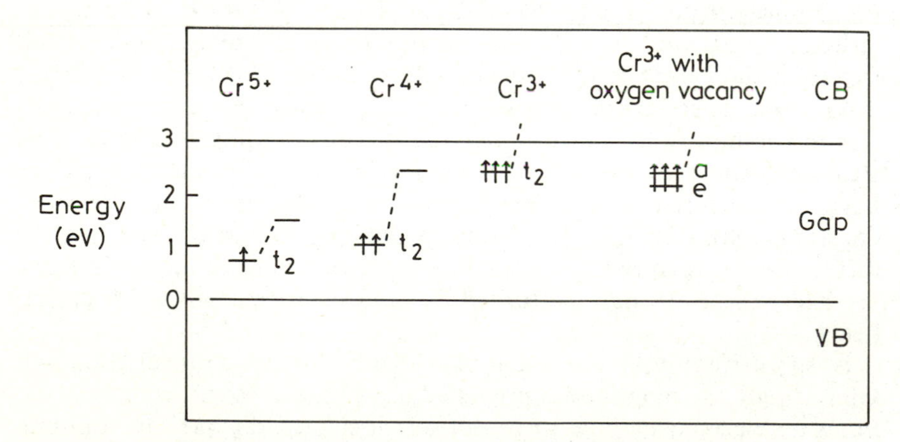

FIG. 2.23 Results of a Green's function impurity-state calculation of chromium in $SrTiO_3$.[95] The spin-up and spin-down orbital energies for different oxidation states of the Cr impurity are shown (compare Fig. 2.10).

exchange effects were allowed for in an approximate way by calculating spin-up and spin-down orbitals separately. Figure 2.23 shows the predicted energies of the defects orbitals within the band gap of $SrTiO_3$, for different oxidation states of chromium. Although the states derive largely from Cr 3*d* orbitals, the calculation shows that they are quite strongly mixed with the bands of the surrounding lattice.

The calculation just described is based on an orbital picture, and in addition to the other approximations required, suffers from the same difficulties as MO cluster models. We recall from Section 2.2 that in accurate calculations we cannot expect to obtain reliable estimates of ionization or excitation energies simply by considering orbital energies. Just as attempts have been made to circumvent this problem in the cluster context by configuration-interaction models which treat *states* directly, so similar ideas have been applied to impurity states. Recently developed **charge-transfer** impurity models are an attempt to extend analogous cluster theories, exactly as are the orbitally based models just described.[96] As with the CI cluster model, the basis states are of the form d^n, $d^{n+1}\underline{L}$, $d^{n+2}\underline{L}^2$, and so on, where now $\underline{L}$ represents a hole in a *band* of oxygen 2*p* states, and so can have a continuous range of energies. Both the input parametrisation and the kind of output obtained are very similar to the CI cluster theories discussed in Section 2.2.2.

2.4.4 Anderson localization

In solids with very high concentrations of defects it is no longer an adequate approximation to think of isolated impurity levels; we can now imagine that these overlap and produce a whole *band* of impurity states. The theoretical difficulties involved in treating such highly disordered materials are immense, but one very important general conclusion can be stated. Although the density of states may resemble that of a regular solid, with continuous bands rather than discrete energy levels, the states making up the bands of a disordered solid may sometimes be *localized*, instead of extending throughout the lattice as in the ordered case.[97,98] This is the phenomenon of **Anderson localization**, and a simple picture of how it can occur is shown in Fig. 2.24.

We imagine an array of atoms which, because of the disorder, have individual energy levels varying randomly from site to site. When the atomic orbitals overlap to form band states, we find that atoms with energies near the average value are likely to have at least some neighbours of similar energy, whereas atoms with energies near the limits of the range are more likely to be 'isolated' and have no similar neighbours. The consequence is that orbitals with energies close to the centre of the band may be able to extend through the solid, but states near the band

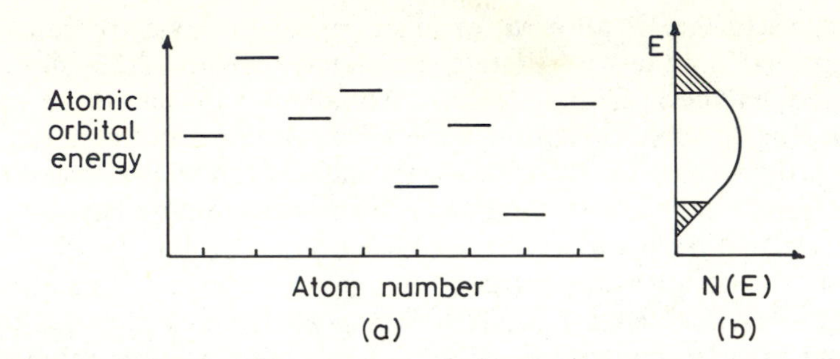

Fig. 2.24 Anderson localization. (a) An array of atoms with randomly varying atomic orbital energies due to disorder. (b) The density of states resulting from their overlap with two kinds of state separated by *mobility edges*: extended states in the band centre, and localized states at the band edges.

edges may be localized near individual atoms. The extent of localization depends on the degree of disorder (i.e. the range of atomic orbital energies) relative to the band width. In the limit of large disorder all states in the band may be localized; when disorder is small only the extreme band tails will be so. A more detailed analysis shows that the boundary between localized and extended states is sharp, and is known as a **mobility edge**; as this name implies the mobility of electrons will drop to zero if the Fermi energy moves into the region of localized states.

Electrons in localized states in a disordered solid can move by a phenomenon known as **variable-range hopping**. Some evidence for this in oxides, and the importance of Anderson localization in highly defective compounds, are discussed in Section 4.4.

2.4.5 *Polarons*

As explained in Section 2.1.1, an isolated carrier in a non-metallic oxide will polarize the surrounding lattice. It was noted that the total polarization can be divided into two parts: that due to purely electronic effects, and that due to the motion of ions as a whole. The former effect is rapid, and gives rise to the high-frequency dielectric constant ε_{opt}, while ionic polarization is slow. Both terms contribute to the static dielectric constant ε_s. Thus by adapting the electrostatic calculation in eqn (2.4), the slow ionic contribution to the polarization energy for a charge e in a sphere of radius R is obtained as

$$\Delta E_{ion} = -e^2/(8\pi\varepsilon_0 R)(1/\varepsilon_{opt} - 1/\varepsilon_s). \quad (2.38)$$

Again, this approximation is hardly suitable for serious numerical predictions, but it suggests that typical values of ΔE_{ion} in a metal oxide may be in the region of 1–2 eV. The importance of this quantity is that when a carrier moves in a band, the ionic polarization is too slow to follow it effectively, and most of the stabilization which this term would otherwise give is lost. On the other hand, the stabilization gained by delocalizing in a band rather than staying on one site is approximately $W/2$, where W is the band width. Thus if the magnitude of the ionic polarization is greater than $W/2$ it will be energetically favourable for the carrier to remain localized.

The situation just described is known as a **small polaron**; that is the trapping of an electron or hole at one site by the local lattice polarization which it causes.[1,99] The simple electrostatic estimate for ΔE_{ion} is very approximate, as there will be a contribution from local bonding effects. Indeed, there is an alternative chemical interpretation of small polarons. An electron or hole corresponds to a change of oxidation state of one atom, and this will cause a change in the lengths of neighbouring bonds, or sometimes a more complicated alteration of coordination geometry. If this local distortion effect is sufficient it can cause *valence trapping*,[100] localizing the carrier at one site. There is always of course the competing tendency to delocalization, depending on the band width.

The estimate given above shows that rather narrow bands are required for small-polaron formation, probably around 3 eV or less. This is often the case for d bands in the $3d$ series, and some of the clearest examples of small-polaron behaviour are found with doped magnetic insulators (see Chapter 4). Even when the bands are somewhat wider, so that small polarons are not formed by free carriers, there is evidence for this kind of behaviour when electrons or holes are bound to an impurity site. In fact it appears that impurity binding and small polaron formation are to some extent *synergic* effects which tend to reinforce each other.

Under most conditions, small polarons must move from site to site by a thermally activated **hopping** process. They also have various spectroscopic signatures, such as an optical absorption band corresponding to a Franck–Condon excitation to a neighbouring atom. To see how these effects can arise, it is convenient to take a simple model, that of just two metal atoms with surrounding oxygens. Figure 2.25 illustrates such a 'dimer', which can be regarded as a small cluster extracted from an oxide lattice.[1] The horizontal coordinate in the diagram represents a vibrational mode which distorts the lattice so that the bond distances around the two metals become unequal. There are two potential curves, corresponding to states where the extra electron resides on one atom or the other. As can be seen, the minimum of each curve occurs at the appropriate distorted configuration, giving the difference of equilibrium bond lengths around the atom with or without the electron. At the symmetrical configuration in the middle, the curves cross, and in the

see p186

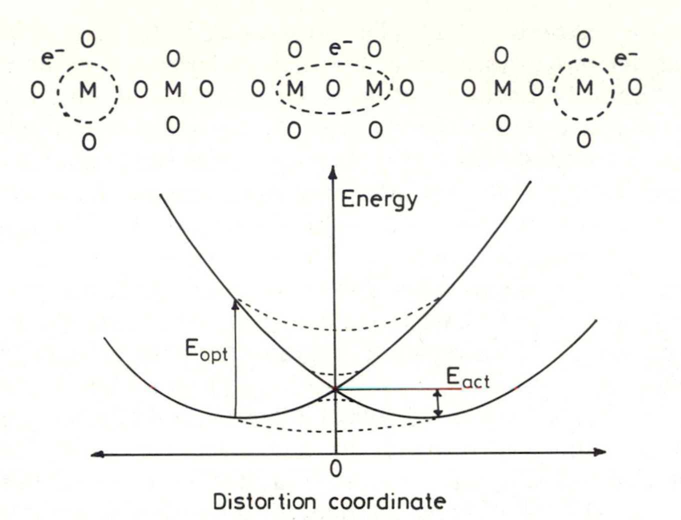

FIG. 2.25 The configuration coordinate model for small polarons. The upper pictures show a portion of an oxide lattice containing two metal atoms. Below, the horizontal coordinate corresponds to a vibrational mode which distorts the bond lengths as shown. The two potential curves correspond to electrons either on one metal or the other. Thermal electron transfer can occur at the crossing point, and requires an activation energy E_{act} to distort the corresponding ground-state configuration. Optical excitation of electrons between the two states takes place vertically, according to the Frank–Condon principle. The dashed curves represent the effect of electronic interaction between the two localized states: when this is sufficient, the symmetrical configuation with a delocalized electron is lowest in energy.

simplest model the system must acquire a sufficient vibrational energy to reach this point before the electron can move between the atoms. If there is some overlap between the two states, a 'bonding' combination of electronic states will be formed near the crossing point, which will lower the energy barrier. The diagram shows what happens if the mixing of the two states is large enough: then the minimum will move to the symmetrical point, and in the ground state the electron will be delocalized equally between the atoms. In the full solid this mixing of states gives the electronic band width: thus we see how the possibility of trapping depends on competition between the band width and the local relaxation accompanying the change of oxidation state.

This picture is known as the **configuration-coordinate** model. It has been widely used to discuss electron-transfer reactions between ions in solution, a process quite analogous to small-polaron motion in a solid.[101] The model emphasizes that the **activation energy** for electron transfer arises from the atomic rearrangement necessary to bring about a crossing of the two energy levels. This in turn depends on the extent of the perturbation involved in changing the oxidation states, and it is not surprising, as we shall see in Chapter 4, that some of the chemical trends found by solution chemists can be carried over to the solid state.

The configuration coordinate model also shows how the optical spectrum of a small polaron can arise. According to the Franck–Condon principle the maximum intensity in an absorption band corresponds to a vertical line on the diagram, as the atoms do not have time to move during the transition.[44] The line drawn in Fig. 2.25 shows the transition of an electron, from its trapped state on one atom, to the upper curve where it is on the other atom while this is the *wrong* geometry. Some of the optical absorptions of carriers in doped oxides (discussed in Section 4.3.1) are almost certainly of this kind.

Even when the bands are wide enough for small polarons to be unfavourable, the local polarization effects of carriers are significant. A **large polaron** is the name given to an electron or hole moving through the lattice with a more extended region of polarization.[13] The theory of large polarons is in many ways more complicated than that of small polarons.[102,103] An important prediction, however, is that although a carrier is not trapped by large-polaron formation, its apparent effective mass will be increased and its mobility reduced. Whereas the thermally activated mobility small polarons increases with temperature, that of large polarons is predicted to *decrease* due to strong scattering by thermally excited dipole-active lattice vibrations.

Polaron formation is probably always important in semiconducting oxides because of the highly polar nature of the oxide lattice. But the distinction between large and small polarons can often be hard to make. The carriers in some reduced oxides show behaviour which is apparently borderline between the two extremes: thus one can find extremely high effective masses, indicating strong electron–lattice interaction effects, but without the activated mobility caharacteristic of small polarons.[18] Some of the problems in this area will be discussed in connection with semiconduction, in Sections 4.1.2 and 4.3.1.

References

1. Cox, P.A. (1987). *The electronic structure and chemistry of solids*. Oxford University Press.
2. Philips, C.S.G. and Williams, R.J.P. (1965). *Inorganic chemistry*. Clarendon Press, Oxford.

3. O'Keefe, M. and Navrotsky, A. (eds) (1981). *Structure and bonding in crystals*, Vol. 1. Academic, New York.
4. Mott, N.F. and Gurney, R.W. (1964). *Electronic processes in ionic crystals*. Dover Publications, New York. First published (1940) by Oxford University Press.
5. Catlow, C.R.A. and Stoneham, A.M. (1983). *J. Phys. C: Solid State Phys.*, **16**, 4321.
6. van Houten, S. (1960). *J. Phys. Chem. Solids*, **17**, 7.
7. Burns, R. G. (1970). *Mineralogical applications of crystal field theory*. Cambridge University Press.
8. Cotton, F.A. (1990). *Chemical applications of group theory*, (3rd edn). Wiley, New York.
9. Van Vleck, J.H. (1939). *J. Chem. Phys.*, **7**, 72.
10. Gerloch, M. and Slade, R.C. (1973). *Ligand field parameters*. Cambridge University Press.
11. Ballhausen, C.J. (1962). *Introduction to ligand field theory*. McGraw-Hill, New York.
12. Figgis, B.N. (1966). *Introduction to ligand fields*. Interscience, New York.
13. Kittel, C. (1976). *Introduction to solid state physics*, (5th edn). Wiley, New York.
14. Gerloch, M. (1986). *Orbitals, terms and states*. Wiley, Chichester.
15. Waddington, T.C. (1959). *Prog. Inorg. Chem. Radiochem.*, **1**, 157.
16. Jahn, H.A. and Teller, E. (1937). *Proc. Roy. Soc. Lond.* A, **161**, 220.
17. Burdett, J.K. (1980) *Molecular shapes*. Wiley, New York.
18. Goodenough, J.B. (1971). *Prog. Solid State Chem.*, **5**, 145.
19. Sturge, M.D. (1967). *Solid State Phys.* **20**, 91.
20. Bersuker, I.D. (1984). *The Jahn–Teller effect and vibronic interactions in modern chemistry*. Plenum, New York.
21. Purcell, K.F. and Kotz, J.C. (1968). *Inorganic chemistry*. W.B. Saunders, Philadelphia.
22. Dunitz, J.D. and Orgel, L.E. (1960). *Adv. Inorg. Chem. Radiochem.*, **2**, 1.
23. Henderson, P. (1982). *Inorganic geochemistry*. Pergamon, Oxford.
24. Wells, A.F. (1984). *Structural inorganic chemistry*, (5th edn). Oxford University Press.
25. McClure, D.S. (1959). *Solid State Phys*, **9**, 399.
26. Lever, A.B.P. (1984). *Inorganic electronic spectroscopy*, (2nd edn). Elsevier, Amsterdam.
27. Low, W. and Offenbacher, E.L. (1965). *Solid State Phys*, **17**, 136.
28. Muller, K.A. (1981). *J. Physique*, **42**, 551.
29. Condon, E.U. and Shortley, G.H. (1935). *The theory of atomic spectra*. Cambridge University Press.
30. Watson, R.E. (1960) *Phys. Rev.*, **118**, 1036; (1960). Ibid., **119**, 1934.
31. Jorgensen, C.K. (1962). *Orbitals in atoms and molecules*. Academic Press, London.
32. Tanabe, Y. and Sugano, S. (1954). *J. Phys. Soc. Japan*, **9**, 753.

33. Griffith, J.S. (1964). *The theory of transition metal ions*. Cambridge University Press.
34. McClure, D.S. (1963). *J. Chem. Phys.*, **38**, 2289.
35. Jost, W. (1933). *J. Chem. Phys.*, **1**, 466.
36. Mott, N.F. and Littleton, M.J. (1938). *Trans. Faraday Soc.*, **34**, 485.
37. Hayes, W. and Stoneham, A.M. (1985). *Defects and defect processes in non-metallic solids*. Wiley, New York.
38. Cox, P.A. (1981). Valence band photoelectron spectroscopy of solids. In *Emission and scattering techniques* (ed. P. Day), NATO ASI Series C, Vol. 73. Reidel, Dordrecht.
39. Mattheis, L.F. (1972). *Phys. Rev.* B, **5**, 290.
40. Kuntz, A.B. (1975). *Phys. Rev.* B, **12**, 5890.
41. Catlow, C.R.A. (1987). *Solid state chemistry: techniques* (ed. A.K. Cheetham and P. Day). Oxford University Press.
42. Mackrodt, W.C. (1984). *Solid State Ionics*, **12**, 175.
43. Stoneham, A.M., Sangster, M.J.L., and Tasker, P.W. (1981). *Phil. Mag.* B, **43**, 609.; (1981). Ibid., B **44**, 603.
44. Atkins, P.W. (1990). *Physical chemistry*, (4th edn). Oxford University Press.
45. March, N.H. (ed.) (1974). *Orbital theories of molecules and solids*. Clarendon Press, Oxford.
46. Murrell, J.N., Kettle, S.F.A., and Tedder, J.M. (1978). *The chemical bond*. Wiley, Chichester.
47. WcWeeny, R. (1979). *Coulson's valence*, (3rd edn). Oxford University Press.
48. Ziman, J.M. (1972). *Principles of the theory of Solids* (2nd edn). Cambridge University Press.
49. Dahl, J.P. and Ballhausen, C.J. (1968). *Adv. Quantum Chem.*, **4**, 170.
50. Larsen, E. and LaMar, G.N. (1974). *J. Chem Educ.*, **51**, 633.
51. Lawley K.P. (ed.) (1987). *Ab-initio methods in quantum chemistry*, Advances in Chemical Physics, Vols 58 and 59. Wiley, New York.
52. Janssen, G.J.M. and Nieuwpoort, W.C. (1987). *Phys. Rev.* B, **38**, 137.
53. Hafid, L., Michel-Calendini, F.M., Chermette, H., and Moretti P. (1987). *Crystal Lattice Defects and Amorphous Materials*, **15**, 97.
54. Gunter, P. (1982). *Phys. Rep.*, **93**, 199.
55. Johnson, K.H. (1973). *Adv. Quantum Chem.*, **7**, 143.
56. Slater, J.C. (1974). *The self consistent field for molecules and solids: quantum theory of molecules and solids*, Vol. 4. McGraw-Hill, New York.
57. Hurley, A.C. (1976). *Electron correlation in small molecules*. Academic, London.
58. Sinanoglu, O. and Bruekner, K.A. (1970). *Three approaches to electron correlation in atoms*. Yale University Press, New Haven.
59. McWeeny, R. and Sutcliffe, B.T. (1969). *Methods of molecular quantum mechanics*. Academic, London.
60. Hubbard, J., Rimmer, D.E., and Hopgood, F.R.A. (1966). *Proc. Phys. Soc. Lond.*, **88**, 13.

61. Davis, L. C. (1982). *Phys. Rev.* B, **25**, 2912; van der Laan, G. (1982). *Solid State Commun.*, **42**, 165.
62. Fujimari, A. and Minami, F. (1984). *Phys. Rev.* B, **30**, 957.
63. Sawatzkty, G. A. (1988). In *Core level spectroscopy in condensed systems,* (ed. J. Kanamori and A. Kotani), Springer Series in Solid State Sciences, Vol. 81. Springer, Berlin.
64. Janssen, G. J. M. and Nieuwpoort, W. C. (1988). *Int. J. Quantum Chem. Symp.*, **22**, 679.
65. Cohen, M. L. and Chelikovsky, J. R. (1985). *Electronic and optical properties of semiconductors,* Springer Series in Solid State Sciences, Vol. 75. Springer, Berlin.
66. Jones, H. (1975) *The theory of Brillouin zones and electronic states in crystals.* (2nd edn). North Holland, Amsterdam.
67. Mattheis, L. F. (1976). *Phys. Rev.* B, **13**, 2433.
68. Slater, J. C. (1964). *Adv. Quantum Chem.*, **1**, 35.
69. Cox, P. A., Goodenough, J. B., Tavener, P. J., Telles, D., and Egdell R. G. (1986). *J. Solid State Chem.*, **62**, 360.
70. Cardona, M. and Ley, L. (eds). (1978). *Photoemission in solids*, Topics in Applied Physics, Vol. 26. Springer, Berlin; (1979). Ibid., Vol. 27.
71. Cox, P. A., Egdell, R. G., Harding, C., Orchard, A. F. Patterson, W. R., and Tavener, P. J. (1982). *Solid State Commun.*, **44**, 837.
72. Graebner, J. E., Greiner, E. S., and Ryder, W. D. (1976). *Phys. Rev.* B, **13**, 2426.
73. Pines, D. and Nozieres, P. (1966). *The theory of quantum fluids.* Benjamin, New York.
74. Fawcett, E. (1988). *Rev. Mod. Phys.*, **60**, 209.
75. Terakura, K., Oguchi, T., Williams, A. R., and Kubler, J. (1984). *Phys. Rev.* B, **30**, 4734.
76. Miller, J. S. and Epstein, A. J. (1976). *Prog. Inorg. Chem.*, **20**, 1.
77. Schlenker, C., Dumas, J., Escribe-Filippini, C., Guyot, H, Marcus, J. and Fourcaudot, G. (1983). *Phil. Mag.* B, **52**, 643.
78. Wilson, J. A., DiSalvo, F. J. D., and Mahajan, S. (1975). *Adv. Phys.,* **24**, 117.
79. Steiner, E. (1976). *The determination and interpretation of molecular wave functions.* Cambridge University Press.
80. Mahan, G. D. (1981). *Many-particle physics.* Plenum, New York.
81. Mott, N. F. (1974). *Metal–insulator transitions.* Taylor and Francis, London.
82. Hubbard, J. (1963). *Proc. Roy. Soc. Lond.* A, **276**, 238.
83. Brandow, B. H. (1977). *Adv. Phys.*, **26**, 651.
84. Tjeng, L. H., Eskes. H., and Sawatzky, G. A. (1989). In ref. 88.
85. Cox, P. A. and Orchard, A. F. to be published.
86. Wilson, J. A. (1972). *Adv. Phys.*, **21**, 143.
87. Lang, J. K., Baer, Y., and Cox, P. A. (1981). *J. phys. F: Met. Phys.,* **11**, 121.
88. Fukuyami, H., Maekawa, S., and Malozemoff, A. P. (1989). *Strong*

correlation and superconductivity, Springer Series in Solid State Sciences, Vol. 89. Springer, Berlin.

89. Bednorz, J.G. and Muller, K.A.(1990). *Superconductivity*, Springer Series in Solid State Sciences, Vol. 90. Springer, Berlin.
90. Baumeister, P.W. (1961). *Phys. Rev.* **121**, 359.
91. Knox, R.S. (1963). *Solid State Phys., Suppl.* 5.
92. Stoneham, A.M. (1985). *Theory of defects in solids*, (2nd edn). Oxford University Press.
93. Catlow, C.R.A., Macrodt, W.C., and Norgett, M.J. (1977). *Phil. Mag.*, **35**, 177.
94. Inkson, J.C. (1984). *Many-body theory of solids.* Plenum, New York.
95. Selme, M.O. and Pecheur, P. (1988). *J. Phys. C: Solid State Phys*, **21**, 1779.
96. Zaanen, J. (1986). The electronic structure of transition metal compounds in the impurity Model. Ph.D. Thesis, University of Groningen.
97. Mott, N.F. and Davis, E.A. (1979). *Electron Processes in non-crystalline solids*, (2nd edn). Oxford University Press.
98. Elliot, S.R. (1983). *Physics of amorphous materials*. Longman, London.
99. Adler, D. (1967). In *Solid state chemistry,* (ed. N.B. Hannay), Vol. 2. Plenum, New York.
100. Robin, M. and Day, P. (1967). *Adv. Inorg. Chem. Radiochem.*, **10**, 247.
101. Cannon, R.D. (1980). *Electron transfer reactions*. Butterworths, London.
102. Appel, J. (1968). *Solid State Phys.,* **21**, 193.
103. Devreese J. (ed.) (1972). *Polarons in ionic crystals and polar Semiconductors.* North Holland, Amsterdam.
104. Cox, P.A. (1984). Chemical bonding in transtion metal oxides. In *Basic Properties of binary oxides*, (ed. A. Dominguez-Rodriguez, J. Castaing, and R. Marquez) University of Seville.

3

INSULATING OXIDES

Insulators have a gap between occupied and empty levels, so that electronic conduction is impossible in the ground state. In transition metal oxides such a gap may arise in various ways. The first two parts of this chapter discuss compounds which may be understood in terms of conventional band theory: in the d^0 oxides treated in Section 3.1 the 'metal *d*' conduction band is empty, while Section 3.2 considers some compounds in which this band is either full, or split by crystal field effects so as to give a closed-shell ground state. The magnetic insulators forming the subject of Section 3.4, on the other hand, have open-shell ground states with unpaired electrons. These compounds are not well described by the band model, as the gap arises largely from electron-repulsion effects. Before discussing 'pure' magnetic insulators, it is appropriate (Section 3.3) to look at the properties of insulating oxides containing transition metal impurities, especially as they show some similar properties.

Many of the compounds described in this chapter are not very good insulators compared with non-transition metal oxides such as MgO or Al_2O_3, but show some degree of semiconduction. *Intrinsic* semiconduction arises from the thermal excitation of electrons and holes across the band gap. In most transition metal oxides the band gap is sufficiently large that intrinsic carriers are usually swamped by those introduced by defects and impurities. For this reason the semiconductive properties are discussed in connection with defects in Chapter 4.

3.1 d^0 compounds

d^0 oxides are formed by the earlier transition elements, as far as manganese in the 3*d* series, and up to ruthenium and osmium in the 4*d* and 5*d* elements (see Fig. 1.2). In many ways, the electronic structures of these compounds are the 'simplest' and least controversial of all classes of transition metal oxides, although by no means are all the details understood. The two following sections discuss the band gaps and optical properties, and the technologically important dielectric properties.

3.1.1 Band gaps and spectroscopic transitions

The nature of the filled and empty electronic states in d^0 oxides can be exemplified by looking in detail at rutile, the most important crystal form of TiO_2. The band structure has been calculated by a variety of methods.[1] Except for the fact that the conduction band is empty, it is essentially similar to the band structure shown in Fig. 2.14 for RuO_2, which has the same crystal structure. The density of states, from a semiempirical calculation, is shown in Fig. 3.1 (a), and compared with the MO levels of an octahedral cluster.[2] As expected from the discussion in Chapter 2, the filled valence band is derived largely from oxygen 2*p* levels and the conduction band from Ti 3*d* orbitals, although considerable mixing of these atomic levels takes place. The crystal field splitting of t_{2g} and e_g orbitals in the conduction band can be seen, and at higher energy, the *d* band merges into one derived from 4*s* and 4*p* states of titanium.

In Fig. 3.1 the theoretical predictions are also compared with some measured spectroscopic transition energies. **X-ray absorption** measures transitions from a sharp core-level into *empty* orbitals.[3] Similar transitions can be seen with **electron energy-loss spectroscopy** (EELS), using incident electrons of high kinetic energy. The strongest peaks appearing in these experiments are those allowed by the electric-dipole selection rule of atomic spectroscopy. That is:

$$\Delta l = \pm 1 \tag{3.1}$$

for the angular momentum *l* of the atomic orbitals involved. In the Ti L_{23} spectrum, where transitions from Ti 2*p* orbitals are observed, the allowed final states are therefore expected to be Ti *d*- and *s*-like. Indeed, the peaks correspond to the two Ti 3*d*-derived bands, and the higher energy 4*s* levels. On the other hand, Ti *K* absorption has a 1*s* initial state, and final states with Ti *p* character are expected. A small forbidden intensity for the 3*d* states does appear, but the major absorption shows Ti 4*p* states at higher energy. The oxygen *K* absorption should give oxygen 2*p* final states; the fact that substantial intensity occurs for the conduction band levels is largely a reflection of the oxygen 2*p* contribution in these.

X-ray emission spectra are excited by removing an electron from a core level by electron or X-ray bombardment. They show transitions of an electron from *filled* valence levels into the vacant core state. The Ti *K* emission shown in Fig. 3.1 has peaks corresponding to the valence band and the lower energy oxygen 2*s* band. As with the oxygen *K* absorption into conduction band states, these emission spectra show

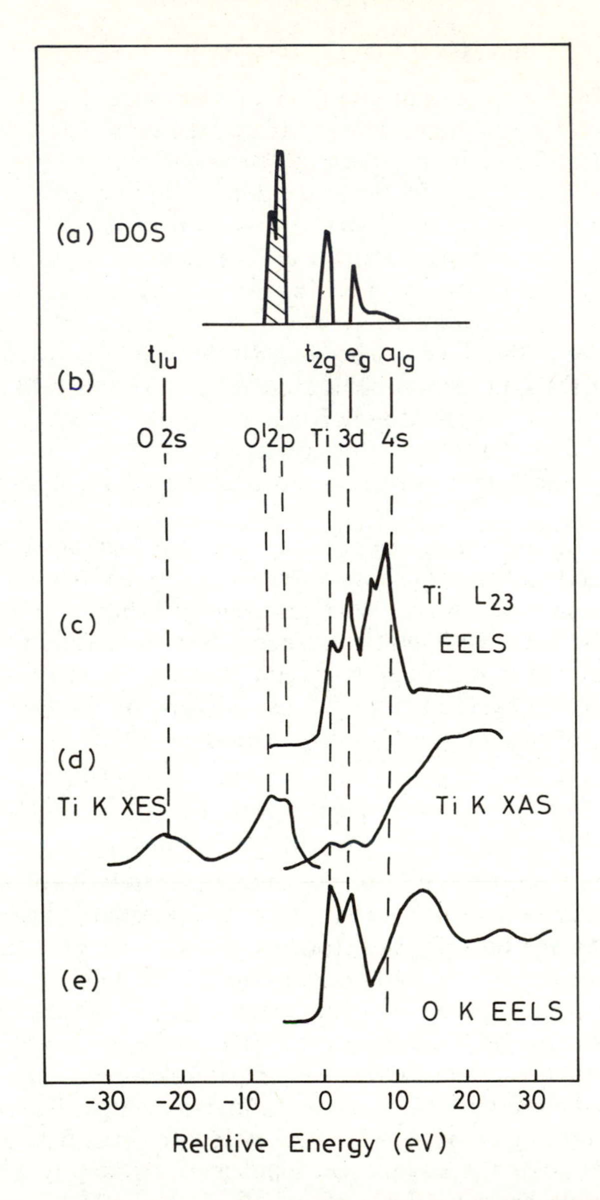

FIG. 3.1 Theoretical and experimental energy levels for TiO_2 (rutile). (a) Density of states from band-structure calculation; (b) empirical cluster molecular orbital energies; (c)–(e) core-level spectra as described in the text. (Redrawn from ref. 2.)

that some orbital mixing occurs, giving a titanium contribution to the predominantly oxygen-derived valence band.

Figure 3.2 shows some measurements of the **optical absorption edge** of rutile at low temperatures. In (a) the photoconductivity and optical absorption spectra are plotted together.[4] The near coincidence of the two edges shows that optical absorption produces free carriers, and thus corresponds to a true band-gap transition without a strongly bound exciton being formed (see Section 2.4.2). However, the spectra in Fig. 3.2(b), measured at a lower temperature and with higher resolution, show more complex structure. The direct absorption spectra are shown, together with ones where a wave length modulation technique has been used to sharpen the features.[5] There is a small exciton peak about 4 meV below the main edge at 3.035 eV, and the edge itself has a second exciton line superimposed. These exciton lines are labelled 1*s* and 2*p* according to the hydrogenic scheme suggested by the *Wannier exciton* model described in Section 2.4.2; the very small binding energies are expected because of the high dielectric constant of rutile (see Table 3.2 later). There is also some **dichroism** at the edge, the lower-energy

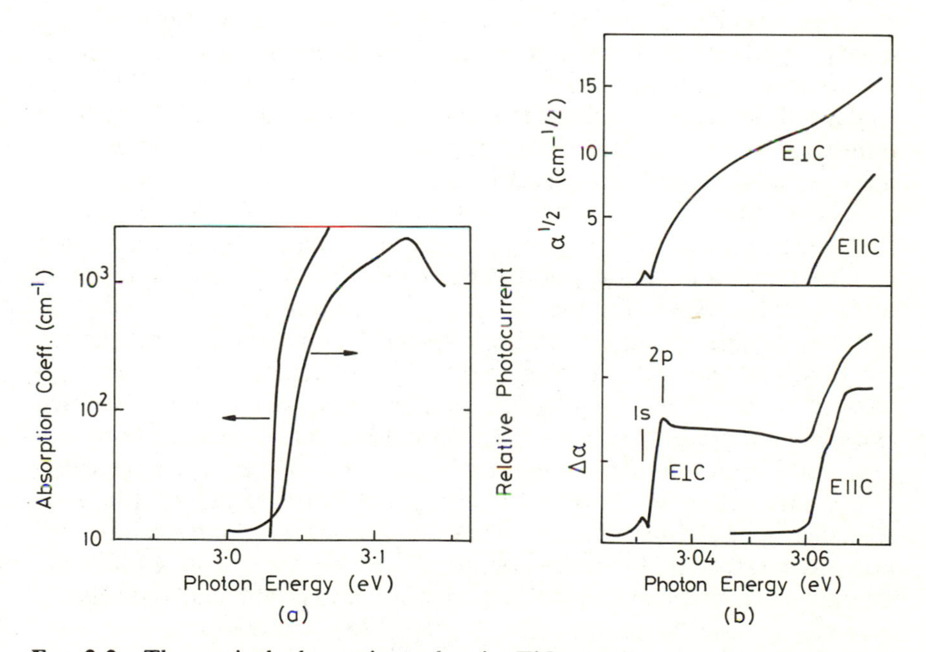

FIG. 3.2 The optical absorption edge in TiO_2 (rutile). (a) Absorption and photoconductivity measured at 4.2 K.[4] (b) Higher resolution spectra measured in two polarizations at 1.6 K; the lower plot shows differential spectra obtained by a modulation technique. 1*s* and 2*p* exciton states are marked.[5]

features being observed only with the electric vector of the radiation polarized perpendicular to the *c* axis.

There is some controversy about the detailed interpretation of this edge structure in rutile. The comparative weakness of the absorption suggests a forbidden transition, which could arise in one of two ways. An **indirect** gap occurs when the top of the valence band is at a different wavevector from the conduction-band minimum. Then the lowest-energy absorption must break the selection rule for electronic states in band-to-band transitions:[3]

$$\Delta \boldsymbol{k} = 0. \tag{3.2}$$

This selection rule depends on perfect lattice periodicity, and like symmetry-based selection rules for localized transitions, can be broken by the intervention of appropriate vibrational modes. Thus $\boldsymbol{k} \neq 0$ phonon modes of the lattice can be combined with the electronic excitation, and the energy of the transition will then include a quantum of the vibrational mode involved. The $E \parallel c$ spectrum of rutile has been interpreted in this way, but the gap measured in the other polarization direction seems to of a different nature. It is thought to be direct (satisfying eqn (3.2)) but forbidden by local symmetry requirements, much as *d–d* transitions are forbidden in centro-symmetric environments.[5] These features of the absorption edge of TiO_2 seem to depend on subtle details of the band structure which are difficult to reproduce in calculations. Some of these give direct gaps; others give indirect gaps for both polarizations.[1]

The example of rutile suggests that the principal features of the electronic structure of d^0 oxides, such as the atomic orbital constitution of the valence and conduction bands, may be fairly readily interpreted using simple models. Details of the band structure, and especially the wavevector and symmetry of the top-filled and lowest empty levels, are harder to ascertain. For the perovskite structure common for ternary d^0 oxides, most calculations predict direct band gaps.[6] Detailed experimental studies such as those just described, however, have not been carried out on many compounds. Because of the strong electron–lattice interaction in these polar compounds, the detail of the edge structure is easily obscured by vibrational excitation, and excitons are not often resolved. Evidence for combined excitonic and vibrational interactions sometimes comes from emission measurements, as discussed below.

Table 3.1 shows band gaps for a selection of d^0 oxides, obtained from optical measurements.[1,7,8] A look at the literature shows many small differences in the values quoted for a single material, reflecting differences in the conditions or the technique employed. As in the

Table 3.1 *Band gaps for some* d^0 *oxides*[a]

Binaries		Ternaries	
Compound	E_g(eV)	Compound	E_g(eV)
(3d⁰)			
TiO_2(rutile)	3.0	$MgTiO_3$	3.7
TiO_2(anatase)	3.2	$SrTiO_3$	3.4
		$BaTiO_3$	3.2
		$La_2Ti_2O_7$	4.0
V_2O_5	2.2		
CrO_3	2		
(4*d*⁰)			
		$SrZrO_3$	5.4
Nb_2O_5	3.9	$LiNbO_3$	3.8
		$KNbO_3$	3.3
MoO_3	3.0		
(5*d*⁰)			
Ta_2O_5	4.2	$LiTaO_3$	3.8
		$NaTaO_3$	3.8
WO_3	2.6		

[a] Data from refs 1, 7, 8.

case of rutile, the optical absorption edge is not sharp, and decreases somewhat with temperature: both features are partly due to the simultaneous excitation of lattice vibrations, i.e. to the influence of electron–phonon interactions. Some attempts to define precise band gaps rely on fitting the optical absorption to a power-law expression of the energy above threshold. Alternatively, the measured edge may be sharpened by employing modulation techniques as in Fig. 3.2(b).

Nevertheless, the uncertainties in individual values are much smaller than the differences between compounds, and Table 3.1 illustrates some important trends. Within each series, band gaps decrease with increasing atomic number: for example $TiO_2 > V_2O_5 > CrO_3$, with similar trends in the 4*d* and 5*d* elements. There is also a clear increase from the 3*d* to heavier elements of the same group: V < Nb ~ Ta, and so on, the difference between 4*d* and 5*d* elements being rather small.

In the *ionic* model (Section 2.1.2) we would expect the band gap of a d^0 oxide containing M^{n+} to depend on the ionization energy corresponding to

$$M^{(n-1)+} = M^{n+} + e^-. \tag{3.3}$$

Atomic ionization energies do indeed follow the same trends mentioned above, although as pointed out in Section 2.1.2, the ionic model cannot be expected to provide *quantitative* estimates of band gaps.

We might also expect a correlation between the band gap and the ease of chemical *reduction* of the corresponding metal ion. A small gap implies low-lying empty orbitals, which can readily accept electrons from some other species. This can be seen with species in aqueous solution, when the ease of reduction, or—what is the same thing—the *oxidizing power*, follows the sequence $Ti^{4+} < V^{5+} < Cr^{6+}$ in line with the falling band gaps.[9] Similar chemical trends are shown with solids, as d^0 oxides can lose oxygen, leaving some metal ions in a lower oxidation state and giving *n*-type semoconducting behaviour. Generally speaking, those oxides with smaller band gaps have more tendency to do this. They are also easier to reduce by the insertion of elements such as hydrogen or alkali metals into the lattice.[10]

Values given for ternary compounds in Table 3.1 are similar to those of the binary oxide containing the same d^0 ion. Empty orbitals of the ternary element, at least when this is from one of the pre-transition groups, are at higher energy than the transition metal *d* orbitals. For example in $SrTiO_3$ the Sr 5*s* orbital contributes to empty orbitals at considerably higher energy than the Ti 3*d*, but not to the lower part of the conduction band.[6] Some of the trends in ternary compounds seem to reflect structural differences. $MgTiO_3$ and $LiNbO_3$, with structures based on corundum, have larger gaps than the corresponding perovskites $CaTiO_3$ and $NaNbO_3$. An ionic picture suggests that lattice potentials could be involved, but the major difference probably lies in the conduction band *width*. Decreasing the band width will lead to an increase in the gap, if the barycentre energy remains the same. According to the qualitative band model of Section 2.3.1, the *d* band width in most oxides comes from indirect metal–oxygen–metal interactions, and therefore depends on the degree of connection between the structural units. Consider the sequence TiO_2 (rutile, with corner- and edge-sharing octahedra), $SrTiO_3$ (perovskite, all corner-sharing), and $MgTiO_3$ (ilmenite, only partial corner sharing). The progressive disconnection of the octahedra leads to a decrease in the interactions that give the conduction band width, and may be the cause of the increase of band gap along this series.

As the bands becomes narrower, the excited state formed in optical absorption should become more localized. It is difficult to observe exciton structure at the absorption edges of most transition metal oxides because of the strong interference with lattice vibrations referred to above. But some evidence of increasing localization comes from studies of **fluorescence** or **luminescence**, coming from the radiative recombination of electrons and holes.[8] In rutile, as in many solids where optical absorption produces free carriers, luminescence appears only from impurity sites where excited carriers are trapped. In titanates with smaller band widths, it is possible to observe direct luminescence from the excited state found in absorption. Furthermore the maximum temperature at which such luminescence can be seen is larger in $MgTiO_3$ than in $SrTiO_3$. These observations suggest that localized excited states are being formed. The fact that such states are not 'simple' excitons, however, but involve vibrational interactions, is shown by the appearance of a **Stokes shift**: that is the luminescence appears at longer wavelength (lower photon energy) than the absorption.

The origin of the Stokes shift can be understood by using the Franck–Condon principle[11] (see Fig. 3.3). When a localized excited state is formed, its equilibrium geometry will be different from the ground state, shown by the horizontal shift of the excited-state potential curve in the figure. Optical transitions occur *vertically* on this diagram; thus the absorption is increased in energy, by the simultaneous excitation of some vibrational energy. However, this excited-state vibrational energy is lost to the lattice very rapidly on the time-scale required for emission (about 10^{-8} s). Thus the emission comes from close to the minimum of the excited-state curve, and is at *lower* energy than the vibrationless (0–0) electronic energy difference. The Stokes shift in the luminescence from titanates is found to increase with the degree of 'disconnection' of Ti–O octahedra referred to above.[8] The larger vibrational relaxation is another sign of the increased localization of the excited state in compounds with narrower conduction bands.

This discussion suggests that the band model becomes less satisfactory for interpreting the optical properties of d^0 compounds when the upper 'conduction band' level becomes narrow. For the later elements in each series, the d^0 oxides become increasingly *molecular* in structure, with metal–oxygen groups completely disconnected from each other. Thus $KMnO_4$ contains discrete tetrahedral MnO_4^- (permanganate) ions. The strong optical absorption of this solid, which covers most of the visible region of the spectrum, is essentially the same as that found with isolated permanganate ions, either in solution or doped as impurities in other solids.[12] Both the energy and the vibrational structure of the absorption indicate an essentially localized excited state. A band structure for a compound such as $KMnO_4$ would presumably show a very narrow 'conduction band', and would be less useful for interpreting the properties than a cluster model based on a discrete permanganate group.

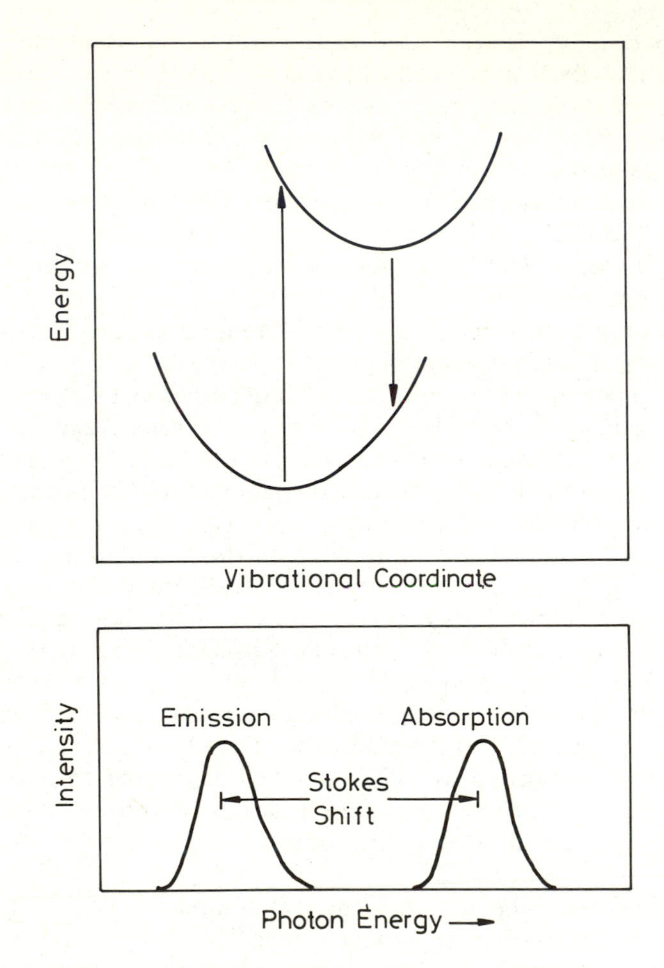

Fig. 3.3 Origin of the Stokes shift in emission. (a) Ground- and excited-state potential curves with a horizontal coordinate corresponding to a local vibrational mode. Vertical lines show the most intense transitions in absorption and emission, expected from the Franck–Condon principle. Typical spectra are shown (b).

3.1.2 Dielectric and non-linear properties

The dielectric properties of a solid represent its response to an applied electric field. They have an important influence on the behaviour of electronic carriers introduced in doping, for example on the binding energies of impurity states and the formation of polarons, discussed in Section 2.4. The interesting dielectric properties of many d^0 oxides also give rise to a number of applications.

The polarization $\boldsymbol{P}$ induced a medium by an electric field $\boldsymbol{E}$ is given by

$$\boldsymbol{P} = \varepsilon_0 (\varepsilon_r - 1) \boldsymbol{E} \tag{3.4}$$

where ε_0 is the permittivity of free space, and ε_r the (relative) dielectric 'constant'. Strictly speaking, as $\boldsymbol{P}$ and $\boldsymbol{E}$ are vector quantities, ε_r is a tensor. It is also far from being constant, but depends on temperature and pressure, as well as on the frequency at which it is measured.[13] Table 3.2 shows a selection of values, making the conventional distinction between dielectric constants determined at *optical frequencies* from the refractive index by

$$n^2 = \varepsilon_{\text{opt}} \tag{3.5}$$

and *static* ε_s values measured at low frequencies. The tensor properties are shown only for the case of TiO_2, where there are two values, depending on the orientation of the electric field with respect to crystal axes. A few examples are also given to show that ε_s may vary strongly with temperature.

The optical dielectric constant is determined by the *electronic* polarizability of the solid, as the ions as a whole are unable to respond at these frequencies. There is a rough inverse correlation between ε_{opt} and the band gap of a solid. Thus in TiO_2, where the band gap (3.0 eV) is just outside the limit of the visible spectrum, the values of ε_{opt} are close to the highest that can be obtained in a solid which is entirely

Table 3.2 *Dielectric constants of some* d^0 *oxides*[a]

Compound	Temperature (K)	ε_{opt}	ε_s
TiO_2(∥c)	300	8.4	170
	0	–	257
(∥a)	300	6.8	86
	0	–	111
$SrTiO_3$	0	6	~30000
$CaTiO_3$	300	6	180
	100	–	330
$PbTiO_3$	300	7	(*see Fig. 3.4*)
$CaZrO_3$	300	–	25
MoO_3(av.)	300	5	20
$KTaO_3$	0	–	~4000
WO_3	300	5	(*v. variable*)

[a] Data from refs 1, 7. Values quoted in the literature are quite variable, especially for ε_s, which is very dependent on sample quality and purity.

transparent to visible light; higher values are normally only found with solids with smaller band gaps, which therefore do not transmit fully. Through its relation to the refractive index (eqn (3.5)) a high ε_{opt} is associated with high *reflectivity*. This is the basis for the use of rutile as a white pigment,[14] as a high reflectivity is required for a finely divided substance to scatter light efficiently; clearly no absorption of visible light can be permitted for this application. It is also interesting that the refractive index of ZrO_2 is very similar to that of diamond. Cubic zirconia (doped with other oxides such as CaO to stabilize the fluorite structure) is used as a cheaper subsitute for diamonds in jewellery.

At low frequencies, the dielectric constant also has contributions from the relative motion of different ions under the influence of an electric field. Values of ε_s for many d^0 oxides are quite exceptionally high, and show striking temperature variations. These properties suggest an unusual degree of lattice polarizability, and are probably related to the marginal stability of many d^0 ions in a regular octahedral geometry (see Section 1.3.1). The most dramatic consequence of such polarizability is the appearance of **ferroelectricity** in some compounds with perovskite-based structures.[7,15] This term implies the appearance of a spontaneous electric polarization, analogous to the spontaneous magnetization appearing in ferromagnets below the Curie temperature T_C. Typical ferroelectrics include $NaNbO_3$, $BaTiO_3$, and $PbTiO_3$, and Fig. 3.4 (a) shows the evolution of the dielectric properties of $PbTiO_3$ with temperature.[16] The large almost discontinuous peak at 500°C in this compound is associated with a structural transition from a regular cubic perovskite at higher temperature, to a distorted form in which both Ti and Pb ions have moved off-centre (Fig. 3.4(b) [17]). Thus the low temperature form has a permanent electric dipole moment in each unit cell; the dipoles are aligned to give the spontaneous polarization in each domain of the ferroelectric form. The structural change in $PbTiO_3$ is rather typical, although some ferroelectrics are more complicated; for example $BaTiO_3$ has a Curie temperature of 120°C associated with a similar distortion, but there are further transitions to lower symmetry structures at lower temperatures.

The large dielectric constants associated with ferroelectricity have important applications, for example in capacitors. Also associated with this behaviour is **piezoelectricity**, which entails either the generation of electric polarization in response to mechanical stress, or the reverse process, of distortion under the application of an electric field.[15] Not all piezoelectric materials are ferroelectric, as only the lack of a centre of symmetry is necessary for piezoelectricity, whereas ferroelectricity also requires that the material can be **poled** by aligning its polarization

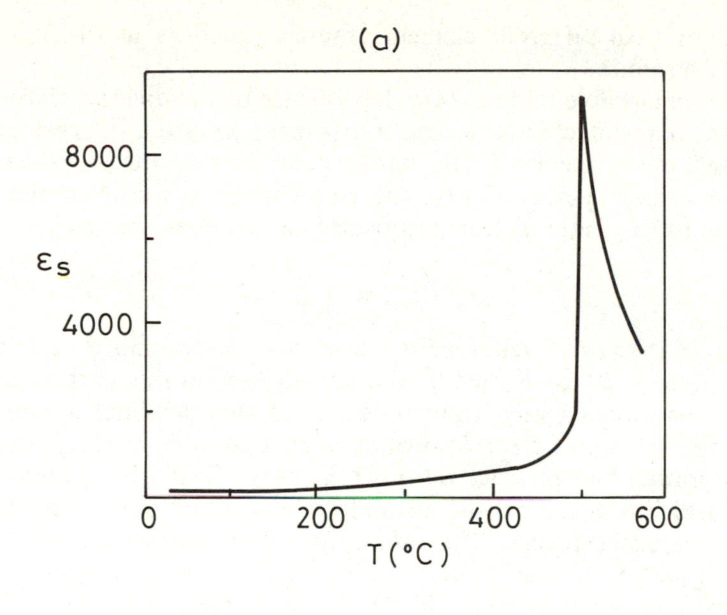

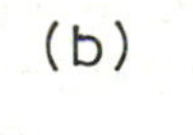

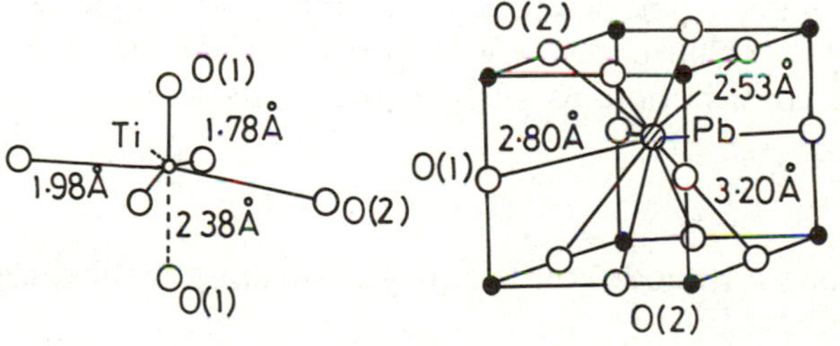

FIG. 3.4 Ferroelectricity in $PbTiO_3$. (a) Static dielectric constant ε_s as a function of temperature, showing the transition at 500 °C.[16] (b) Structure of the low-temperature phase, showing the different Ti–O (left) and Pb–O (right) distances in the distorted perovskite.[17]

in an applied field. For example quartz (SiO_2) is piezoelectric, but the asymmetry is 'locked into' the crystal structure and cannot be changed in direction, and so it is not ferroelectric. Piezoelectric materials have important applications in transducers which convert electrical into mechanical signals and vice versa. Among transition metal oxides, the relevant properties can be 'tuned' to a large extent by making solid solutions with different elements present, such as in $Pb(Zr_{1-x}Ti_x)O_3$ (PZT) ceramics.

The remarkable temperature dependence of the dielectric properties, and the associated ferroelectric transitions, are often interpreted using the **soft-mode** theory.[18] The background to this theory is based on two important ideas. Firstly, the contribution made by lattice polarization to the static dielectric constant can be expressed as

$$\varepsilon_s - \varepsilon_{opt} = \sum_i \Omega_i^2 / \omega_i^2 \tag{3.6}$$

where ω_i is the frequency of a vibrational mode, and Ω_i is related to the *effective charge* carried by the ions moving in this particular vibration. The modes which contribute to this sum are ones giving rise to a long-range dipole field, known as **transverse-optic** modes.[19] Secondly, the vibrating ions interact via this long range field, giving an *attractive* term which acts against the normal force constant coming from short-range repulsive forces. The long-range electrostatic force may lower the frequency of one mode, to give a very large contribution to the dielectric constant in eqn (3.6). In the limit, the frequency of a particular mode may even *soften* to zero; then the static dielectric constant will diverge, and the appropriate atomic motion will change from a vibrational mode into a *static* distortion of the lattice. This is one way of describing a ferroelectric transition. The temperature dependence arises from anharmonic effects in the short-range forces. These give a positive contribution to the frequency of a mode which is roughly given by

$$\omega^2 = C(T - T_0). \tag{3.7}$$

If only one soft mode contributes significantly to the dielectric constant, we then have

$$\varepsilon_s = C' / (T - T_0). \tag{3.8}$$

This is the same as the **Curie–Weiss law** for interacting magnetic systems (see Section 3.4.1). In many solids the constant T_0 is less than zero, so that a complete mode-softening does not take place, and there is no strong dielectric anomaly. It appears that in some oxides T_0 may be only just negative: thus in $KNbO_3$ and $SrTiO_3$ the dielectric constants become very large at low temperatures, although they do not become truly ferroelectric. When T_0 is greater than zero, then it is the Curie temperature below which the ferroelectric phase becomes stable.

The predictions of the simple form of the soft-mode theory given above are: (i) that the vibrational spectrum of the solid should show a transverse optic mode which softens to zero frequency at the Curie

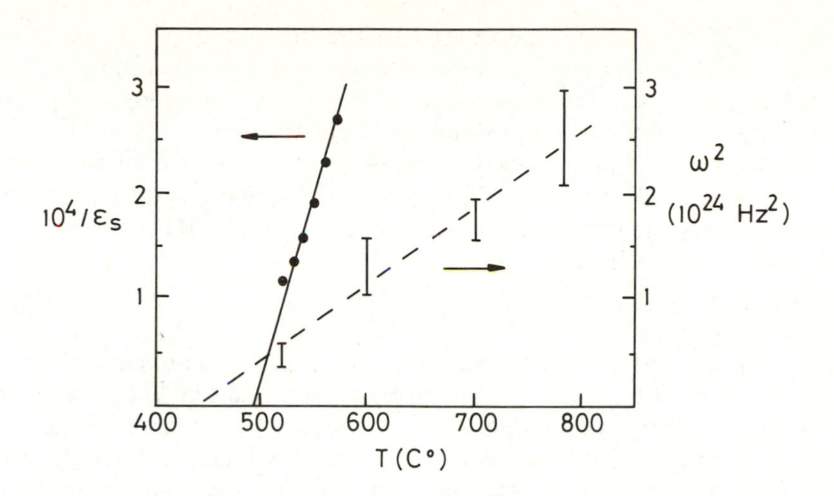

FIG. 3.5 Temperature dependence of ω^2 for the lowest energy transverse-optic-phonon mode and $1/\varepsilon_s$ for $PbTiO_3$. (Based on refs. 16,21.)

temperature according to eqn (3.7); and (ii) that the static dielectric constant should follow the Curie–Weiss form of eqn (3.8). There are some approximations in this theory: thus the damping of the modes is neglected, and anharmonicity is only roughly taken into account. Also this is a *mean-field* theory, and like similar theories of cooperative magnetism does not include the effect of fluctuations which are particularly important near the transition temperature.[20] In spite of these limitations the predictions are roughly borne out by many ferroelectric oxides. For example, Fig. 3.5 shows how the square of the frequency of the lowest TO mode and $1/\varepsilon_s$ vary with temperature in $PbTiO_3$.[16,21] Approximate straight lines are found, although they extrapolate to zero at temperatures rather lower than the Curie temperature.

The soft-mode theory emphasizes the importance of long-range electrostatic forces in the ferroelectric transition. However, it is clear that local bonding forces must also play a role. The effect of elemental substitutions on T_C throws some interesting light on these. For example, replacing some Ti^{4+} in $BaTiO_3$ by the larger Zr^{4+} ion lowers T_C. Size is therefore important, the smaller ions giving rise to easier distortion of the metal–oxygen octahedra. However, Sn^{4+}, which is smaller than Zr^{4+}, affects T_C even more; thus the d^0 configuration itself seems to be crucial.[22] There is some interesting ESR evidence which suggests that the d^5 impurity ion Fe^{3+}, substituting for Ti^{4+} in $BaTiO_3$, does *not* follow the ferroelectric distortion but remains almost central in the octahedron in spite of its similarity in size to Ti^{4+}. The peculiar features of d^0 ions may be related to the absence of electrons in the antibonding 'd' orbitals. Thus the stronger bonding (especially π type) resulting from the closer approach of one or two oxygens is not cancelled by a corresponding antibonding contribution.

Another important property of some d^0 oxides is that of **non-linear optics.**[15] In principle all materials show a non-linear response to electric fields, so that the polarization given by eqn (3.3) should really be given by a series in the field strength:

$$P = \chi_1 E + \chi_2 E^2 + \chi_3 E^3 + \dots \tag{3.9}$$

The most important non-linear term for practical applications is the second-order one χ_2, and simple arguments show that this is zero for centrosymmetric crystals. Thus many useful non-linear materials are solids without a centre of symmetry, and this includes ferroelectrics. When the symmetry conditions are met, the strongest non-linear effects are generally shown by solids with a high *linear* polarizability; that is ones with large dielectric constants. This is another reason why d^0 oxides are important in this field. Applications of non-linear optics include *frequency doubling*, for example to extend the range of frequencies obtainable from lasers, and various *electro-optical* effects in which optical properties can be coupled to low-frequency electric fields.[23] Research into possible future *optical computers* makes use of these properties.

In order to see in a simple way how non-linear optical properties arise, we may look briefly at two important materials, lithium niobate $LiNbO_3$, and potassium titanyl phosphate, $KTiO(PO_4)$ (KTP). Figure 3.6 shows part of the crystal structures of these compounds, emphasizing the local metal–oxygen coordination in each case.[24] The essential point is that this coordination is strongly *asymmetric*. In $LiNbO_3$ the asymmetry comes from the way in which octahedral sites are occupied

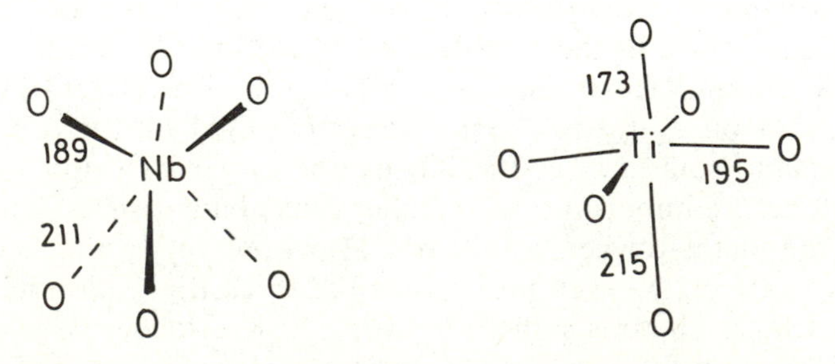

FIG. 3.6 Metal–oxygen coordination in (a) $LiNbO_3$ and (b) $KTiO(PO_4)$ (KTP), showing the asymmetry important for non-linear optical properties. (Distances in pm; from ref. 24.)

in the structure, which is based on corundum (see Figs 1.7 and 1.8). Each Nb^{5+} has Li^{+} in one neighbouring site, and an empty site opposite. Thus Li^{+}–Nb^{5+} repulsion causes a strong distortion of the octahedral coordination, with the unequal bond lengths shown. In KTP it is remarkable that there is one very short Ti–O bond, a feature often found in the ferroelectric titanates discussed above. Calculations show that the shortest metal–oxygen bonds contribute most to the electronic polarizability, so that the asymmetry of the coordination leads to an asymmetric polarizability.[25] In order for the crystal as a whole to have a non-zero value of χ_2 it is also necessary that the individual asymmetric groups in the crystal structure be aligned so that their non-linear effects do not cancel out.

3.2 Other closed-shell oxides

In addition to d^0 there are some other electron configurations that can give insulating compounds with a closed-shell diamagnetic ground state. A few examples are listed in Table 3.3.

The d^{10} configuration is found in Cu_2O, where the band gap involves the filled Cu $3d$ levels and an empty band of mostly Cu $4s$ character. Many oxides of *post-transition metals* effectively have a d^{10} configuration, as with Zn^{2+} in ZnO and Sn^{4+} in SnO_2. Increasing the nuclear charge beyond the copper group causes a rapid stabilization of the d shell, however, so that for later elements the d orbitals should be regarded as core levels.[9] In ZnO and SnO_2 the valence bands are composed essentially of oxygen $2p$ orbitals, as with d^0 oxides.

For the other electron configurations listed in Table 3.3, the band gap is a result of the crystal-field splitting of the d levels, and for this

Table 3.3 *Some oxides with closed-shell* d^n *ground states*

Compound	Electron configuration	Metal coordination[a]	Band gap (eV)[b]
Cu_2O	$3d^{10}$	2 (linear)	2.16
Ag_2O_2	$4d^{10}$	2 (linear)	
	$4d^8$	4 (square planar)	
PdO	$4d^8$	4 (square planar)	1
$LaCoO_3$	$3d^6$	6 (octahedral)	0.1
$LaRhO_3$	$4d^6$	6 (octahedral)	1.6

[a] Geometry of only the *closest* oxygens is shown.
[b] Data from refs 1, 26.

reason the metal coordination geometry has also been shown in the table. An octahedral site, as in the perovskite compounds $LaCoO_3$ and $LaRhO_3$, gives a lower t_{2g} set of three orbitals, and a closed-shell state is obtained when they are filled in *low-spin* d^6 compounds. The band gap in these compounds is presumably between a filled t_{2g} band and an empty e_g band, both composed of metal d with some antibonding oxygen admixture. As many other chemical examples show,[9] Co^{3+} is the only common ion of the 3d series with this configuration, but it appears to be only just stable in $LaCoO_3$, as the band gap is very small. Above room temperature $LaCoO_3$ undergoes a transition to a metallic form.[26] Both the experimental and theoretical details are still disputed. It has been claimed that there is a quite sudden, first-order transition, but more recent work suggests a more gradual increase in conductivity, associated with a lattice expansion.[27] It has also been suggested that a *localized* view of the d^n configuration might be more appropriate than the band model; we should note, however, that as far as the ground state is concerned it makes no difference whether we think of a localized $(t_{2g})^6$ configuration or a filled t_{2g} band. These pictures correspond to the *same* wavefunction, just as we can also think of a filled oxygen 2p band in terms of localized ions (see the introduction to Section 2.4). If metallic conduction occurs at higher temperatures the band model would seem more appropriate. Indeed, a metallic state is also formed by removing some Co 3d electrons, in $La_{1-x}Sr_xCoO_3$.[28] The transition in $LaCoO_3$ probably involves band-crossing, similar to that in Ti_2O_3 which will be discussed in Section 5.3.2. As the temperature increases, lattice expansion reduces the t_{2g}–e_g splitting, and facilitates the thermal excitation of carriers into the e_g band. Populating this more antibonding level will itself increase the lattice spacing, and thus the gap will disappear. It is interesting that the corresponding 4d^6 compound $LaRhO_3$ is reported to have a significantly larger gap,[26] and does not seem to have a transition. Crystal field splittings are always larger in the lower transition series, due to the stronger overlap of metal d orbitals with surrounding atoms. This may be the principal factor in giving the difference between the Co and Rh compounds.

Low spin Co^{3+} occurs in a number of other oxides, and an interesting contrast with $LaCoO_3$ is found in the compound $LiCoO_2$, which has a structure based on rock-salt, but with alternate layers of Li^+ and Co^{3+}. Photoelectron spectroscopy and other experiments suggest that the filled t_{2g} band in $LiCoO_2$ is appreciably narrower than in $LaCoO_3$. For example, in the former compound it is possible to observe electronic excitations corresponding to the crystal field transitions expected for a *localized* low-spin d^6 configuration.[29] The difference in band width may be understood using the qualitative band picture of Section 2.3.1. There it was shown how bands are broadened by the indirect metal–oxygen–metal interaction (see Fig. 2.12). The linear M–O–M geometry found in the perovskite structure is optimal for this interaction. In the structure of $LiCoO_2$ the bond angle is close to 90°, and in consequence two neighbouring Co^{3+} ions cannot overlap with the same *p* orbital of their bridging oxygen. Thus the main source of *d* band width is removed.

The other electron configuration found in the examples in Table 3.3 is d^8. In octahedral geometry this would always give a state with two e_g electrons, and thus by Hund's rule a triplet ($S = 1$) state. Square planar coordination of four close oxygens is a common alternative geometry with this configuration. As shown in Fig. 2.2 the crystal-field splitting now leaves one *d* orbital significantly higher than the others, and if this remains empty, a closed-shell configuration is obtained. As with d^6, this low-spin arrangement is more common with 4*d* and 5*d* elements. The electronic details of PdO shown in the table are uncertain; as with most of the compounds here the intrinsic band gap is hard to measure due to problems of non-stoichiometry.[1] Both the top filled and bottom empty levels are probably of Pd 4*d* character, but the filled oxygen 2*p* orbitals may be of similar energy.

The compound AgO is especially interesting, as it illustrates that a simple chemical formula may not give a good guide to the oxidation states and electronic configurations of the elements present. Whereas CuO is a magnetic insulator containing the d^9 ion Cu^{2+}, both the structure and properties of AgO suggest a *mixed-valency* compound best formulated in terms of $Ag^+(d^{10})$ and $Ag^{3+}(d^8)$.[30] The latter ion has the tetragonal four-coordination expected for low-spin d^8, and Ag^+ has two close oxygens in a linear configuration, as with Cu^+ in Cu_2O. AgO is black and semiconducting; the intrinsic band gap is probably between the filled 4*d* orbitals on Ag^+ and the empty one in Ag^{3+}, but because of non-stoichiometry it is hard to characterize properly.

The chemical *disproportionation* of one oxidation state (Ag^{2+}) into two others (Ag^+ and Ag^{3+}) is rather surprising in the light of the Hubbard model discussed in Section 2.4.1, as it indicates *negative Hubbard U*. This behaviour is common in the post-transition elements, which have stable oxidation states differing by two units, corresponding to the ionic d^{10} and $d^{10}s^2$ configurations.[9,30] Disproportionation is rarer among transition elements, although it appears to occur at low temperatures in some Fe^{4+} compounds. It is probably no coincidence that disproportionation is found with the same electron configurations (d^4 and d^9) that under other circumstances are associated with strong Jahn–Teller distortions. Both effects must be related to the strongly antibonding nature of the octahedral e_g orbitals, although as discussed in Section 5.3.3, the behaviour of Fe^{4+} compounds is not well understood.

3.3 Transition metal impurities

In this section we shall look at the properties of d^n ions as dilute impurities, either in non-transition metal oxides such as MgO and Al_2O_3, or in d^0 compounds like TiO_2. Impurity ions are ubiquitous in oxide minerals, and even supposedly 'pure' oxides may be very hard to obtain free from common elements such as iron or chromium. Small concentrations of these give rise to optical properties, and thus contribute to the colours of gemstones such as ruby and sapphire.[31] From a more fundamental point of view, these materials offer the possibility of studying the electronic levels of isolated d^n ions in oxide lattices. Such a study ought, in principle, to provide many clues to the behaviour of compounds with more concentrated d^n ions, such as magnetic insulators. In spite of the apparent simplicity of these dilute systems, there remain a number of important uncertainties regarding the electronic levels.

3.3.1 Spectroscopic properties of d^n *states*

The two most important techniques for studying d^n impurities are optical spectroscopy, which measures the energies of crystal field and charge-transfer transitions, and electron spin resonance (ESR or EPR), which is a sensitive probe of the magnetic properties of the ground state. The basis of the crystal field (CF) model was explained in Section 2.1.1; in this section we shall look at some further applications of this model, especially to the important Cr^{3+} system. Charge-transfer transitions between the transition metal ion and the oxide lattice will be considered in the following section.

Figures 2.3 and 2.4 showed the absorption spectrum of ruby (Al_2O_3 with Cr^{3+}) and its interpretation in terms of the CF levels of a d^3 ion in an octahedral site. The more intense transitions are spin-allowed ones from the $^4A_2$ ground state to $^4T_1$ and $^4T_2$ excited states; weak spin-forbidden transitions to ^2E and $^2T_2$ states can also be seen. Figure 3.7(a) shows a more detailed view of these spin-forbidden bands, together with part of the *emission* spectrum produced in ruby by illumination of the spin-allowed bands.[32] Both doublet states have a small splitting, which is a consequence of the true site symmetry of Cr^{3+} being not quite octahedral. Emission from ^2E, known as the 'R' lines,

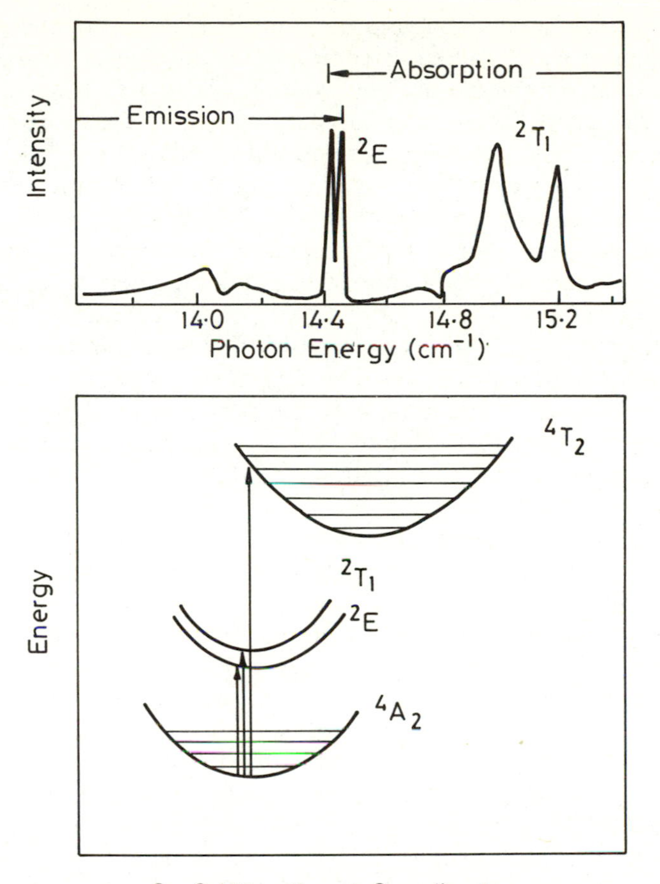

FIG. 3.7 Part of the absorption and emission spectra of ruby (Al_2O_3 with Cr^{3+} impurities), showing the spin-forbidden transitions under high resolution.[32] (b) Potential curves for ground and excited crystal field states for octahedral Cr^{3+} (d^3). The horizontal coordinate represents a local Cr-O stretching mode. Vertical lines show the band maxima for absorption and emission processes, as predicted by the Franck-Condon principle.

forms the basis of the ruby laser, historically interesting as the first system in which optical laser operation was achieved.[33]

In order to see how this emission occurs, and also to understand the widths of the different transitions, it is necessary to consider the way in which the equilibrium geometry surrounding the Cr^{3+} ion changes in the various electronic states. In Fig. 3.7(b) the horizontal coordinate represents a local vibrational mode, leading to an alteration of the Cr–O distance. In the Tanabe–Sugano diagram in Fig. 2.3 the energy of the excited quartet states increases rapidly with the crystal field splitting, Δ; this is understandable as, for example, $^4T_2$ arises from the configuration t^2e^1 where an electron has been excited to the more antibonding e orbital. Thus we expect the equilibrium bond length to *increase* in this state, as indicated in the figure. On the other hand, the ^2E and $^2T_2$ states are almost horizontal in the Tanabe–Sugano diagram, as they come largely from a *rearrangement* of electrons within the t_2 levels, and do not involve a change in bonding. Their potential curves are therefore almost directly above that for the $^4A_2$ ground state. The vertical lines in Fig. 3.7(b) show transitions taking place according to the Franck–Condon principle. The transition to $^4T_2$ is accompanied by vibrational excitation, and it is the partly unresolved vibrational structure which broadens the line. On the other hand, very little vibrational excitation should accompany the doublet transitions, which are therefore sharp.

Figure 3.7(b) also shows that the $^4T_2$ surface crosses the doublet ones; thus before direct emission from $^4T_2$ can take place, there is a high probability of non-radiative crossing into the doublets. Most emission in ruby comes from the lower of these, ^2E. The spectrum shows that there is a vibrational mode at 375 cm^{-1} weakly excited in the ^2E. This appears at higher energy in absorption, as one quantum of this mode is produced in the excited state, but at *lower* energy in emission, which occurs primarily from the vibrationally relaxed upper level, and leaves one quantum of vibration in the ground state. The 'mirror-image' form of absorption and emission is quite common in the spectra of localized states (including molecules).[11]

The absorption and emission of Cr^{3+} has been studied in other host crystals, and shows some interesting variations. In rutile (TiO_2), emission is harder to observe, but can be seen at 4 K to come mostly from the $^4T_2$ level, giving an approximate mirror image to the broad absorption to this state.[34] The absorption band now peaks at 14 000 cm^{-1}, much lower than the value of 18 000 cm^{-1} in ruby. The CF splitting, Δ, must be correspondingly smaller, and reference to the Tanabe–Sugano diagram suggests why the emission is different. As Δ is decreased, the $^4T_2$ state drops below the doublets. The minimum in its potential surface is lower, and crossing to the doublets impossible. In MgO, emission from both $^4T_2$ and ^2E can be seen, with a relative intensity depending on temperature. Presumably now the two states have almost the same energy.

Optical spectroscopy offers one way of identifying the elements and their oxidation states present as impurities in oxides. However, ESR has proved to be even more powerful in this respect.[35] This technique depends on the *ground state* electronic structure, and specifically on its spin and orbtial degeneracy. The basic idea to is measure the **Zeeman splitting** in a magnetic field $\boldsymbol{B}$; thus a state with spin S will split into $(2S+1)$ components labelled by their quantum number M_S according to:

$$\Delta E = g\beta \boldsymbol{B} M_S. \tag{3.10}$$

β is the *Bohr magneton* and the **g value** will in general be different from the free-spin value 2.0023. Many further effects appear, due to small energy splittings with the same order of magnitude as the Zeeman effect. For example **hyperfine splitting** comes from interaction between the electron spin and those of nearby nuclei, and **zero-field splittings** often appear in ions with $S > \frac{1}{2}$; these latter effects are very sensitive to the local symmetry and thus can give important information about the environment of an impurity ion.[22]

The ESR parameters determined from the spectrum form a combination which is generally highly characteristic of a particular ion. Table 3.4 gives an idea of the range of 3*d* ions and their apparent charge states that have been identified in this way in MgO and $SrTiO_3$ hosts.[36] Not all transition metal ions can be seen in ESR. For example, low-spin Co^{3+} has the closed-shell t_{2g}^6 configuration with $S = 0$. In many ions with $S > \frac{1}{2}$ the spectra are broadened by relaxation effects. Sometimes these ions can be studied at low temperatures, but there remain a number of cases, such as Mn^{3+}, where relaxation seems to prevent a spectrum from being observed. For these reasons, Table 3.4 is certainly not a complete list of all the states that occur.

ESR and optical studies indicate that most transition metal impurities

Table 3.4 *Oxidation states of some* 3d *impurities found in MgO and* $SrTiO_3$[a]

MgO				$SrTiO_3$			
Ti^{+}							
	V^{2+}	V^{3+}				V^{4+}	
	Cr^{2+}	Cr^{3+}			Cr^{3+}		Cr^{5+}
	Mn^{2+}		Mn^{4+}	Mn^{2+}		Mn^{4+}	
Fe^{+}	Fe^{2+}	Fe^{3+}		Fe^{2+}	Fe^{3+}	Fe^{4+}	Fe^{5+}
Co^{+}	Co^{2+}				Co^{3+}		
Ni^{+}	Ni^{2+}	Ni^{3+}		Ni^{2+}	Ni^{3+}	Ni^{4+}	

[a] From ref. 36. Most identifications made by ESR, but Co^{3+} (probably low spin and hence diamagnetic) by Mössbauer. Some others not seen (e.g. Mn^{3+}) probably exist but have very broad ESR lines.

occupy *substitutional* sites in an oxide lattice, for example replacing Mg^{2+} in MgO, and Ti^{4+} in TiO_2. When an impurity ion substitutes for one of different charge, additional defects must be present to preserve overall charge neutrality in the crystal. With Cr^{3+} replacing Mg^{2+} in MgO for example, it would be expected that some Mg^{2+} vacancies would be present. Replacing Ti^{4+} on the other hand, oxygen vacancies are likely. It is electrostatically favourable for the different defects to associate, so that for example a Cr^{3+} substitutional site and an Mg^{2+} vacancy may be at neighbouring sites rather than randomly distributed. Such neighbouring defects perturb the crystal field levels of the transition metal impurity, and in some cases can be detected in spectroscopic measurements. ESR studies show that substitutional impurities in $SrTiO_3$ are sometimes associated with an oxygen vacancy at a neighbouring site.[22] High-resolution laser spectroscopy can also show a range of slightly different absorption energies for many transition metal impurities, corresponding to perturbed states where other defects are nearby. Sometimes it is possible to use these spectroscopic methods to study the equilibrium between isolated defects and various defect associations. For Cr^{3+} in MgO it has been shown that isolated Cr^{3+} ions (with compensating vacancies sufficiently distant not to give a perturbation) are present only at very low impurity concentrations; the spectra become dominated by (Cr^{3+})–(Mg^{2+} vacancy) pairs above a Cr^{3+} level of 0.02 per cent.[37] The presence of defect interactions such as this is one of the factors that makes both the theoretical and experimental study of defect electronic structure in oxides very difficult.

3.3.2 Charge transfer and impurity-state energies

The crystal field model, together with suitable refinements necessary for the small energy splittings seen in ESR, provides a very satisfactory interpretation of the energy levels of the d^n configuration appropriate for an ion in a fixed charge state. Unfortunately, it has proved much harder to measure and to understand the energies of the different d^n states with respect to the bands of the host crystal. This is an important goal for several reasons. One would like to be able to interpret the energies of *charge-transfer* transitions, in which an electron is excited either from the impurity into the conduction band, or from the valence band to the impurity. This type of transition is intrinsically much more intense than the *d–d* kind treated by CF theory, and may often be important in the optical properties of solids. Charge transfer can generate free carriers, which are important in semiconduction (when thermally excited), and photoconduction (if produced optically). Optically generated carriers can influence other properties: for example some non-linear materials such as $LiNbO_3$ show a *photorefractive effect* in which the optical properties change under illumination. Although this effect has important potential applications, it is a nuisance in many situations. It is probably caused by carriers which are excited from Fe^{2+} and Fe^{3+} impurities, and then migrate in the solid and set up an electric field.[38]

Another important problem related to charge transfer is how to understand the remarkable range of charge states found in oxide lattices, and displayed in Table 3.4.[39] For example, the stability of a low-charge state such as Fe^{+} shows that the process

$$Fe^{+} = Fe^{2+} + e^{-} \tag{3.11}$$

must be energetically unfavourable, if e^{-} is an electron in the conduction band. The stability Fe^{5+} on the other hand, is limited by:

$$Fe^{5+} = Fe^{4+} + h^{+} \tag{3.12}$$

generating a hole h^{+} in the valence band. The energy levels of different charge states of an element are also related to the Hubbard U parameter, important in the theory of magnetic insulators.

As an example of a charge-transfer transition, Fig. 3.8 shows part of the optical spectrum of MgO doped with Ni^{2+}.[40] The weak band around 3 eV energy is a *d–d* transition, and is plotted on a different absorbance scale from the stronger transitions beginning at 6 eV. Since

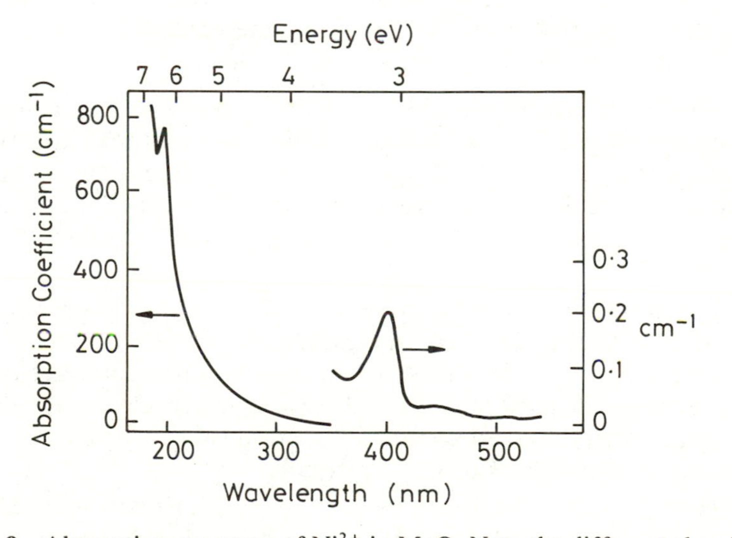

FIG. 3.8 Absorption spectrum of Ni^{2+} in MgO. Note the different absorbance scale for the CF transition around 3 eV, and the much stronger charge-transfer band around 6 eV.[40]

the band gap of MgO is 7.5 eV, this transition must also be due to the impurity, and is assigned to the charge-transfer process

$$Ni^{2+} = Ni^{+} + \hbar^{+}. \quad (3.13)$$

Although this assignment is probably correct, there is often great difficulty in identifying the precise transition involved. For example, we might also expect a transition corresponding to

$$Ni^{2+} = Ni^{3+} + e^{-} \quad (3.14)$$

and it may be very hard to distinguish these two processes. As we have seen above, many transition metal impurities are associated with other defects in order to provide charge compensation. Often the nature of these is unknown, and they may also give rise to absorption bands. Because of all these problems, the number of unambiguously assigned charge-transfer energies is rather small.

Partly because experimental assignments are difficult, there have been many attempts to calculate the energies of transition metal *d* levels relative to the band edges of oxide lattices. Some of the methods used were discussed in Chapter 2. Unfortunately it is not easy to assess the reliability of these calculations, as they all involve some kind of approximation. An especially serious problem is that the precise atomic positions which are an essential input to a reliable electronic calculation are generally unknown. Another difficulty in this area is to know how to display the energy levels of impurities in a satisfactory way. Figure 3.9 shows some different ways of doing this; the limitations of the representations labelled (a) to (g) are discussed below.

(a) Shows the *d*-orbital energies of the impurity, with a crystal field splitting appropriate to an octahedral site. No electrons are shown, and unless this represents a d^0 ion, it is unsatisfactory: as was emphasized in Section 2.2.1, even if we accept the usefulness of simple orbital pictures, we must recognize that orbital energies depend on their occupancy.

(b) Specifies the orbital occupancy, in this case a d^6 high-spin configuration. Unfortunately we still cannot expect the orbital energies to give a good representation of the energies of different spectroscopic *states*, as these depend also on the way in which electron repulsion changes as electrons are excited. This problem was discussed in the context of crystal field theory in Section 2.1.1.

(c) Is often used in the context of *different orbitals for different spins* calculations, where the exchange splitting between up- and down-spin electrons is included. We have seen this type of representation earlier (see Figs. 2.10 and 2.23). The simple approximation to electron

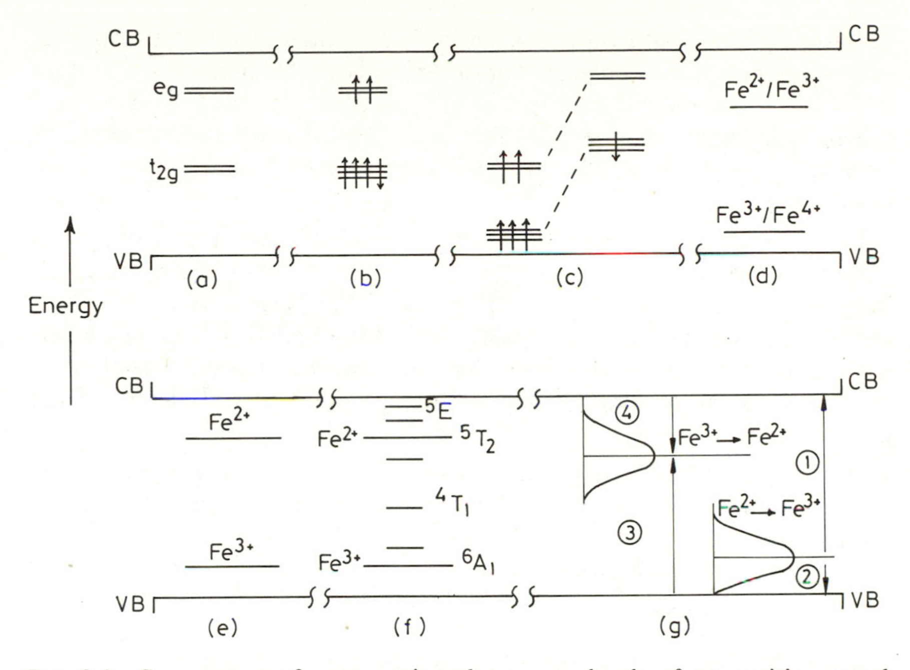

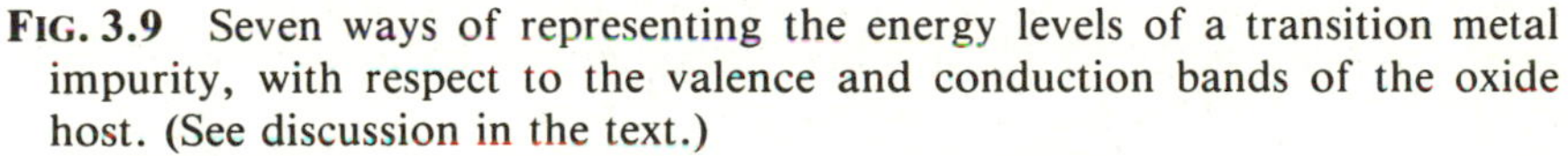

FIG. 3.9 Seven ways of representing the energy levels of a transition metal impurity, with respect to the valence and conduction bands of the oxide host. (See discussion in the text.)

repulsion terms that is used (or implied) in this plot is incompatible with the treatment of these in CF theory. Although it might be satisfactory to use such a representation to read off charge-transfer energies (but see below), we should not expect to be able to obtain the energies of CF transitions from the orbital energies shown.

(d) Avoids the orbital approximation altogether, by only specifying the energies of *states*. It is essential to specify the actual transition involved. Thus the difference between the Fe^{2+}/Fe^{3+} level and the conduction band gives the energy of the process

$$Fe^{2+} = Fe^{3+} + e^{-}. \tag{3.15}$$

The same level measured with respect to the valence band shows the energy required for

$$Fe^{3+} = Fe^{2+} + h^{+}. \tag{3.16}$$

In each case, it is assumed that the ground states of the different d^n configurations are involved. This is the form of representation used to show the calculated defect levels in Fig. 2.7. It is a satisfactory way of showing the equilibrium energies of the states involved in *thermal* excitation processes, i.e. the donor and acceptor ionization energies important in semiconduction, and discussed in Chapter 4. However, the *optical* excitation energies measured for the same processes are expected to be different, because of the Franck–Condon principle (see below).

(e) Is a representation often used instead of (d). It is ambiguous without more information, as the transitions are not fully specified. Does Fe^{3+} refer to the energy of the Fe^{3+}/Fe^{2+} transition, or that of Fe^{4+}/Fe^{3+}? Normally, it is the latter which is implied; i.e. all the levels are regarded as potential *donors*, rather than *acceptors*. Nevertheless, this could be misleading, and the more complete specification of (d) is preferable.

(f) Shows not only the ground states of the different configurations, but also various excited crystal field states above them. It is often used for transition metal impurities in covalent semiconductors, and since vibrational excitation effects are less of a problem in the spectra of these solids, it may be a valuable way of summarizing a wide range of transition energies.[41] Even so it is rather confusing, as the different types of transition involved (CF and charge transfer) are not very clearly distinguished. It cannot be recommended for oxides, especially in the light of the comments below on method (g).

(g) Recognizes that optical transitions are accompanied by vibrational excitation, especially in charge transfer where the geometries of the two states involved may be very different. The Gaussian curves show the spectral distribution expected from the Franck–Condon principle. The band maxima for the Fe^{2+} to Fe^{3+} transition and the reverse process are *not* the same; the difference between these comes from the different equilibrium geometry of the two states, and is essentially the same as the *Stokes shift* illustrated in Fig. 3.3.

The four possible transitions shown in Fig. 3.9 (g) are

(1) $Fe^{2+} = Fe^{3+} + e^-$ (optical absorption); (3.17a)

(2) $Fe^{2+} + h^+ = Fe^{3+}$ (with emission of light); (3.17b)

(3) $Fe^{3+} = Fe^{2+} + h^+$ (absorption); (3.17c)

(4) $Fe^{3+} + e^- = Fe^{2+}$ (emission). (3.17d)

It is clear that if the diagram gives a correct representation of these, the following relations must hold between the energies of the corresponding band maxima:

(1) + (2) = (3) + (4) = band gap;

(1) − (4) = (3) − (2) = Stokes shift.

Unfortunately the data available to test this are rather limited. Table 3.5 shows some values that have been assigned to the appropriate Fe^{2+}/Fe^{3+} transitions in MgO, together with the results of an ionic calculation using the shell model parametrization.[39] It can be seen the experimental assignments taken together are not compatible with the band gap of 7.5 eV, or with the results of the calculation. This may indicate that the assignments are wrong, but it must also be said that there is no clear consensus on the importance of Stokes-shift effects in this type of charge transfer. Certainly, some data indicate that thermal transfer energies are different from those measured optically: thus in electron transfer to Fe^{4+} and Fe^{5+} from the valence band in $SrTiO_3$, the thermal activation energies for producing carriers are about 1 eV less than the optical energies.[42] However, the *shapes* of these same absorption bands have not been interpreted as Franck–Condon envelopes as in Fig. 3.9 (g), but rather in terms of the *electronic* density of states in the valence band. This indeed shows one limitation of the model, as it is assumed in Fig. 3.9 that transitions occur from the band *edges*; in fact of course both valence and conduction bands have a width which may be 5 eV. One ought therefore to perform some kind of convolution between the density of states (modulated by an electronic matrix element) and the Franck–Condon distribution. None of the existing interpretations has attempted to do this.

In view of the various problems just discussed, attempts to show systematic trends in defect levels should be regarded with caution. Figure 3.10 shows one such scheme proposed for 3*d* ions in TiO_2.[43] The notation used is that of Fig. 3.9 (d), and these values should probably be regarded as equilibrium rather than spectroscopic levels, although the authors do not make a clear distinction between them. Even if the precise values shown in Fig. 3.10 can be disputed, the general trends displayed here are probably correct. Levels with a fixed

Table 3.5 *Energies of optical Fe^{2+}/Fe^{3+} transitions in MgO*[a]

Transition[b]	Experimental energy (eV)	Calculated energy (eV)
1	6.7, 4.9	4.1
2	1.9	3.7
3	4.3, 5.7	7.2
4	3.3	0.6

[a] Based on data from ref. 39.

[b] Labelled according to eqns (3.17a)–(3.17d) in the text and Fig. 3.9 (g).

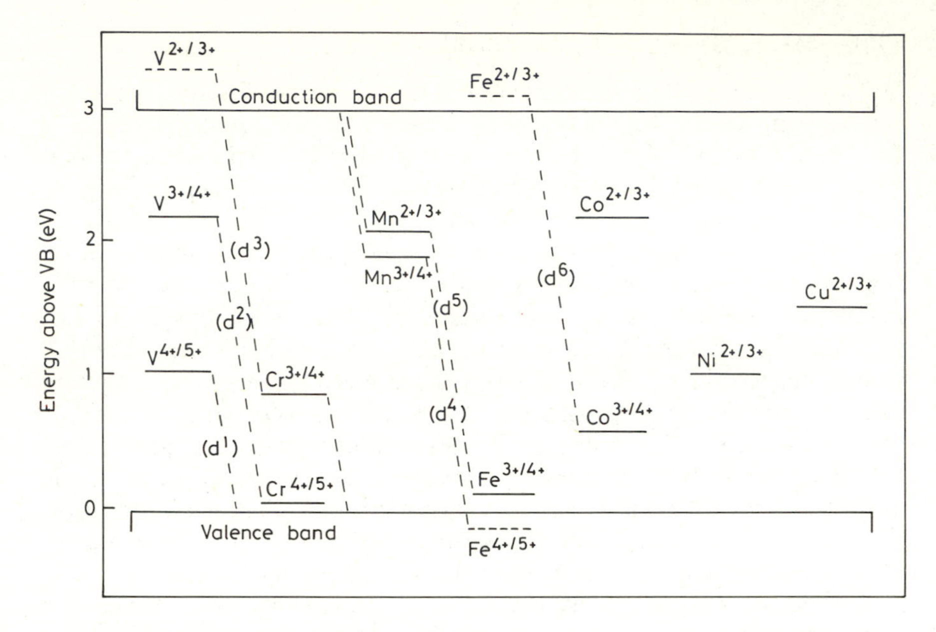

FIG. 3.10 Proposed impurity energies of some $3d$ ions in rutile, TiO_2.[43]

d electron configuration show a progressive stabilization of about 2 eV as the nuclear charge increases from one element to the next. The irregular changes as d electrons are added are a consequence of crystal field and exchange terms: for example there is a particularly large increase of energy on going from d^5 to d^6, associated with a loss of exchange energy for electron configurations beyond the half-filled d shell in these high-spin ions.

There is an important connection between the energy levels illustrated in Fig. 3.10 and the Hubbard model described in Section 2.4.1. The increase in energy as an electron is added to the d shell is a result of the extra electron repulsion. Energy differences between successive charge states of the same element therefore provide one measure of the Hubbard U parameter. The energies in Fig. 3.10, and in other plots shown in the previous chapter (for example Fig. 2.23), suggest U values between 1 and 3 eV, and thus enormously reduced from the gas-phase values given in Fig. 2.22. We can argue that even if the precise energies in Fig. 3.10 are incorrect, the very existence of a large range of charge states found for many ions, shown in Table 3.4, implies very small U values for transition metal ions in oxides. On the other hand, some of the models of magnetic insulators and high-T_c superconductors discussed later (see Sections 3.4.2 and 5.4.2) suggest substantially larger values, in the range 5–9 eV.

The difficulty in obtaining unique estimates of U has already been mentioned in Section 2.4.1. It arises partly from the different ways in which various models interpret the reduction from gas-phase to solid-state values. From an ionic viewpoint, this reduction must be ascribed entirely to *polarization* effects.[39] On the other hand, cluster and impurity-state calculations show that the d orbitals are mixed with the bands of the host, and that repulsions between 'd' electrons are reduced by their appreciable *delocalization*.[44] It may not be possible to make a meaningful separation between polarization and covalency; for example the parametrization of ionic model calculations ensures that polarization terms incorporate many of the effects which appear due to covalent mixing in other theories.[45] The best 'compromise' model in this area may be a cluster calculation which treats the local electronic structure in a proper quantum-mechanical way, embedded in a lattice of polarizable ions to take account of longer-range effects. An example of such a calculation was discussed in Section 2.2.1.

There are probably two reasons why the U values suggested by defect energies are much lower than ones used in semiempirical models of magnetic insulators and high-T_c superconductors. In the first place, covalent mixing of orbitals is treated explicitly in these theories, and so the screening effect of 'delocalization' of the d orbitals is not included in U itself. The other reason is that using the *thermodynamic* stability of different charge states gives values appropriate to fully relaxed equilibrium geometries. They therefore include the effect of slow atomic-relaxation terms which do not play a role in purely electronic processes. As discussed above, there is some disagreement about the importance of this distinction, but ionic relaxation could easily make a difference of 2 eV or more.

3.3.3 Interaction between impurities

As the impurity concentration increases, one expects on a statistical basis to find a proportion of transition metal ions at neighbouring lattice sites. Because of the large number of different possible defect configurations involved, a detailed study of such systems is difficult. Nevertheless some interesting effects can be observed, a few examples of which are discussed in this section.

Both ESR and high-resolution optical studies on ruby show the effects of interaction between Cr^{3+} ions at neighbouring sites in the corundum lattice.[46] It has been possible to identify some of the signals as coming from configurations of impurities, and Fig. 3.11 shows the

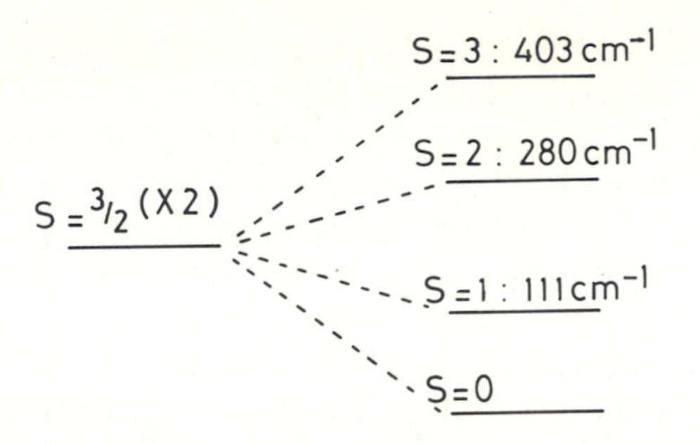

FIG. 3.11 Low-lying energy levels measured for near-neighbour Cr^{3+} pairs in ruby.[46] The four states labelled by their total spin values arise from antiferromagnetic coupling of the two $^4A_{2g}$ ($S = 3/2$) states.

ground-state energy-level structure found for nearest-neighbour pairs.[47] These levels show the result of **antiferromagnetic coupling** between the spins of neighbouring ions. Two $^4A_2$ ground states, each with $S = \frac{3}{2}$, give states with total spin values ranging from 0 to 4, with $S = 0$ being lowest in energy. This type of effect is known as **exchange splitting**. It is the same interaction that gives rise to cooperative magnetic effects in solids, and its origin is discussed in Section 3.4.4. The most significant fact is the small energy involved. The splittings, which are quite typical for this situation, are of the order of $100\,\text{cm}^{-1}$. This is 10^{-2} eV and thus *two orders of magnitude* less than the other energies (CF splitting and charge-transfer transitions) characteristic of the electronic states of d^n ions. One must conclude that even at neighbouring lattice sites the interaction between transition metal ions gives a very small perturbation. A 'delocalized orbital' model, which attempted to describe the electronic structure in terms of molecular orbitals spread equally over the two ions, would be quite inappropriate here. By far the best picture is that of highly localized states with a weak interaction. A good molecular analogy is the H_2 molecule at a very large internuclear separation.[48] It is well known that the single-configuration MO wavefunction describes this very poorly; the best picture is that of the Heiter–London wavefunction, representing one localized electron on each atom, coupled to form a singlet $S = 0$ ground state. Analysis of the H_2 case shows that, as in the Hubbard model, it is their mutual repulsion which forces electrons into localized states. The importance of electron-repulsion effects in transition metal pairs is a strong argument in favour of the Mott–Hubbard model for the gap in magnetic insulators.

Another feature of transition-metal pairs is the possibility of direct charge transfer between them.[31] These may be of lower energy than transitions into the bands of the host crystal, and many minerals may be coloured by such transitions, such as

$$Fe^{2+} + Fe^{3+} = Fe^{3+} + Fe^{2+}. \tag{3.18}$$

This process requires energy (typically 1–2 eV) because the two charge states have different environments. Fe^{2+} and Fe^{3+} may be at different types of lattice site, but even if this is not the case, their equilibrium bond lengths will be different. In the latter case, the optical transition resembles that expected from *small polarons*, as explained in Section 2.4.5.

Neighbouring pairs of Fe ions may have a small exchange interaction similar to that described for chromium. The different energies resulting from the interaction cannot be resolved in the charge-transfer transitions, as these are quite broad due to vibrational interaction. Nevertheless, an exchange splitting in the ground state may give rise to an unusual temperature dependence of the absorption intensity.[49] This happens because the matrix element for electron transfer depends on the relative spin direction in the two ions, which is different for the various levels arising from the splitting. (A similar effect is described in Section 5.2.2, in connection with the *metallic* transfer of electrons between ions with non-zero spin.) Changing the temperature will alter the Boltzmann distribution within the exchange-split ground state, and thus change the charge-transfer intensity.

As might be expected from the difficulties in interpreting charge-transfer energies of *single* transition metal ions, detailed assignments of transitions involving pairs have been much disputed. However, the transition

$$Fe^{2+} + Ti^{4+} = Fe^{3+} + Ti^{3+} \tag{3.19}$$

between Fe–Ti impurity pairs in Al_2O_3 is thought to be responsible for the absorption band between 10 000 and 15 000 cm^{-1} which gives the blue colour of sapphire.[31]

3.4 Magnetic insulators

The term *magnetic insulator* implies the simultaneous existence of an energy gap, and of unpaired electrons which give rise to magnetic properties. The importance of this class of solid extends far beyond the field of transition metal oxides; a large majority of compounds of lanthanides (rare earths) and many of those of the actinides fall within it, as do most transition metal halides. They are also known as *Mott insulators*, as it was Mott's suggestion that electron repulsion is responsible for a breakdown of normal band properties for the d electrons, which was later formalized in the Hubbard model.[50,51] At a qualitative level, this model provides a straightforward interpretation of some general systematics of magnetic insulator versus metallic properties. As described in Section 2.4.1, it is predicted that insulators should be formed by stoichiometric compounds when the bands are narrow. Chemical experience, reinforced by calculations of the radial distribution and overlap properties of different atomic orbitals, suggests that the interaction between orbitals that provides band width in a solid should follow the sequence

$$4f < 5f < 3d < 4d \sim 5d.$$

Magnetic insulators are indeed more prevalent among $3d$-series oxides, although a few are known from the later series.[26] Halides also appear to have smaller band widths than oxides, partly because of the smaller degree of covalency expected with more electronegative elements, and partly because of their stoichiometries ($CrCl_3$ compared with Cr_2O_3 for example), which lead to structures with a lower degree of connectivity between metal coordination polyhedra. Compounds with less electronegative elements such as sulphides and phosphides are more often metallic. Thus oxides appear to be in an intermediate position, with behaviour sometimes close to the borderline between insulating and metallic character; we shall look at some examples of this in Chapter 5. In spite of the qualitative success of the Hubbard model, it is not universally accepted as the best way of describing magnetic insulators; some of the difficulties and controversy surrounding these materials will be discussed below (Section 3.4.2).

3.4.1 Survey of properties

The most characteristic features of a magnetic insulator are:

(1) non-metallic character suggesting the presence of a band-gap;

(2) magnetic properties showing the existence of unpaired electrons;

(3) crystal field transitions characteristic of open-shell d^n ions.

Some of these properties are summarized for a selection of binary and ternary compounds in Table 3.6.

Unfortunately the precise magnitude of the band gap is often difficult to measure. In very pure stoichiometric materials it is possible to observe an absorption edge in the spectrum, although this is often complicated by the presence of d–d transitions at lower energy. It may also be possible to obtain an estimate from the activation energy for conductivity. As with other types of insulator, however, both these measurements are frequently subverted by the presence of defects and non-stoichiometry.[1,52] The most obvious consequence is that the *intrinsic* conduction is swamped by carriers provided by defect levels, so that a serious underestimate of the gap is obtained from conductivity measurements. Defects also tend to give a strong background in the absorption spectrum, so that optical determinations are also difficult. Some other types of spectroscopy are less sensitive to the presence of defects. In principle, a combination of X-ray absorption and emission measurements, like those shown for TiO_2, should give an estimate of the band gap. Unfortunately, the core hole which is always involved in these techniques can give a serious perturbation.[53] A more promising method is the combination of photoelectron spectroscopy (looking at occupied levels) and inverse photoelectron (or bremsstrahlung isochromat) spectroscopy, which studies empty states. Figure 3.12 shows spectra of this kind for NiO.[54] With the photoelectron and inverse spectrum plotted on the same energy scale, the gap between filled and empty states is measured to be about 4 eV, in agreement with that (3.8 eV) found from the optical absorption edge. Unfortunately, measurements on non-metallic solids are complicated by charging effects, and not many results have been reported for magnetic insulators.

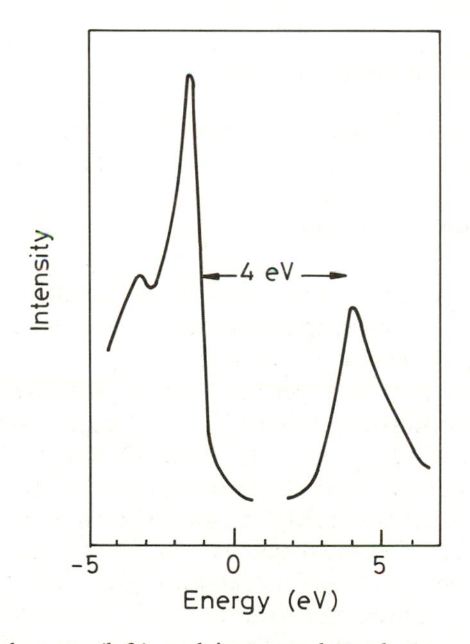

FIG. 3.12 Photoelectron (left) and inverse photoelectron spectra (right) for NiO.[54] The former shows ionization from filled levels; the inverse spectrum shows empty levels. The band gap, ~4 eV, is marked.

Table 3.6 *Properties of some magnetic insulators*[a]

Compound	Band gap (eV)	$d-d$ transitions (eV)	Magnetic properties[b]		
			S	Θ (K)	T_N (K)
Binaries					
Cr_2O_3	3.3	2.1, 2.6	$\frac{3}{2}$	−550	308
MnO	3.6	2.2, 2.9, 3.5	$\frac{5}{2}$	−417	118
MnO_2	-	-	$\frac{3}{2}$	−1000	92
FeO	2.4	1.2	2	−500	198
α-Fe_2O_3	1.9	1.4, 2.1, 2.6, 2.9	$\frac{5}{2}$	−3000	950[c]
CoO	2.6	1.1, (2.0), 2.3	$\frac{3}{2}$	−300	292
NiO	3.8	1.1, 1.8, 3.2	1	−2000	523
CuO	1.4	None (?)	$\frac{1}{2}$	-	226
Ternaries					
$LaVO_3$	-	-	1	−300	137
$LaCrO_3$	-	-	$\frac{3}{2}$	−600	300
$LaMnO_3$	-	-	2	-	100
$LaFeO_3$	-	-	$\frac{5}{2}$	−2000	750
$Y_3Fe_5O_{12}$	3	Many	$\frac{5}{2}$	-	559[d]

[a] Based on data selected from refs 1, 78.
[b] Weiss constants, Θ, are normally *very approximate*, as many of these compounds do not obey the Curie–Weiss law well.
[c] α-Fe_2O_3 (haematite) shows *weak ferromagnetism* between 265 K and 950 K.
[d] Yttrium–iron garnet (YIG) is *ferrimagnetic* due to the opposed spins of three tetrahedral and two octahedral Fe^{3+} ions.

Table 3.6 shows band gaps for some binary oxides, normally derived from optical measurements; in a few cases values have been confirmed by conductivity or photoelectron studies.[1]

The existence of strong defect absorptions also makes the measurement of *d–d* transitions quite hard in many cases. When they can be seen, as for example in pure samples of Cr_2O_3 and NiO, they appear at energies very similar to those of the same ion in doped compounds.[55,56] Studies of halide magnetic insulators have shown many effects of small interactions, similar to those discussed above in connection with impurity pairs in oxides.[57] These interactions give rise, for example, to splittings of sharp bands, and to unusual temperature effects on transition intensities. Although some small splitting effects have been observed in Cr_2O_3,[55] very little work of this kind has been reported on oxides. Generally it is difficult to obtain suitable defect-free specimens, especially in sections thin enough for good absorption measurements.

Recently it has been shown that *d–d* transitions in metal oxides can also be observed by electron energy-loss spectroscopy, which measures the losses occuring as a beam of mono-energetic electrons is scattered from the surface.[58] Although this technique, like photoelectron spectroscopy, is in principle expected to show surface effects, it appears that bulk transitions are in fact being measured. When a comparison is possible, the energies observed coincide with those found by optical measurements. An example of such an energy-loss spectrum for NiO is shown in Fig. 3.13, with an assignment of the crystal field excitations expected from the $^3A_{2g}$ ground state for octahedral Ni^{2+}. There is a background caused by defect transitions, but in a scattering technique these are less serious than in optical absorption, and spectra can be measured from samples where optical measurements would be impossible.

Some crystal field transition energies are given in Table 3.6. They can be interpreted in the same way as the transitions shown by impurity ions, and octahedral splitting parameters, Δ found from such fits were given previously in Table 2.2.

The magnetic properties of these compounds also show the presence of ions with the expected d^n configurations. As in the case of the impurity pairs discussed above, there are perturbations produced by interactions between magnetic ions. Although small in terms of energy these interactions have an important effect, giving rise to **cooperative magnetic ordering** below a certain temperature.[19,59] Above the ordering temperature, the paramagnetic susceptibility follows approximately the **Curie–Weiss law**:

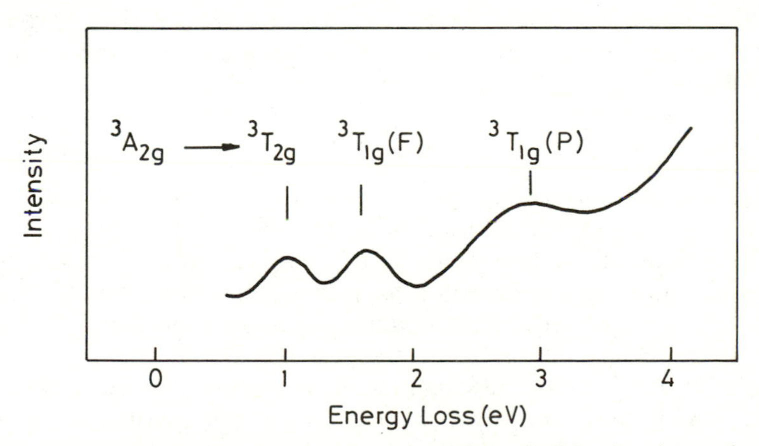

FIG. 3.13 Electron energy-loss spectrum (EELS) of NiO, showing assignments of the losses to crystal field excitations of octahedral Ni^{2+}.[58]

$$\chi_p = C/(T - \theta) \tag{3.20}$$

where the Weiss constant θ is negative for the most common situation of antiferromagnetic order. The Curie constant is given by

$$C = N\mu_0\mu_{\text{eff}}^2/(3k) \tag{3.21}$$

where N is the number of magnetic ions present, k is the Boltzmann constant, and the effective magnetic moment μ_{eff} for many d^n ions is roughly given by the **spin-only** formula:

$$\mu_{\text{eff}} = 2\,[S(S+1)]^{1/2}\beta. \tag{3.22}$$

Unlike ions of the lanthanide series, orbital contributions to the moments are largely quenched by the crystal field splittings, although spin–orbit coupling does cause some deviations from the spin-only formula. In particular, ions with T_1 and T_2 ground states may have temperature-dependent moments because of spin-orbit coupling effects.[60]

The ordering of spins is opposed by their thermal randomization. The ordering transitions have a specific-heat anomaly associated with them. The entropy changes associated with spin disorder can thus be estimated, and are often found to be close to the expected value of $k\ln(2S+1)$. The ordering temperatures shown in Table 3.6 are of the order of 100 K, suggesting that the interaction energies between ions are typically around 0.01 eV. This is the same order of magnitude as is found by spectroscopic measurements of pairs of impurity ions.

We shall discuss magnetic ordering phenomena, and the nature of the interactions which produce them, in more detail below. The important point to note here is that they are *weak* effects in terms of the overall electronic structure. Magnetic as well as spectroscopic properties are most easily interpreted in terms of highly *localized* configurations of d electrons.

3.4.2 The nature of the band gap

There have been two main approaches to understanding the electronic structure of magnetic insulators. The Hubbard model suggests that we should think of localized states of d^n ions, with itinerant (band-like) behaviour inhibited by electron-repulsion effects. On the other hand, it has been claimed that these materials can be understood in band-theory terms, when the correct ground-state magnetic order is included. We briefly described the idea behind this approach in Section 2.3.3. Figure 3.14 shows in more detail the origin of the energy gap in antiferromagnetically ordered MnO and NiO, as predicted by the spin-dependent band model.[61] The crucial terms in this diagram are the

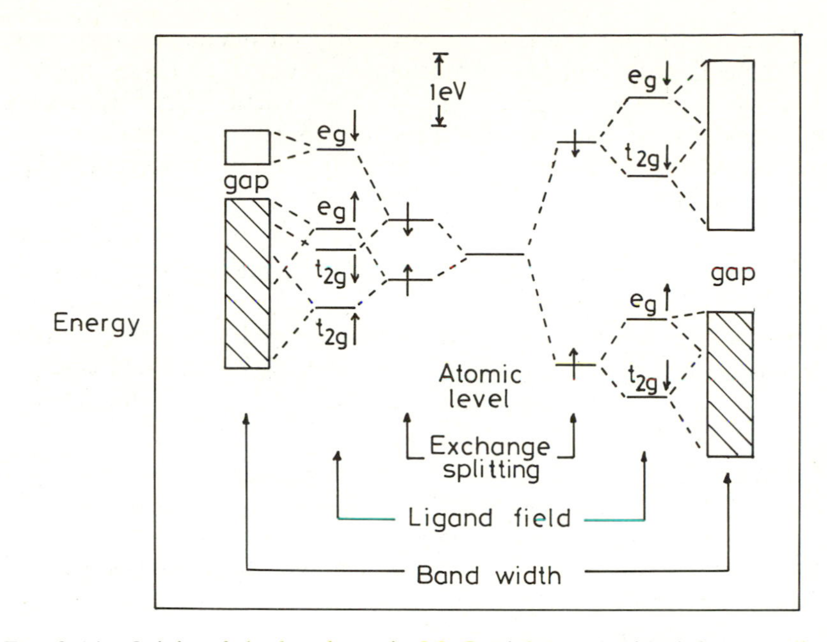

Fig. 3.14 Origin of the band gap in MnO (right) and NiO (left) according to spin-dependent band calculations in the local-density formalism.[61]

exchange splitting of the *d* orbitals, giving an energy difference between spin-up and spin-down electrons on a given ion, and the crystal field splitting from the octahedral environment. Interaction between the ions gives rise to bands with a width of 1–2 eV, but in MnO a gap remains between the occupied (up-spin) and empty (down-spin) levels; in NiO the empty level is the down-spin e_g, split off from t_{2g} by crystal field effects.

The very different pictures of the band gap provided by these two approaches have given rise to a great deal of controversy.[62] Since most estimates of the Hubbard U are at least 3 eV, and therefore larger than the predicted band width, the localized approach would seem preferable. There are a number of other arguments against the band model, listed below.

1. The gap predicted by band calculations is much too small (0.5 eV compared with 3–4 eV found experimentally).

2. Apart from the detailed *values*, the interactions treated by band calculations (exchange and crystal field effects which are experimentally only around 1 eV) seem by their *nature* to be too small to give the observed gaps.

3. The band-theory gap depends explicitly on magnetic order, and indeed the correct type of antiferromagnetic order (type II: see Section 3.4.3) must be present to give a gap in these calculations.[61] But most experiments suggest that the gap does not change appreciably at the magnetic-ordering temperature, and certainly does *not* disappear in the paramagnetic phase.

4. The prediction of a gap in MnO and NiO depends on their having a *half-filled* set of orbitals, so as to give filled and empty states when the spin direction is included. But many magnetic insulators do not have this feature. For example CoO, with the ground-state configuration $t_{2g}^5 e_g^2$, has neither a completely nor a half-filled t_{2g} band. Band calculations predict a *metal*, although the experimental gap is similar to that in NiO and MnO.

5. Band theory cannot account for the existence of crystal field excitations. As was suggested earlier (Section 2.4.2) these are best regarded as *Frenkel excitons*. They are localized by essentially the same Coulomb interaction that appears in the Hubbard model, and their existence confirms that the Hubbard U is considerably larger than the band width.

6. The *spin-waves* discussed later are another type of low-energy excitation that cannot be treated by band theory.

It is possible that some of these objections could be overcome, as they are partly a feature of the *local-density approximation* used in band calculations, rather than of the band model as such. The local-density method is designed to give a reasonable approximation to the electronic ground state of a system, and many ground-state features do indeed seem to be satisfactorily treated: for example the trends in lattice parameters of the 3*d* monoxides, and the magnetic ordering of the insulators, are correctly predicted.[63] It is the *excitations* of magnetic insulators that are especially poorly described by the local-density calculations. An underestimate of the band gap is a very general result of this method, although not often to the degree found here. In a *Hartree–Fock* calculation, the energy of empty states includes the contribution of an extra Coulomb term, so that the effect of a Hubbard U could appear. Model calculations, designed to represent some of the features of copper oxides important for superconducting properties, suggest that a band approach of this kind can be a useful alternative to the localized approach, although low-energy excitations such as spin-waves must be treated by an extension to the theory, using many-body approaches such as the random-phase approximation.[64] Attempts are also being made to modify the local-density approximation, so that Hubbard energies may be incorporated.[65] A satisfactory band approach to compounds such as MnO, CoO, and NiO would be valuable for a number of reasons. In particular, recent measurements of angular-resolved photoelectron spectra show that there is some dispersion in the *d* bands.[66] At present, the localized cluster and impurity models discussed below offer the best interpretation of photoelectron spectra; these are, however, unable to treat band-structure effects, and there is a need for a new approach that can combine the band model with local correlation effects.

If we accept that the existence of a band gap in magnetic insulators is primarily due to a Hubbard splitting, there still remain several possibilities for the nature of the top-filled and bottom-empty levels. Four possible energy-level schemes are shown in Fig. 3.15 (a) to (d). The split *d* states are represented by the levels showing the energies of processes

$$(1)\ d^n \rightarrow d^{n-1} + e^- \tag{3.23}$$

$$(2)\ d^n + e^- \rightarrow d^{n+1} \tag{3.24}$$

corresponding respectively to 'occupied' and 'empty' *d* states. Both the width of the *d* band and its mixing with other orbitals have been ignored. The other levels are the broad bands arising from the filled oxygen 2*p* orbitals, and the empty metal *s* and *p*. In possibility (a) the band gap *is* the Hubbard gap, corresponding to the transition of a *d* electron from one ion to another:

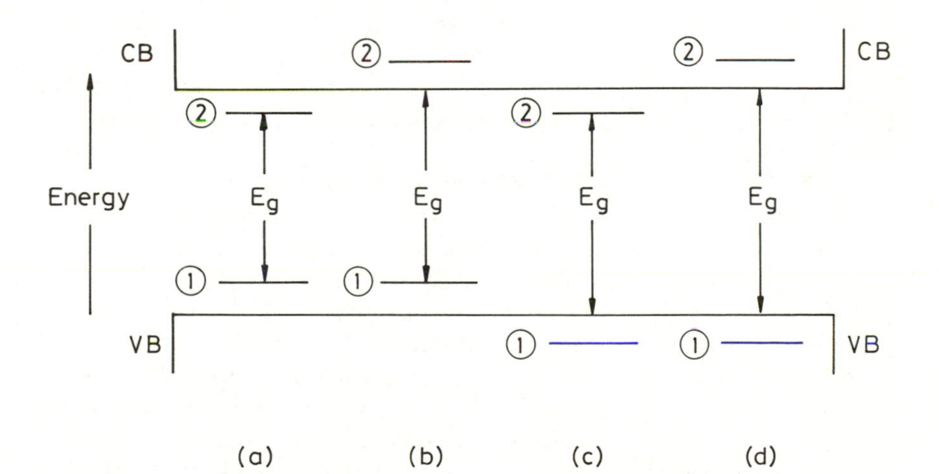

FIG. 3.15 Four different possible orderings for the energy levels in a magnetic insulator. The band edges correspond to the top of the oxygen 2*p* valence band, and the bottom of the empty metal ($s + p$) band. The Hubbard-split *d* levels are labelled (1) and (2), with band-width neglected.

$$\text{(a)}\quad d^n + d^n \rightarrow d^{n+1} + d^{n-1}. \tag{3.25}$$

This is the band-gap excitation predicted for NiO by the ionic model, as shown in Fig. 2.6 (Section 2.1.2). The other possible transitions are

$$\text{(b)}\quad d \rightarrow \mathrm{M}(s + p); \tag{3.26}$$

$$\text{(c)}\quad \mathrm{O}\ 2p \rightarrow d; \tag{3.27}$$

$$\text{(d)}\quad \mathrm{O}\ 2p \rightarrow \mathrm{M}(s + p). \tag{3.28}$$

All these assignments have at one time or another been suggested for band-gap transitions, and it is not easy to decide between them.[1] It is important to realize that the existence of *d–d* excitations within the band gap is *not*, as might be thought, an argument for always preferring (a). As we have emphasized, these are *excitonic* transitions of a different nature from the charge-transfer ones involved in band gap processes. Their energy is strongly modified by the Coulomb attraction between electrons and holes within the *d* shell of a single transition-metal ion, and is not simply related to any of the energy differences between the levels in Fig. 3.15. On the other hand, modulated reflectance spectroscopy of the monoxides shows the existence of strongly allowed excitations occurring at about 6 eV.[67] The most obvious assignment of these is to transition (d). If this is a typical magnitude of the oxygen $2p$ to metal $(s + p)$ gap, it is too large to account for the actual band gaps, and so must be ruled out. The appearance of the absorption spectrum at the gap threshold has been offered as evidence for one process or another. For example MnO and Fe_2O_3 seem to have forbidden transitions, whereas that in NiO is apparently allowed.[1] This may suggest a transition from oxygen $2p$ in NiO, as opposed to one between metal *d* levels in the other compounds. But it is hard to be certain about this, as there seems to be little theoretical guide on what to expect for optical transitions between the Hubbard sub-bands.

One important source of information on filled levels in solids is photoelectron spectroscopy, and as mentioned above this can be complemented in a few cases by measurements on empty states, using the inverse PES technique. PES of most d^n oxides shows that the first ionization comes from a level above the top of the oxygen $2p$ band. In metallic oxides (see Chapter 5), this is assigned to a *d* band; in magnetic insulators these spectra have also been taken as evidence that the occupied d^n level is above the top of the O $2p$ band, as in Fig. 3.15 (a) and (b). Spectra of NiO were shown in Fig. 3.11, and Fig. 3.16 shows the PES of Cr_2O_3, measured with both He-I (21.2 eV) and He-II (40.1 eV) radiation. The increase of intensity of the 'Cr $3d$' feature with higher photon energy is expected from the change in relative ionization cross-sections of *d* and *p* atomic orbitals. Similar features

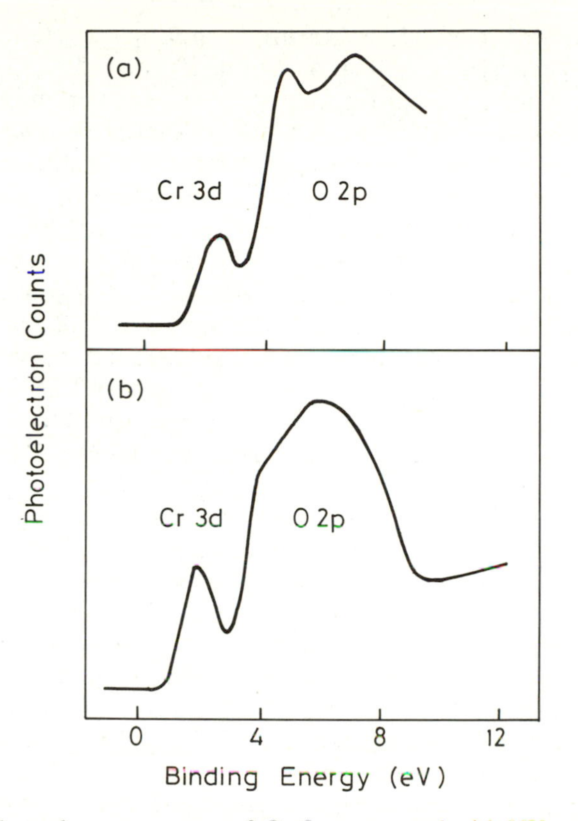

Fig. 3.16 Photoelectron spectra of Cr_2O_3, measured with UV radiation at (a) 21.2 and (b) 40.8 eV.

are found in PES of a wide variety of magnetic insulators, although in many cases the structure of the '*d*' ionizations is more complicated, as a variety of crystal field states may be formed in ionization.[68,69]

Recent experimental and theoretical work has led to some re-interpretations of the nature of the levels observed in PES. In NiO and in many other oxides of elements late in the 3*d* series, *satellite* peaks appear which cannot be interpreted in terms of any single-electron ionization process.[70] In addition, the detailed changes in band intensities with photon energy, especially when *resonance* effects occur due to the interference of core-level excitations, appear to be quite anomalous, and inconsistent with the simple *d* character assumed for the top-filled level.[71] So far, the only satisfactory interpretation of all these features is based on cluster and impurity models using the *configuration-interaction* formalism explained in Section 2.2.2. These are semi-empirical methods, requiring as input parameters the energies of filled and empty levels such as O 2*p*, d^n, and d^{n+1}, together with some off-diagonal terms corresponding to the covalent mixing between them. A fully consistent set of parameters should be able to interpret transitions seen in the absorption spectrum, as well as those observed by photoelectron techniques. The result of one such calculation on NiO was shown in Fig. 2.11. The model suggests that for NiO and CuO it is Fig. 3.15 (c) which most closely represents the band gap; that is the top-filled level has more oxygen 2*p* than metal *d* character. There is, however, quite a strong interaction between these orbitals, which leads to a mixed *impurity state* above the top of the main valence band.[72] It is this state that appears as the apparent '*d* band' in PES. For elements earlier in the series, such as Mn and Fe, the positions of the unperturbed oxygen 2*p* and metal *d* levels are reversed, so that Fig. 3.15 (a) is more appropriate.

The prediction that in some magnetic insulators the top-filled level has more ligand (oxygen) *p* than metal *d* character, has led to the suggestion that these should be called **charge-transfer insulators**, rather then Mott insulators where the gap involves the Hubbard-split *d* levels.[73] This may be a useful distinction, but it must not obscure one essential point: irrespective of the detailed interpretation of the band gap, the *presence* of the Hubbard splitting in Fig. 3.15 is crucial. It is clear that in all the level orderings shown, if the splitting between the the two *d* levels (1) and (2) were eliminated, these would meet and overlap in energy; the result would be a partially filled *d* band of itinerant, metallic electrons. It may seem, therefore, that the precise order of levels is a rather academic problem, of interest only to spectroscopists. This is not true, because electrons and holes can be introduced into magnetic insulators in other ways, for example by doping to give semiconductors, metals, and indeed high-T_c superconductors. It is important to know the nature of the hole states involved in these solids if their properties are to be understood. Spectroscopically deduced levels are not *necessarily* a reliable guide to the orbitals occupied by carriers in thermal equilibrium, because as we have seen there is an important difference between states accompanied by full lattice relaxation and those arising spectroscopically by Franck–Condon transitions. Nevertheless, there is strong evidence that holes in doped nickel and copper oxides do have substantial oxygen 2*p* character, in agreement with the models just discussed. Some of the evidence regarding hole states will be explained in Sections 4.3.1 and 5.4.2.

3.4.3 Magnetic ordering

Magnetic ordering is a phase change associated with a discontinuity in the susceptibility; it may also involve a small change in lattice spacing

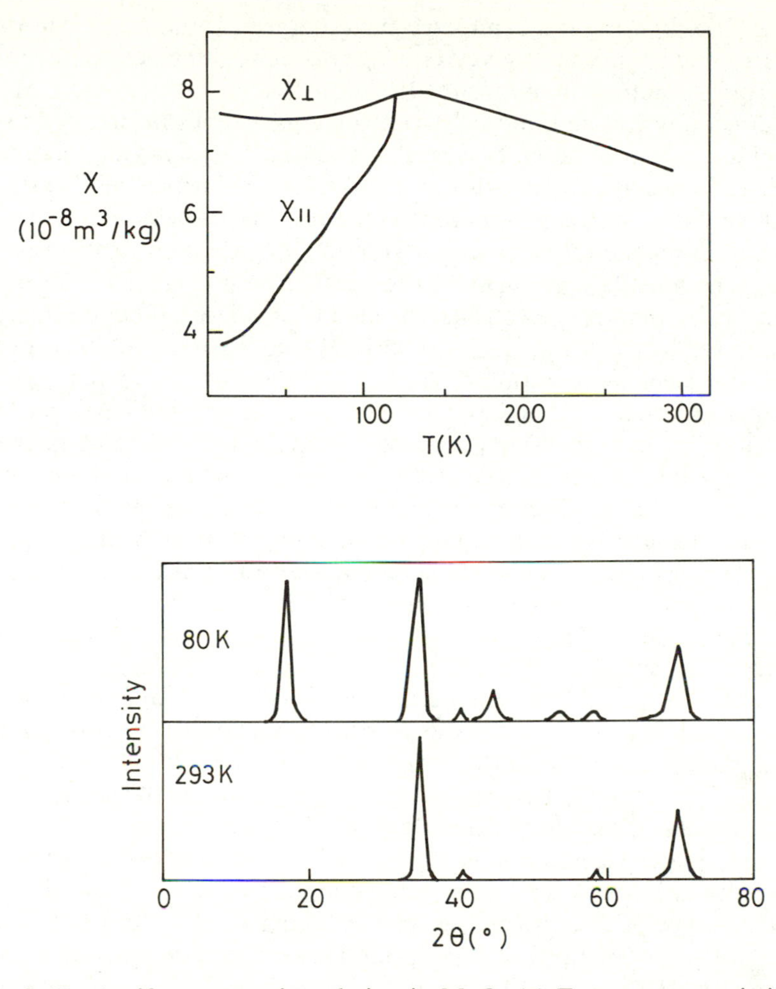

FIG. 3.17 Antiferromagnetic ordering in MnO. (a) Temperature variation in the magnetic susceptibility, showing its discontinuity at the Néel temperature, and its anisotropy below this temperature.[74] (b) Neutron diffraction peaks above and below T_N, with the extra peaks showing the enlarged magnetic unit cell.[76]

and often in crystal symmetry. The commonest behaviour in magnetic insulators is **antiferromagnetism**, where the susceptibility drops below the **Néel temperature**. Figure 3.17 (a) illustrates this for the case of MnO.[74] It can also be seen that below T_N at 120 K the susceptibility becomes anisotropic, and depends on whether the field is applied parallel or perpendicular to the [111] crystal direction.

Magnetic ordering is a **symmetry-breaking** transition. In the simplest antiferromagnetic case it gives rise to two **sub-lattices** of metal ions, with spins pointing in opposite directions. Some information about spin directions can be obtained from the anisotropies in magnetic susceptibility, but the most powerful technique for studying magnetic structure is **neutron diffraction**.[75] Unlike X-rays, which are scattered by all electrons, neutrons are unaffected by closed shells, and conventional (non-magnetic) neutron diffraction depends on scattering by nuclei. The magnetic moment of the neutron, however, gives an interaction with unpaired electrons in magnetic solids. The diffraction patterns for MnO in Fig. 3.17 (b) show the appearance of extra peaks below the Néel temperature.[76] Indexing these on the normal crystal lattice gives fractional indices such as $(\frac{1}{2}, \frac{1}{2}, \frac{1}{2})$, and indicates the formation of an enlarged **magnetic unit cell** in the antiferromagnetic phase. The intensities of the peaks, especially when combined with polarization studies where the orientation of the crystal with respect to the neutron beam is changed, give detailed information both about the spin directions and the **sub-lattice magnetization**, indicating the degree of spin alignment. Sub-lattice magnetizations may also be measured for some elements (especially Fe) by Mössbauer spectroscopy, from the hyperfine splittings that appear in magnetically ordered solids.[77]

Figure 3.18 shows the magnetic structures found in a number of oxides.[1,78] In each case, only the transition metal ion and its relative spin direction is shown. In the monoxides, MnO, FeO, CoO, and NiO, the spins form parallel sheets within (111) planes in the lattice, with each adjacent sheet antiparallel. The actual spin direction relative to the (111) plane varies, and is controlled by quite subtle features of crystalline anisotropy and spin–orbit coupling. But the basic pattern, known as **type II** f.c.c. ordering, can be understood in terms of a predominantly antiferromagnetic coupling between *next-nearest-neighbour* ions in the [100] lattice directions.[79] These metal ions are separated in the NaCl structure by an oxide ion. Although nearest-neighbour interactions (in the [110] directions) are not zero, they cannot be responsible for this type of ordering. This is because out of the twelve nearest neighbours to a given metal ion, six have parallel and six have opposed spins: thus the net ordering force between near neighbours is zero.

Associated with the antiferromagnetic ordering in the monoxides is a small crystal distortion. For example MnO becomes rhombohedral below its Néel temperature, with the crystal axes inclined at about 0.62° from the ideal cubic directions.[80] This type of distortion is known as **magnetostriction**, and arises because the magnetic interaction between metal ions gives a slightly 'attractive' or 'repulsive' force according to their relative spin direction. Thus at low temperatures in MnO, the

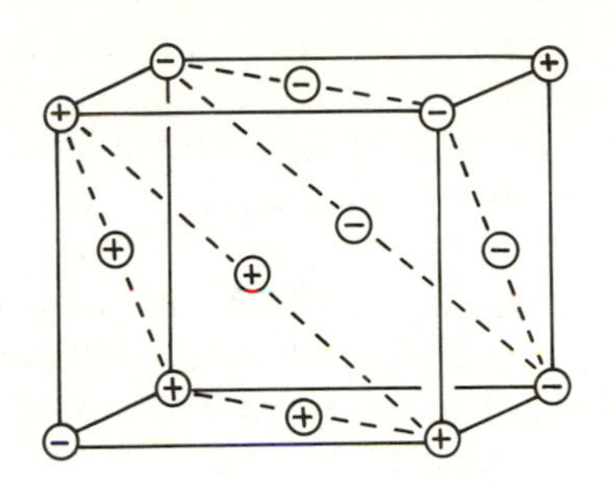

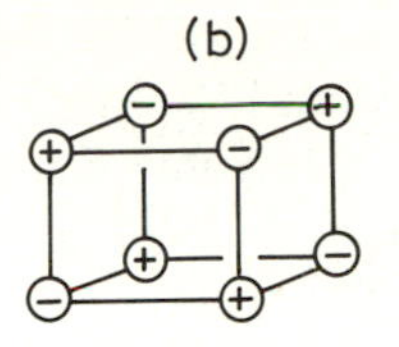

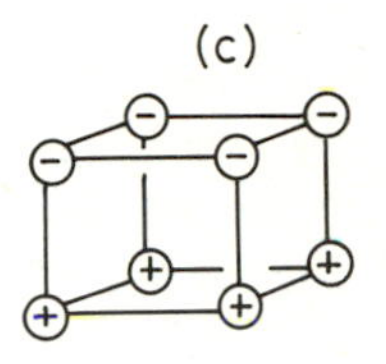

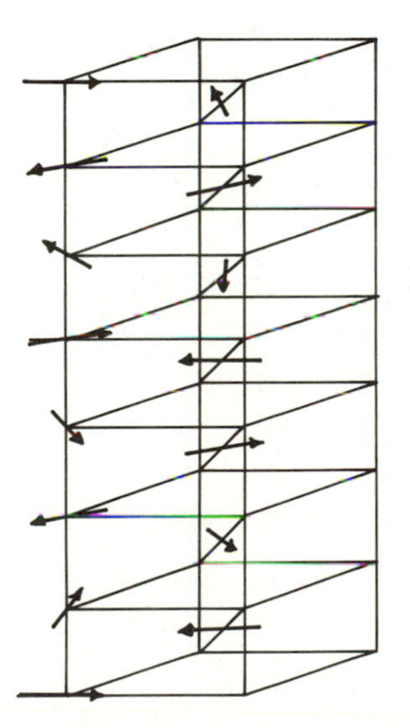

FIG. 3.18 Some magnetic-ordering patterns in antiferromagnetic oxides. (a) Type II order of rock-salt monoxides; (b) type G and (c) type A order in perovskites; (d) spiral-spin structure of MnO_2 (rutile structure).[81] Only metal atoms and their relative spin directions are shown.

near-neighbour Mn–Mn distances are 331.4 pm for pairs with parallel spin, and 331.1 pm for ones with antiparallel spins.

Antiferromagnetic coupling via a bridging oxygen also gives rise to the commonest magnetic ordering pattern, **type G**, in perovskites.[78]

This is found for example in $LaVO_3$, $LaCrO_3$, and $LaFeO_3$. On the other hand, $LaMnO_3$ has the different **type A** pattern, with alternately aligned ferromagnetic (100) planes. A more complex form of ordering is found in MnO_2, where as the figure shows there is a magnetic unit cell comprising a stack of seven cells of the rutile structure (see Fig. 1.7). The spin directions form a *spiral* which rotates by an angle $2\pi/7$ between each cell and the next.[81] Such ordering patterns must be the result of competition between different interactions; in MnO_2 the coupling between a corner and a body-centred ion, and that between neighbours along the *c* axis, are both antiferromagnetic; the spiral configuration seems to be the one that minimizes the combined energy.

The degree of spin alignment increases progressively as the temperature decreases below T_N. Figure 3.19 shows how the sub-lattice magnetization varies with temperature for the case of Cr_2O_3.[82] A 'classical' picture of totally aligned spins predicts a **saturation moment** equal to $2S\mu_B$, different from the effective moment $2[S(S+1)]^{1/2}\mu_B$ appropriate to the paramagnetic susceptibility (see eqn (3.22)). Even at absolute zero, however, antiferromagnets show a sub-lattice magnetization slightly less than this classical value, as the ideal spin alignment is upset by quantum-mechanical 'zero-point' fluctuations. *Mean-field* theories of magnetism cannot predict the detailed form of sub-lattice

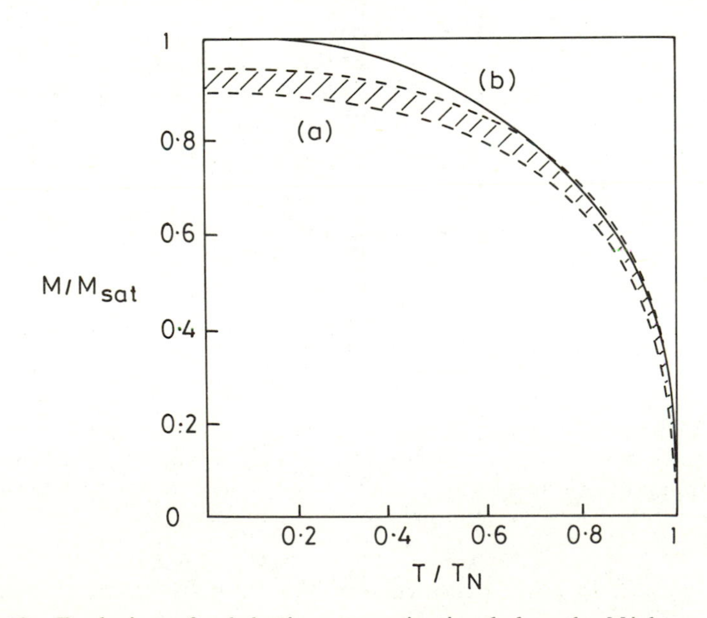

FIG. 3.19 Evolution of sub-lattice magnetization below the Néel temperature in Cr_2O_3. The experimental curve from neutron diffraction (a) is compared with the prediction of mean-field theory (b).[82]

magnetization curves such as that in Fig. 3.19, although the deviations are not large for this case with $S = \frac{3}{2}$. The errors in classical theories are more significant for smaller S values, and also in 'low-dimensional' solids, where magnetic interactions occur predominantly in one or two dimensions only.[83] The sub-lattice magnetization in antiferromagnets may also be reduced by covalent-bonding effects, essentially of the same kind that give the *superexchange* interaction between spins, described below.

While elastic scattering (diffraction) of neutrons shows the ground-state structure, it is also possible to perform inelastic scattering experiments which give information about low-energy excitations. The most characteristic of these are **spin-waves** or **magnons**. A spin-wave can be thought of as an oscillation of the spin about its equilibrium direction. The same interactions that give rise to magnetic ordering cause such oscillations to propagate in the crystal like vibrational waves (phonons).[84] Since momentum as well as energy transfer can be determined in inelastic neutron scattering, it is possible to measure a complete dispersion spectrum, showing how the energy depends on the wavevector $\boldsymbol{k}$ of the excitation. Such a spectrum for MnO is shown in Fig. 3.20; as with a band-structure plot, the horizontal axis is split into sections showing various non-equivalent $\boldsymbol{k}$ directions within the first Brillouin zone.[85]

The energies of spin-waves are typically of the order of 100 cm^{-1}, or 0.01 eV, which once again give an indication of the magnitude of magnetic interactions. Theoretical fits to spin-wave spectra in fact provide the most accurate source of information about these. The interaction between each pair of ions (i and j) is often represented by a **Heisenberg exchange Hamiltonian**:

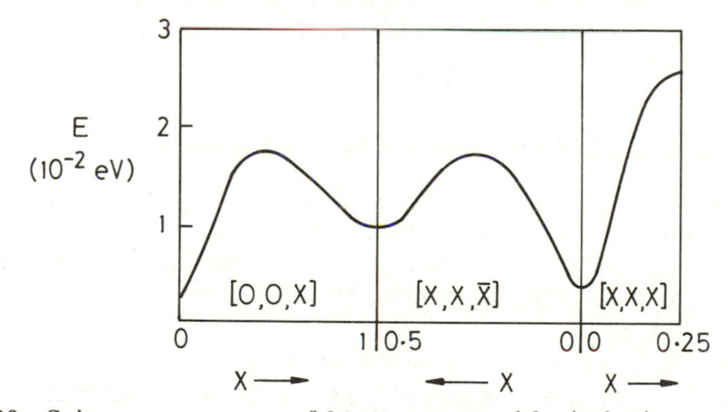

Fig. 3.20 Spin-wave spectrum of MnO, measured by inelastic neutron scattering at 4.2 K.[85] The curves show the spin-wave energies as a function of wavevector in three different directions.

$$H_{ij} = -J_{ij}(S_i \cdot S_j) \tag{3.29}$$

where the exchange parameter J is positive for ferromagnetic, and negative for antiferromagnetic, interactions. For MnO it is found that interactions between nearest neighbours are as large as those involving next-nearest Mn^{2+} ions, although as pointed out above, it is the latter interaction that is responsible for the actual magnetic ordering. The Heisenberg Hamiltonian only gives a first approximation to the interaction, and often must be supplemented by higher-order terms in each spin, which may also incorporate anisotropic effects.

Other forms of magnetic ordering are possible. True ferromagnetism is not common in insulators, a rare oxide example being $LiNiO_2$ with a Curie temperature of 6 K.[91] **Weak ferromagnetism** can result from a slight **canting** of the spins in a basically antiferromagnetic structure, as happens in haematite ($\alpha-Fe_2O_3$) between 256 K (below which it becomes truly antiferromagnetic) and the ordering temperature of 905 K.[86] Permanent magnetism in insulating oxides is more often the result of **ferrimagnetism**, where the opposed alignment of ions with *different* spins does not give a net cancellation. The example shown in Table 3.6 is yttrium–iron garnet (YIG) $Y_3Fe_5O_{12}$. The Fe^{3+} ions are not crystallographically equivalent, each formula unit containing three in tetrahedral sites and two in an octahedral environment. The total magnetization below the ordering temperature 559 K comes from the opposition of these two types of Fe, giving a net saturation value equivalent to one ion with $S = \frac{5}{2}$. The large class of magnetic materials known as **ferrites**, containing magnetically inequivalent ions in crystal structures based on spinel, derive their name from their ferrimagnetic properties. Many of these compounds are technologically important.[87,88] For example it is desirable to have an insulating solid for high-frequency applications such as inductance cores, where a metallic material would give rise to large eddy-current losses.

3.4.4 Origin of exchange interactions

The direct through-space magnetic dipole interaction of spins is far too small to account for most magnetic-ordering phenomena. One of the interesting aspects of exchange interactions, in fact, is that they depend on *chemical bonding effects*, and can, in principle, give information about these.[79,89]

Suppose that we have two weakly-interacting atoms, each with one unpaired electron. The energy difference between the singlet ($S = 0$) and triplet ($S = 1$) states of the combined system can be expressed as

$$\Delta E = 2K - 4St \tag{3.30}$$

where the two terms have a different origin. The **potential exchange** term $2K$ represents the difference in electron repulsion in the two states, arising from the Pauli-principle requirement that the overall wavefunction be antisymmetric with respect to interchange of electrons. This is the exchange effect that operates between two electrons in different orbitals in one atom, and leads to *Hund's rule*, and to the occurrence of high-spin ground states (see Section 2.1.1). It is always *ferromagnetic* in its effect, although its contribution to the coupling between ions is small when they are well separated. The second part of eqn (3.30) is known as the **kinetic exchange** term. It depends on the **overlap integral** S of the two orbitals, and the **hopping** or **resonance integral** t between them. In a molecular orbital picture, t would be the interaction giving rise to the energy difference between bonding and antibonding orbitals. In a band calculation on a solid, it represents the effective interaction between neighbouring atoms which determines the band width. The kinetic exchange term therefore depends directly on chemical bonding. It is *antiferromagnetic* in sign, and arises from the fact that electrons in overlapping orbitals can obtain some bonding stabilization when their spins are opposed, but not when they are parallel.

Equation (3.30) is not often used explicitly to treat exchange interactions in extended solids, but it is important in showing how both ferromagnetic and antiferromagnetic interactions can arise in different circumstances.[89] When the orbitals on different magnetic ions are orthogonal, only the potential exchange term contributes, and ferromagnetic coupling is expected. If there is an appreciable overlap, however, the kinetic exchange term usually dominates, and the coupling is antiferromagnetic. An alternative expression for this antiferromagnetic interaction is frequently quoted, based on a perturbation approximation to the Hubbard model.[90] The Heisenberg exchange parameter J is given approximately by

$$J = -2t^2/U. \tag{3.31}$$

As noted above, the hopping integral t is proportional to the band width that would be calculated in a simple tight-binding approach. We have seen in Section 2.3.1 that although there may be some direct overlap of metal d orbitals, in a majority of metal oxides the band width is mostly a result of *indirect* interactions, involving two metal atoms with an intervening oxygen. The same indirect overlap is also generally responsible for magnetic interactions in insulators, and the resulting effect is known as **superexchange**.

In many metal oxides the most significant interaction is that between two metal ions and an oxygen in a linear configuration. The way in

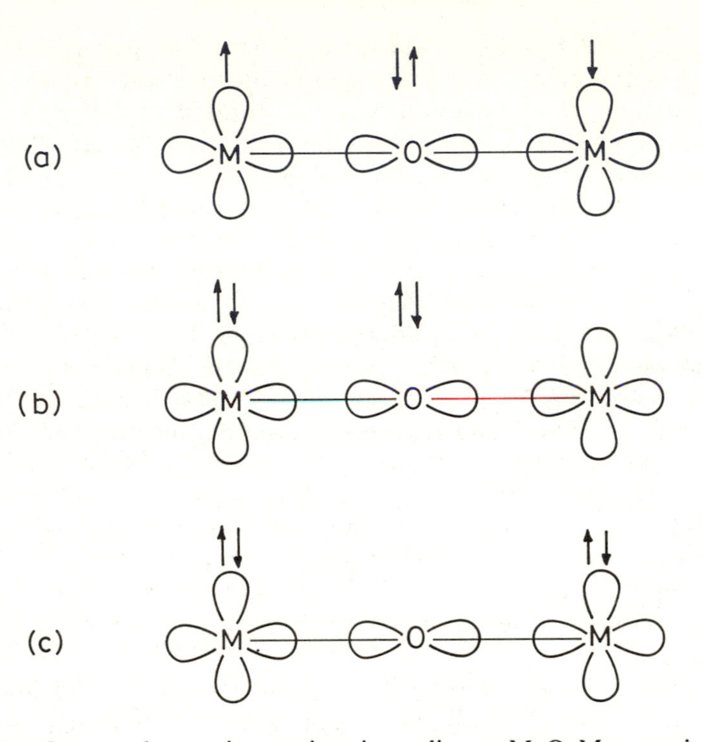

FIG. 3.21 Superexchange interaction in a linear M–O–M grouping. The ground-state ionic electron configuration (a) can mix with two types of excited configuration (b) and (c), so long as the spins are antiparallel as shown. With parallel spins this interaction is prevented by the exclusion principle.

which superexchange operates in this group is shown in Fig. 3.21. The antiferromagnetic ground state (a) can mix with the two types of charge-transfer configuration shown in (b) and (c), thus lowering its energy. The same mixing is prevented by the exclusion principle, however, if the spins are parallel. It can be seen that whereas one of the excited configurations, that of Fig. 3.22 (b), involves only the Hubbard energy U; the other (c) depends on the *charge-transfer gap* between the valence (O 2p) band and the empty d level. Thus the perturbation estimate of J should really include this charge-transfer term as well. Writing the energy of this state as E_{CT} gives the modified expression:

$$J = -2t^2[1/U + 1/E_{CT}] \qquad (3.32)$$

With this extended model it is possible to obtain reasonable estimates of the next-nearest-neighbour exchange terms in monoxides, using parameters derived independently from spectroscopic data.[72] The increase in J along the series MnO, FeO, CoO, NiO, which is reflected in the increase in the Néel temperatures noted in Table 3.6, comes largely from a decrease the charge-transfer gap E_{CT} as the nuclear charge on the metal increases. A simple interpretation of the trend in T_N is that the progressive reduction in the gap leads to more covalent mixing between metal and oxygen for the later elements, and so to a larger effective overlap (via the common oxygen) between the metal ions.

Ferromagnetic exchange interactions dominate only if the normal superexchange bonding contribution is small. The electrons involved must still come close to each other, however, to give a significant potential exchange term. One way in which this can happen is when a ligand atom forms a 90° bridge between two metals.[79] In a simple bonding picture each metal then interacts with a *different* p orbital on the ligand, as shown in Fig. 3.22. The fact that these are orthogonal guarantees that the kinetic term is zero, and the ferromagnetic interaction comes predominantly from the Hund's rule coupling of electrons in the two ligand orbitals: in other words, the excited configuration shown in Fig. 3.22 (b) has lower energy than the one formed with antiparallel spins.

In the layer structures adopted by many halides, 90° bridges are common, and often give rise to ferromagnetic ordering within each layer. These structures are not common in oxides, but 90° bridges between transition metal ions are found in the compounds of formula AMO_2, which have structures based on rock-salt with the metals A and M occupying alternate layers. It was noted in Section 3.2 that the low-spin d^6 compound $LiCoO_2$ apparently has a very narrow d band

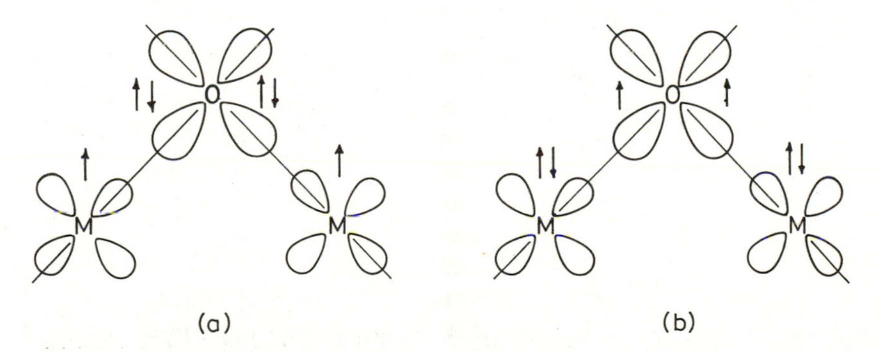

FIG. 3.22 Ferromagnetic coupling in a 90° M–O–M group. The ground-state ionic configuration in (a) mixes with the excited configuration (b). The state shown, with parallel spins in the two oxygen $2p$ orbitals, has lower energy than the antiparallel configuration which would be reached from an antiferromagnetically coupled ground state.

as a result of this geometry. $LiNiO_2$, with Ni^{3+} in the low-spin d^7 configuration, appears to be a rare example of a ferromagnetic insulating oxide.[91]

Ferromagnetic interactions can also arise when the superexchange 180° pathway couples an occupied orbital on one atom with an *empty* one on a neighbour.[79] An example is found in $LaMnO_3$, where the d^4 ion Mn^{3+} gives rise to a co-operative distortion of the Jahn–Teller kind, as described in Section 2.1.1. Within the *a–b* plane of the distorted perovskite structure there are alternate long and short Mn–O bonds.[78] Figure 3.23 shows this, and the orientation of the occupied e_g-type orbitals (one per Mn) which results. The orbitals on adjacent Mn^{3+} ions are differently oriented, and the superexchange leads to the interaction of an occupied orbital on one centre with an empty one on another. Some spin-up electron density from one ion will therefore be transferred into empty e_g orbitals of its neighbours. The Hund's

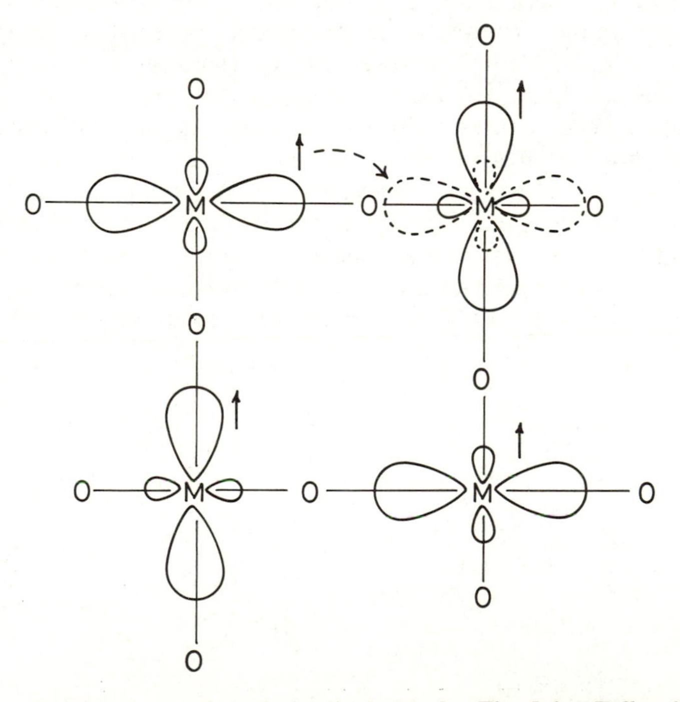

FIG. 3.23 Origin of type A ordering in $LaMnO_3$. The Jahn–Teller distortion of the d^4 Mn^{3+} gives alternate long and short Mn–O bonds within the *a–b* plane. The occupied e_g type *d* orbitals are shown by full lines, with different orientations on adjacent Mn. Thus an occupied orbital on one Mn interacts most strongly with an empty one on a neighbour (shown by broken line), giving ferromagnetic interaction within the plane.

rule coupling between different orbitals on one ion gives an overall ferromagnetic effect. Between the planes in $LaMnO_3$, the normal antiferromagnetic superexchange interaction operates, thus giving the type A order which was shown in Fig. 3.18.

In some mixed-valency oxides such as Fe_3O_4 and $La_{1-x}Sr_xMnO_3$, ferromagnetism is associated with high electronic conductivity through a different type of mechanism known as *double exchange*. This is described in Section 5.2.3.

References

1. Hamnett, A. and Goodenough, J.B. (1984). Binary transition metal oxides. In *Physics of non-tetrahedrally bonded compounds*, (ed. O. Madelung), Landolt-Bornstein New Series, Vol. 17g. Springer, Berlin.
2. Grunes, L.A., Leapman, R.D., Wilker, C.D., Hoffmann, R., and Kunz, A.B. (1982). *Phys. Rev.* B, **25**, 7157.
3. Cox, P.A. (1987). *The electronic structure and chemistry of solids*. Oxford University Press.
4. Cronemeyer, D.C. (1952). *Phys. Rev.*, **87**, 876.
5. Pascual, J., Camassel, J., and Mathieu, H. (1977). *Phys. Rev. Lett.*, **39**, 1490.
6. Mattheis, L.F. (1972). *Phys. Rev.* B, **6**, 449.
7. Mitsui, T. (1969, 1975). *Ferroelectric and antiferroelectric oxides*, Landolt-Bornstein New Series III, Vols 3, 9, and 16a. Springer, Berlin.
8. de Haart, L.G.D., de Vreis, A.J., and Blasse, G. (1985). *J. Solid State Chem.*, **59**, 291.
9. Phillips, C.S.G. and Williams, R.J.P. (1965). *Inorganic chemistry*. Clarendon Press, Oxford.
10. Dickens, P.G. and Wiseman, P.J. (1975). Oxide bronzes. In *Solid state chemistry*, MTP International Review of Science, Inorganic Chemistry Series 2, Vol. 10. Butterworths, London.
11. Atkins, P.W. (1990). *Physical chemistry*, (4th edn). Oxford University Press.
12. Cox, P.A., Robbins, D.J., and Day, P. (1975). *Mol. Phys.*, **30**, 405.
13. Burfoot, J.C. and Taylor, G.W. (1979). *Polar dielectrics and their applications*. Macmillan, London.
14. Darby, R.S. and Leighton, J. (1977). In *The modern inorganic chemicals industry*, Special Publication No 31. The Chemical Society, London.
15. Lines, M.E. and Glass, A.M. (1977). *Principles and applications of ferroelectrics and related materials*. Clarendon Press, Oxford.
16. Reide, V. (1970). *Phys. Status Solidi* (a), **2**, K193.
17. Shirane, G., Pepinsky, R., and Frazer, C.B. (1956). *Acta Cryst.*, **9**, 131.
18. Samara, G.A. and Peercy, P.S. (1981). *Solid State Phys.*, **36**, 1.
19. Kittel, C. (1976). *Introduction to solid state physics*, (5th edn). Wiley, New York.

20. Huang, K. (1987). *Statistical mechanics* (2nd edn). Wiley, New York.
21. Shirane, G., Axe, J.D., Harada, J., and Remeika, J.P. (1970). *Phys. Rev.* B, **2**, 155.
22. Muller, K.A. (1981). *J. Physique*, **42**, 551.
23. Chen, C. and Lin, G. (1986). *Ann. Rev. Mater. Sci.*, **16**, 203.
24. Abrahams, S.C., Levistein, H.J., and Reddym, J.M. (1966). *J. Phys. Chem. Solids*, **27**, 1019; Trudjman, J., Maesi, R., and Guitel, J.C. (1974). *Z. Kryst.*, **139**, 103.
25. Levin, B.F. (1973). *Phys. Rev.* B, **7**, 2600.
26. Goodenough, J.B. (1971). *Prog. Solid State Chem.*, **5**, 145.
27. Thornton, G., Morrison, F.C., Partington, S., Tofield, B.C., and Williams, D.E. (1988). *J. Phys. C: Solid State Phys.*, **21**, 2871.
28. Kemp, J.P., Beal, D.J., and Cox, P.A. (1990). *J. Solid State Chem.*, **86**, 50; (1990). *J. Phys.: Condensed Matter*, **2**, 3767.
29. Kemp, J.P. and Cox, P.A. (1990). *J. Phys.: Condensed Matter*, **2**, 9653.
30. Robin, M. and Day, P. (1967). *Add. Inorg. Chem. Radiochem.*, **10**, 247.
31. Burns, R.G. (1985). Electronic spectra of minerals. In *Chemical Bonding and Spectroscopy in Mineral Chemistry* (ed. F.J. Berry, and D.J. Vaughan). Chapman and Hall, London.
32. Nelson, D.F. and Sturge, M.D. (1965). *Phys. Rev.* A, **137**, 1117.
33. Svelto, O. (1989). *Principles of lasers*, (3rd edn). Plenum, New York.
34. Grabner, L., Stokowski, S.E., and Browser, W.S. (1970). *Phys. Rev.* B, **2**, 590.
35. Low, W. and Offenbacher, E.L. (1965). *Solid State Phys.*, **17**, 136.
36. Hayes, W. and Stoneham, A.M. (1989). *Defects and defect processes in non-metallic solids*. Wiley, New York.
37. Poliak, J.R., Noon, K.R., and Wright, J.C. (1989). *J. Solid State Chem.*, **78**, 232.
38. Gunter, P. (1982). *Phys. Rep.*, **93**, 190.
39. Stoneham, A.M., Sangster, M.J.L., and Tasker, P. (1981). *Phil Mag.* B, **43**, 609; (1981). Ibid., B, **44**, 603.
40. Blazey, K.W. (1977). *Physica*, **89B**, 47.
41. Zunger, A. (1986). *Solid State Phys.*, **39**, 275.
42. Blazey, K.W., Schirmer, O., Berlinger, W., and Muller, K.A. (1975). *Solid State Commun.*, **16**, 589; Blazey, K.W., Aguilar, M., Bednorz, J.G., and Muller, K.A. (1983). *Phys. Rev.* B, **27**, 5836.
43. Mizushima, M., Takanaka, M., Asai, A., Iida, S., and Goodenough, J.B. (1979). *J. Phys. Chem. Solids*, **40**, 1129.
44. Haldane, F.D.M. and Anderson, P.W. (1976). *Phys. Rev.* B, **13**, 2553; Zunger, A. and Lindfelt, U. (1983). *Phys. Rev.* B, **27**, 1191.
45. Catlow, C.R.A. and Stoneham, A.M. (1983). *J. Phys. C: Solid State Phys.*, **16**, 4321.
46. Powell, R.C. and DiBartolo, B. (1972). *Phys. Status Solidi* (a) **10**, 315.
47. van der Ziel, J.P. (1974). *Phys. Rev.* B, **9**, 2846.
48. McWeeny, R. (1975). *Coulson's valence*, (3rd edn). Oxford University Press.
49. Cox, P.A. (1980). *Chem. Phys. Lett.*, **69**, 340.

50. Mott, N.F. (1949). *Proc. Phys. Soc. Lond.* A, **62**, 416.
51. Hubbard, J. (1963). *Proc. Roy. Soc. Lond.* A, **276**, 238.
52. Honig, J.M. (1984). Properties of first row transition metal monoxides at low temperatures. In *Basic properties of binary oxides*, (ed. A. Dominuez-Rodriguez, J. Castaing, and R. Marquez). University of Seville.
53. Kanamori, J. and Kotani, A. (eds) (1988). *Core-level spectroscopy in condensed systems*, Springer Series in Solid State Sciences, Vol. 81. Springer, Berlin.
54. Sawatzky, G.A. and Allen, J.W. (1984). *Phys. Rev. Lett.*, **53**, 2339.
55. McClure, D.S. (1963). *J. Chem. Phys.*, **28**, 2289.
56. Newman, R. and Chrenko, R.M. (1959). *Phys. Rev.*, **147**, 1507.
57. Cottam, M.G and Lockwood, D.J. (1986). *Light scattering in magnetic solids*. Wiley, New York.
58. Cox, P.A. and Williams, A.A. (1985). *Surf. Sci.*, **152–3**, 791; Kemp, J.P., Davies, S.T.P., and Cox, P.A. (1989). *J. Phys.: Condensed Matter*, **1**, 5313.
59. West, A.R. (1984). *Solid state chemistry and its applications*. Wiley, Chichester.
60. Figgis, B.N. (1966). *Introduction to ligand fields*. Interscience, New York.
61. Terakura, K., Oguchi, T., Williams, A.R., and Kubler, J. (1984). *Phys. Rev.* B, **30**, 4734.
62. For opposing views, see the adjacent papers in *J. Magn. Magn. Mater.*, **54–57** (1986): Kubler, J. and Williams, A.R., p. 603; and Allen, J.W., Sawatzky, G.A., and Zaanen, J. p. 607.
63. Andersen, O.K. and Satapathy, S. (1984). Calculation of the electronic bandstructure of the 3*d* monoxides and the vacancy compound, Nb_3O_3. In *Basic properties of binary oxides*, (ed. A. Dominuez-Rodriguez, J. Castaing, and R. Marquez). University of Seville.
64. Schrieffer, J.R., Wen, X.G., and Zhang, S.C. (1989). *Phys. Rev.* B, **39**, 11663.
65. Ansimov, V.I., Zaanen, J. and Andersen, O.K. *Phys. Rev.* B, (submitted).
66. Shen, Z.-X., Shih, C.K., Jepsen, O., Spicer, W.E., Lindau, I., and Allen, J.W. (1990). *Phys. Rev. Lett.*, **64**, 2442.
67. Messick, L., Walker, W.C. and Glosser, R. (1972). *Phys. Rev.* B, **6**, 3941.
68. Hufner, S. (1979). In *Photoemission in solids II*, (ed. L. Ley and M. Cardona), Topics in Applied Physics, Vol. 27. Springer, Berlin.
69. Cox, P.A. (1975). *Structure and Bonding*, **24**, 59.
70. Sawatzky, G.A. (1988). In ref. 53.
71. Fujimori, A., Saeki, M., Kimizuka, N., Taniguchi, M., and Suga, S. (1986). *Phys. Rev.* B, **34**, 7318.
72. Zaanen, J. (1986). The electronic structure of transition metal compounds in the impurity model. Ph.D. Thesis, University of Groningen.
73. Sawatzky, G.A. and Allen, J.W. (1984). *Phys. Rev. Lett.*, **53**, 2239.
74. Jagadesh, M.S. and Seehra, M.S. (1981). *Phys. Rev.* B, **23**, 1185.
75. Bacon, G.E. (1962). *Neutron diffraction*, (2nd edn). Clarendon Press, Oxford.

76. Shull, C.G., Strauser, W.A., and Wollen, E.O. (1951). *Phys. Rev.*, **83**, 333.
77. Greenwood, N.N. and Gibb, T.C. (1971). *Mössbauer spectroscopy*. Chapman and Hall, London.
78. Goodenough, J.B. and Longo, J.M. (1970). Crystallographic and magnetic properties of perovskites and related compounds. In *Magnetic and other properties of oxides*, (ed. K.-H. Hellwege), Landolt-Bornstein New Series III, Vol. 4. Springer, Berlin.
79. Goodenough, J.B. (1963). *Magnetism and the chemical bond*. Wiley, New York.
80. Block, D., Cherbit, P., and Georges, R. (1968). *Comptes Rendues*, **266B**, 430.
81. Yoshinori, A. (1959). *J. Phys. Soc. Japan*, **14**, 807.
82. Samuelson, E.J., Hutchings, M.I., and Shirane, G. (1970). *Physica*, **48**, 13.
83. See ref. 20, especially Chapter 20.
84. Kittel, C. (1967). *Quantum theory of solids*. Wiley, New York.
85. Pepy, G. (1974). *J. Phys. Chem. Solids*, **35**, 433.
86. Tasaki, A. and Iida, S. (1963). *J. Phys. Soc. Japan*, **18**, 1148.
87. Standley, K.J. (1970). *Oxide magnetic materials*. Clarendon Press, Oxford.
88. Craik, D.J. (ed.) (1975). *Magnetic oxides*. Wiley, New York.
89. Kahn, O. (1987). *Structure and Bonding*, **68**, 89.
90. Anderson, P.W. (1959). *Phys. Rev.*, **115**, 2.
91. Kemp, J.P., Cox, P.A., and Hodby, J.W. (1990). *J. Phys.: Condensed Matter*, **2**, 6699.

4

DEFECTS AND SEMICONDUCTION

Although all the 'insulating' oxides described in Chapter 3 show some degree of intrinsic semiconduction, the transport properties of most of these compounds are in practice dominated by impurities and non-stoichiometry. The first two parts of the present chapter describe the transport properties of the electronic carriers in semiconductors, and how the carrier concentrations are controlled by chemical equilibria involving defects. Section 4.3 gives a brief discussion of the spectroscopic properties of electronic carriers in semiconductors, and the theoretical interpretation of their energy levels. The final Section, 4.4, looks at more heavily doped compounds, and the transition from semiconducting to metallic behaviour which is often found.

4.1 Electronic carrier properties

One of the main problems is to understand the electronic conductivity of semiconducting oxides, and especially how it varies with temperature and chemical composition. Firstly we shall consider the effect of temperature. Before discussing conductivity itself and some related transport properties, it is useful to review briefly some conventional semiconductor theory, concerning the thermal distribution of carriers between bound defect states and the bands of the solid.[1-3]

4.1.1 Free carriers in thermal equilibrium

The distribution of electrons in thermal equilibrium is described by the **Fermi–Dirac** function:

$$f(E) = [1 + \exp\{(E - E_F)/kT\}]^{-1}. \qquad (4.1)$$

Here $f(E)$ is the fractional occupation of the available levels at energy E and E_F the Fermi level, which may also be regarded as the *chemical potential* for electrons. If E_c is the energy of the conduction-band edge and N_c the effective density of states in the band, the concentration of electrons in the conduction band is therefore

$$n = N_c[1 + \exp\{(E_c - E_F)/kT\}]^{-1}. \qquad (4.2)$$

Most of the oxides we shall deal with in the earlier part of this chapter are **non-degenerate semiconductors**, which means that

$$E_c - E_F \gg kT. \tag{4.3}$$

Then the exponential term in eqn (4.2) is very much larger than unity, and the Fermi–Dirac equation may be replaced by a Boltzmann-like form:

$$n = N_c \exp\{-(E_c - E_F)/kT\}. \tag{4.4}$$

If the electrons in the conduction band can be treated as free carriers with an effective mass m_e^*, the effective density of states N_c is given by

$$N_c = (8\pi m_e^* kT/h^2)^{3/2}. \tag{4.5}$$

As a rough order of magnitude, N_c at room temperature is around 10^{20} per cm^3, or one carrier per hundred atoms, if m_e^* is the unmodified free-electron mass.

A similar treatment for the hole concentration, p, gives

$$p = N_v \exp\{-(E_F - E_v/kT\} \tag{4.6}$$

where

$$N_v = (8\pi m_h^* kT/h^2)^{3/2}. \tag{4.7}$$

A general condition may therefore be written for electrons and holes simultaneously present in equilibrium:

$$np = N_c N_v \exp(-E_g/kT) \tag{4.8}$$

where E_g is the band gap ($= E_c - E_v$).

For an *intrinsic semiconductor* where electrons and holes are formed only by excitation across the band gap, we can obtain the familiar results

$$E_F \approx (E_c + E_v)/2 \tag{4.9}$$

and

$$n = p = (N_c N_v)^{1/2} \exp(-E_g/2kT). \tag{4.10}$$

These conditions rarely apply to transition metal oxides: except in very pure stoichiometric samples or at high temperatures, the intrinsic carriers are swamped by ones introduced by defects or impurities.[4–6] n-type behaviour, where $n \gg p$, can be produced by slight reduction of an insulator (as in TiO_{2-x}), by substitutional doping (for example of Ti by Nb), or by insertion of an electropositive element (as in $Li_xV_2O_5$). In all these cases the lattice defects present must carry a *net positive charge* to provide overall electroneutrality in the presence of the extra carriers. A natural consequence of this positive charge is to produce a bound *impurity state* for each electron. The energy required to excite the electron from the impurity state into the conduction band is known as the **donor ionization energy** E_d. We described some models for impurity states in Section 2.4.3, and will further discuss the application of these later (see Section 4.3.2). For the moment we may note that typical values of E_d in oxides are in the region of 0.1–0.5 eV (corresponding to 10–50 kJ mol^{-1}), although much lower values are occasionally found.

Writing the donor ionization process as

$$D = D^+ + e^- \tag{4.11}$$

the following equilibrium condition, similar to (4.8), can be written for an *n*-type material:

$$n[D^+] = [D]N_c \exp(-E_d/kT) \tag{4.12}$$

where [D] and $[D^+]$ are the concentrations of neutral and ionized donors. If carriers are provided *only* by ionization of this one type of donor, we must have

$$n = [D^+]. \tag{4.13}$$

Neglecting the $T^{3/2}$ dependence of N_c (eqn(4.5)) in comparison with that of the exponential Boltzmann term, we find that at temperatures where only a small fraction of donors are ionized (i.e. well below the *exhaustion regime*):

$$n \propto \exp(-E_d/2kT). \tag{4.14}$$

In place of (4.10), the Fermi level now lies roughly half way between the donor level and the conduction-band edge.

Similar considerations apply to *p*-type semiconductors, where electrons are removed by oxidation of a stoichiometric compound or by appropriate doping. For example $Ni_{1-x}O$ and $Li_xNi_{1-x}O$ show *p*-type behaviour, as the average oxidation state of nickel is higher than the value 2+ found in stoichiometric NiO, so that there must be some holes in the top-filled level of NiO. The defect forms an **acceptor** centre, having a net negative charge with respect to the perfect lattice, and can trap the hole in an impurity state similar to that found with donors. With only one type of defect present with an **acceptor ionization energy** E_a, we would have

$$p \propto \exp(-E_a/2kT). \tag{4.15}$$

This theory is applicable when only one type of defect is present, which is not often the case in transition metal oxides. A *p*-type compound with metal vacancies may well have some oxygen vacancies too. These provide some degree of **compensation**, by trapping many of the holes produced by the majority defect.

For an n-type oxide, strong compensation by acceptor centres will give a large number of ionized donors, even if no free carriers are present. Thus

$$n \ll [\mathrm{D}^+] \tag{4.16}$$

and under these conditions,

$$n \propto \exp(-E_\mathrm{d}/kT) \tag{4.17}$$

the activation energy for carrier production being equal to E_d, rather than half this value. The Fermi level will now be **pinned** close to the donor level, at E_d below the conduction-band edge. Similar results apply in compensated p-type compounds. It is often assumed that strong compensation is present in oxide semiconductors, even though the precise nature of the compensating defects (or even in many cases, of the majority defect which provides carriers) is uncertain.[7] In principle, one could test these models by comparing the experimental activation energy with an independent estimate of the impurity ionization energy. Unfortunately this is rarely possible, as both experimental and theoretical estimates of ionization energies are beset by uncertainties. Furthermore, it cannot in general be assumed that the activation energy for *conduction* is the same as that for generation of carriers. We shall return to these various problems below.

Whether or not compensation is present, as the temperature is increased a point is reached where nearly all donor or acceptor states are ionized. This is the **exhaustion regime**, where the carrier concentration becomes essentially independent of temperature. The temperature required for this depends on both the carrier ionization energy and the doping level. In some lightly reduced d^0 oxides, exhaustion may be complete at room temperature. In the majority of non-stoichiometric oxides, however, much higher temperatures are required. Eventually, as the temperature is increased, we reach the **intrinsic regime**, where carriers excited across the band gap come to dominate over those introduced by doping.

4.1.2 Transport properties

Both the measurement and the interpretation of transport properties of oxides are beset with enormous uncertainties. We shall be concerned mostly with the problems of interpretation, but it is important to be aware that there are often severe *experimental* difficulties in making reliable measurements of conductivity and other properties. Contacts between a semiconducting sample and metal leads may give rise to spurious effects. When powdered samples are used, conductivity can be determined by contacts between the grains; this can lead to problems, especially as semiconductor surfaces may have very different electronic properties from the bulk materials. When carriers are provided by defects, even the bulk properties can be very dependent on sample preparation conditions and thermal history. For these reasons, enormous variations can be found in the literature data.

A selection of data derived from transport measurements is given in Table 4.1. For the reasons just discussed, and in the light of the problems of interpretation explained below, the critical reader should regard the precise values with some scepticism. Any survey of the literature on binary compounds shows just how variable the numbers quoted for the same compound can be.[6]

The conductivity of a semiconductor is given by

$$\sigma = ne\mu_e + pe\mu_h \tag{4.18}$$

where μ_e and μ_h are the carrier **mobilities**, defined as their average drift velocity in an electric field of unit strength. It may sometimes be possible to estimate directly the concentration of the dominant carrier. For example in the exhaustion regime when all defects are ionized their concentration can be obtained, in principle, from the precise chemical composition of the oxide. In this case, the drift mobility of the carriers can then be determined directly from a conductivity measurement. Some of the mobility values in Table 4.1 have been obtained in this way. Often, however, the types of defect present and the number of carriers produced by each defect centre are unknown. Also, it may not be possible to make measurements under exhaustion conditions. For these reasons an independent measurement of the carrier concentration may be necessary in order to calculate the mobility. Some ways in which this can be done are explained later in this Section.

The variation of conductivity with temperature is frequently interpreted using the **Arrhenius equation** familiar in chemical kinetics:

$$\sigma = A\exp(-E_\sigma/kT). \tag{4.19}$$

E_σ is the **activation energy** for conduction. This behaviour is expected from equations such as (4.14) if the mobility varies only slowly with temperature. It is tempting therefore to identify E_σ with the energy required to produce free carriers. Even when this is correct, the discussion in the previous section shows that it may not be obvious whether E_σ is equal to the donor ionization energy E_d or to half this value. Another serious complication is that the mobility may not vary slowly

Table 4.1 *Carrier properties of some semiconducting oxides*[a]

	Carrier properties			Activation energy (eV)[d]	
Compound[b]	Type	Mobility[c] ($cm^2 V^{-1} s^{-1}$)	Effective mass (m^*/m_0)	E_a	E_μ
TiO_{2-x}	*n*	0.1	20	0.03 (var.)	0
$Ti_{1-x}Nb_xO_2$	*n*	0.1	–	0.14	0
$SrTiO_{3-x}$	*n*	3	12	0	0
$BaTiO_{3-x}$	*n*	–	–	–	0.15
$La_{1-x}Sr_xVO_3$	*p*	0.15	17	0.1	0
$Li_xV_2O_5$	*n*	0.03	–	0.1	0.15
$La_{1-x}Sr_xCrO_3$	*p*	–	–	0	0.11
$Mn_{1-x}O$	*p*	10^{-5} (var.)	–	0.5	0.4
	n	10	–	1.5	–
$Li_xMn_{1-x}O$	*p*	10^{-5} (drift)	–	0.4	0.3
		10^{-2} (Hall)			
$Li_xCo_{1-x}O$	*p*	0.4	–	0.3	0 (?)
$Li_xNi_{1-x}O$	*p*	–	6	0.3	0
$LiNbO_{3-x}$	*n*	–	–	–	0.4
$KTaO_{3-x}$	*n*	30	0.8	0	0

[a] Based on data from refs 6–13. Values given in the literature are *very variable*.
[b] Properties given are appropriate to *small* values of x and are often different for greater deviations from stoichiometry.
[c] At room temperature.
[d] E_a and E_μ are the activation energies associated with the production and mobility of carriers respectively. The activation energy E_σ for conduction is the sum of these two values.

with temperature. In fact, if the carriers are small polarons (see Section 2.4.5) we may expect something like

$$\mu = B\exp(-E_\mu/kT) \tag{4.20}$$

where E_μ is the activation energy for carrier 'hopping' from site to site. In this case the overall activation energy for conduction will be a sum of two terms:

$$E_\sigma = E_\mu + E_a \tag{4.21}$$

where E_a is the energy associated with carrier generation.

It is obvious that independent measurements of the carrier concentration, and its variation with temperature, may be essential in order to disentangle the various parameters discussed above.[3] One traditional method for semiconductors uses the **Hall effect**, explained later in this section.[1] Its interpretation is not always straightforward for oxides, especially when magnetic ions are present. More generally useful is the **Seebeck effect** or **thermopower**. This is the effect used in thermocouples, and measures the potential difference generated across a sample in response to a temperature gradient. Seebeck coefficients of semiconductors are generally much larger than those of metals, and often easier to interpret.[2,14]

For non-degenerate semiconductors, the Seebeck coefficient is given by

$$\alpha = -(k/e)[(E_c - E_F)/kT + A] = -(k/e)[\ln(N_c/n) + A] \tag{4.22}$$

in the n-type case, and

$$\alpha = +(k/e)[(E_F - E_v)/kT + A] = +(k/e)[\ln(N_v/p) + A] \tag{4.23}$$

if p-type conduction dominates. The term A in each case depends on the way in which carrier mobility varies with energy in the bands, and is not easy to estimate. Fortunately it is usually small (of the order of unity) and may often be neglected in comparison with the other term.

One simple consequence of these equations is that the sign of a thermopower measurement should indicate the dominant carrier type. For example, Fig. 4.1 shows the variation of Seebeck coefficient with composition in the mixed spinel $Co_{3-x}Fe_xO_4$.[15] Clearly a change occurs from p-type conduction when $x < 2$ to n-type when $x > 2$. The change in conduction mechanism is also shown by a marked drop in the activation energy; the intepretation of these data is discussed below. MnO also shows a change in sign of the thermopower according to the conditions.[16] p-type behaviour is normally found, and as explained in Section

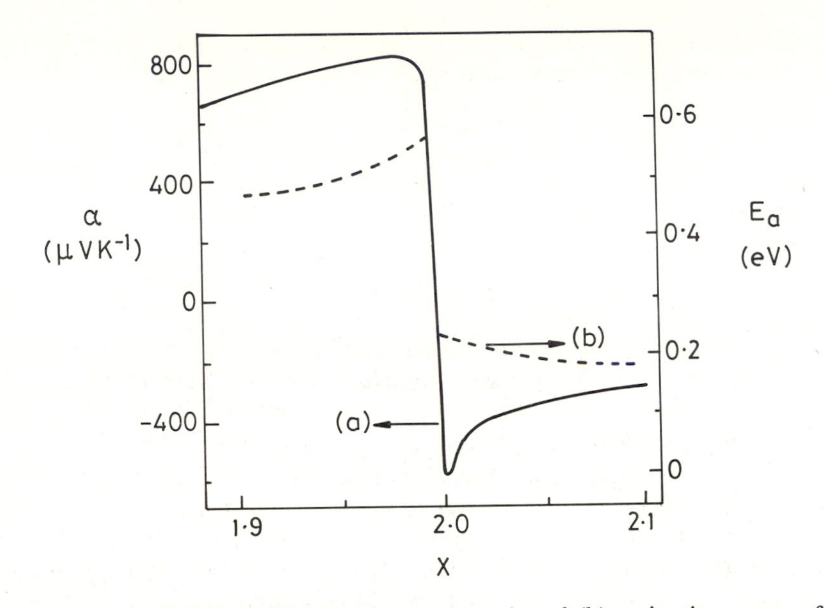

Fig. 4.1 (a) Seebeck coefficient (thermopower) and (b) activation energy for conduction in $Co_{3-x}Fe_xO_4$ as a function of x.[15] The change in sign of the thermopower indicates a change from p-type to n-type behaviour at $x = 2$.

4.2.2, holes are associated with a small natural Mn deficiency. But at temperatures above 1200 °C and oxygen pressures below 10^{-10} bar (10^{-5} Pa) the Seebeck coefficient shows that n-type conduction predominates. This has been interpreted as an indication of *intrinsic* semiconduction in MnO. Although the concentration of electrons is small compared with that of holes even at high temperatures, the electrons come to dominate the transport because their mobilities are much higher.

If a separate estimate of n or p is available, for example in a compound in the exhaustion regime, eqns (4.22) or (4.33) may be used to determine the band properties of the carriers, through the N_c or N_v terms which are given by eqns (4.5) and (4.7). Carrier effective masses may thus be deduced, and some values are shown in Table 4.1.

The way in which the Seebeck coefficient varies with temperature is especially informative. We can see by comparing eqns (4.22) and (4.4) that the slope of a plot of (αe) against $1/T$ should give the activation energy for the production of free carriers. This is the term written E_a in eqn (4.21), and if the activation energy for conduction E_σ is also known, it is clearly possible to determine the mobility activation energy E_μ. Since E_σ is the slope of a plot of $(-k \ln \sigma)$ against $1/T$ (eqn (4.19)), it is useful to give this quantity together with (αe) on the same graph.

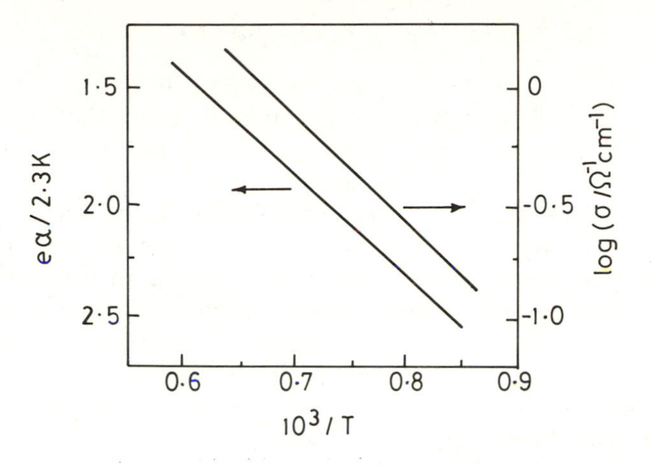

FIG. 4.2 Comparison of log (conductivity) and 'reduced thermopower' ($\alpha e/k \ln 10$) as a function of reciprocal temperature for *p*-type NiO. The fact that the two curves are parallel indicates that the temperature dependence for conduction is entirely due to the activation energy for carrier production, and that mobility is not activated.[17]

Figure 4.2 shows such a plot for NiO in a mathematically equivalent form, with log σ on one scale and ($-\alpha e/k \ln 10$) on the other.[17] In this example the two lines have the same slope, which suggests that the activation energy for conduction is the same as that for carrier activation: thus the mobility in this compound is not activated. In many compounds, such as in *p*-type MnO, the thermopower variation indicates a smaller E_a than the value of E_σ obtained from the conductivity, so that activated small-polaron mobility is suggested. Separate values of E_a and E_μ given in Table 4.1 have often been obtained in this way; we should point out that, as with other values in the table, many of the details are still disputed.

As described in Section 2.4.5, small polarons are formed when the local polarization produced by carriers is sufficient to trap them at one lattice site. A chemical picture of localized valencies is more appropriate than a band model in this situation.[2] We can think of a mixed-valency compound containing an element in two different oxidation states, for example Mn^{2+} and Mn^{3+} in $Mn_{1-x}O$. Conduction arises by the activated hopping of carriers between such neighbouring ions.

In some small-polaron materials it is assumed that the activation energy for conduction is almost entirely due to the mobility term, the number of carriers remaining essentially constant with temperature. A formula for the conductivity under these conditions is [18]

$$\sigma = c(1-c)(e^2a^2\nu/kT)\exp(-E_\mu/kT). \qquad (4.24)$$

Here c and $(1-c)$ are the relative proportions of the two different oxidation states present. Then $c(1-c)$ is the probability that a jump can take place between neighbouring sites, assuming that the two states are randomly distributed. a is the lattice constant and ν a vibrational frequency, representing the maximum frequency at which a carrier 'jump' can take place. Thus $(e^2a^2\nu/kT)$ is the conductivity estimated from diffuson theory.[19] The final exponentional factor is the probability that a given site will possess the thermal energy necessary for the jump. Equation (4.24) has been applied to a number of oxides. For example in $La_{1-x}Sr_xCrO_3$ it seems to fit the variation of conductivity with composition, x, quite well, with a vibrational frequency $\nu \sim 10^{13}$ Hz and an activation energy of 0.11 eV.[8]

Under the conditions where hopping conductivity is the appropriate model a useful alternative expression for the thermopower is the **Heikes formula**:[14]

$$\alpha = (k/e)[\ln\{c/(1-c)\} + A]. \qquad (4.25)$$

It is necessary to specify here that c and $1 - c$ are the relative amounts of the *lower* (M^{n+}) and the *higher* ($M^{(n+1)+}$) oxidation states respectively. The thermopower predicted by the Heikes formula then has the same signs for *n*- and *p*-type compounds as in the previous equations ((4.21) and (4.22)). For example, an *n*-type material results when the higher oxidation state is the predominant one, as conduction then arises from transfer of the extra *electron* present in the ion of lower oxidation state.

The Heikes formula has been widely used for interpreting thermopower data on transition metal oxides. In the $La_{1-x}Sr_xCrO_3$ series mentioned above, the hole concentration $(1-c)$ deduced from the thermopower is found to correlate well with the strontium content x.[8] More interesting examples are provided by vanadium bronzes such as $Li_xV_2O_5$ and $Cu_xV_2O_5$.[20] These are *n*-type conductors for small x due to the presence of V^{4+}. In order to interpret the thermopower variation with x, it is necessary, however, to use values of c and $1-c$ that are not simply given by the stoichiometric ratio of V^{4+} to V^{5+}; rather it appears that electron hopping takes place only among a *subset* of the available vanadium sites. The favoured sites are probably those closer to the dopant ion, where the potential is more favourable for occupation by an extra electron.

According to the discussion in Section 2.4.5, the small polaron activation energy E_μ is a measure of the energy required to reorganize the lattice locally, so that an electron can transfer between sites at the same energy. Precisely the same effect is involved in the **Marcus–Hush** theory of electron-transfer reactions between ions in aqueous solution.[21] A chemical picture suggests that the most important factor determining E_μ is the difference in bond lengths of the two oxidation states involved (see for example the configuration coordinate diagram of Fig. 2.24). Because of the similarity betweent the two processes, it is interesting to see whether some of the chemical trends found in solution chemistry can be carried over to the solid state. An example is provided by the compound $Co_{3-x}Fe_xO_4$, for which some data were shown in Fig. 4.1.[15] This compound has the spinel structure; the stoichiometric compound with $x = 2$ may be formulated $Fe^{3+}[Co^{2+}Fe^{3+}]O_4$, the square brackets showing the occupation of the equivalent octahedral sites. When $x < 2$ there is some cobalt excess, giving $Fe^{3+}[Co^{2+}Co^{3+}_x Fe^{3+}_{1-x}]O_4$. Electron transfer can then take place between Co^{2+} and Co^{3+}, the small concentration of the higher oxidation state giving p-type behaviour as the thermopower shows. On the other hand with $x > 2$ iron is in excess, and in n-type $Fe^{3+}[Co^{2+}_{1-x}Fe^{2+}_x Fe^{3+}]O_4$, electrons transfer between Fe^{2+} and Fe^{3+}. Figure 4.1 shows that the activation energy for conduction is much less for electron transfer between Fe^{2+} and Fe^{3+} than between Co^{2+} and Co^{3+}, the same trend being found for these ions in aqueous solution.[22] The explanation probably lies in the electron configurations involved. Fe^{2+} and Fe^{3+} are both high-spin ions, with configurations $(t_{2g}{}^4e_g^2)$ and $(t_{2g}^3e_g^2)$ respectively. The change in occupancy of a weakly non-bonding t_{2g} electron does not entail a very large change in bond length, so that the activation energy for transfer is small. On the other hand, with cobalt, the electron configurations change from $(t_{2g}^5e_g^2)$ to low-spin (t_{2g}^6). The removal of two strongly antibonding e_g electrons gives a large bond-length change, and consequently a high activation energy for electron transfer.

The other important transport property is the **Hall effect.**[1,2] This involves the measurement of a potential difference V_y across a sample, in response to a current j_x passing in the presence of a magnetic field B_z; x, y, and z here represent a right-handed set of orthogonal axes. The Hall coefficient is defined as

$$R_H = V_y/(B_z j_x) \tag{4.26}$$

although in anisotropic materials this equation should really be expressed in tensor form. Simple theory shows that in an n-type semiconductor,

$$R_H = -1/(ne) \tag{4.27}$$

whereas for the p-type case,

$$R_H = +1/(pe). \tag{4.28}$$

Thus a measurement of the Hall coefficient should give the carrier type (n or p) and its concentration. By combining this with a conductivity measurement, we obtain

$$\sigma R_H = -\mu_e \qquad (n\text{-type}) \tag{4.29}$$

$$= +\mu_h \qquad (p\text{-type}). \tag{4.30}$$

Figure 4.3 shows the **Hall mobility** determined in this way for reduced TiO_2 as a function of temperature.[23] Hall mobilities may differ considerably from values obtained in other ways. This is partly because of averaging approximations, it being assumed in simple theories that all carriers have the same mobility. However, there seem to be some peculiar difficulties in interpreting Hall measurements on many transition-metal compounds. Anomalies occur in magnetic materials; thus for example the Hall coefficient changes sign at the Néel temperature in NiO,[24] although other techniques show that conduction is still predominantly p-type. It also appears that in small-polaron conductors the activation energy for the Hall mobility may be different from that of the drift mobility: one theory suggests a factor of three less for the Hall mobility, which is reported to be found experimentally in reduced $LiNbO_3$.[13]

It can be seen from Table 4.1 that carrier mobilities are mostly very low in comparison with ones found in elemental metals and semiconductors. Carrier scattering must therefore be very rapid, and in many cases the mean-free paths are comparable with the lattice spacing.

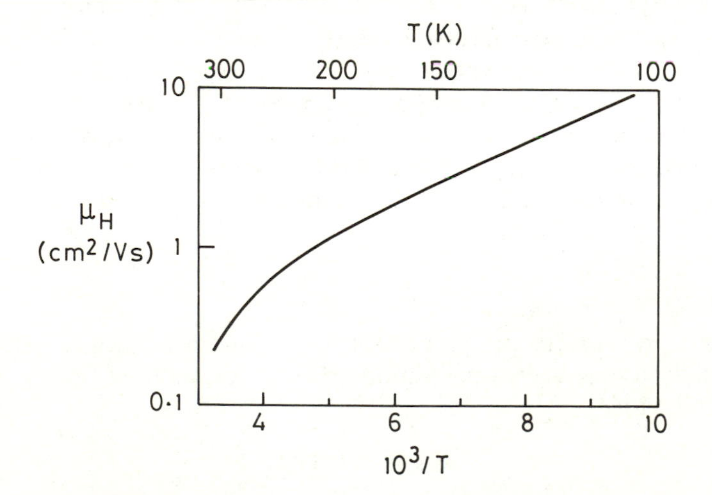

FIG. 4.3 Hall mobility in reduced TiO_2 as a function of inverse temperature.[23]

When carrier mobility is not activated, rather high effective masses are usually found, probably larger than can be explained purely by the electronic band width. These observations indicate a strong interaction of carriers with defects and with lattice vibrations. Effective masses will be enhanced by large-polaron formation. When no activation energy is observed, mobilities decline considerably with temperature, and various mechanisms can be proposed for carrier scattering.[3] Impurity scattering, either by charged or neutral defects, is probably only dominant at low temperatures except in very highly defective compounds. Scattering by polar vibrations probably dominates in the d^0 compounds under normal conditions. Conventional theory suggests that the mobility should then vary as $T^{-3/2}$, but in fact this is rarely found, and some d^0 compounds show a T^{-3} variation in the Hall mobility.[10,11] In magnetic solids, the presence of spins may also provide an important scattering mechanism, which is described in Section 5.2.2.

Another interesting conclusion is that the borderline between activated (small polaron) and non-activated mobility is rather fine. This is illustrated by the comparisons between $SrTiO_3$ and $BaTiO_3$, $LaVO_3$ and $LaCrO_3$, and NiO and MnO, where in each case the former compound apparently shows no activation energy for mobility, while the latter one does. Small variations in band width and/or dielectric properties must be sufficient to tip the balance. We shall also see in Section 4.3 that there is often evidence for small-polaron behaviour in *bound* carriers, even when no activation energy appears in the mobility.

4.2 The point-defect model

At a qualitative level it is often possible to understand the connection between chemical composition and semiconduction without having in mind any particular model for the defects present in a non-stoichiometric oxide. For example, some oxygen loss from a d^0 insulator, as in TiO_{2-x}, inevitably introduces extra electrons and leads to n-type behaviour. In order to treat this relationship in a more quantitative way, however, and to make a connection between carrier concentrations and chemical equilibria in non-stoichiometric oxides, it is necessary to be more specific about the types of defect present.

As explained in Section 1.3.3, the simplest models for non-stoichiometry are based on the idea of **point defects**. We also pointed out there some of the limitations of this picture, and in particular the various ways in which defects can interact and form clusters. Nevertheless the simple point-defect model, in conjuction with equally elementary notions of chemical equilibria, forms the basis for most chemical theories of non-stoichiometric solids.[25] We shall first look at the applications of the model; some of its limitations will be discussed later.

4.2.1 The Kröger–Vink notation

The most appropriate notation for defects in solids depends on what properties we wish to treat.[26] Consider for example a Cr^{3+} impurity in an oxide. It is convenient to specify the normal ionic charge in this way if we wish to understand the ligand-field spectrum where the d^3 electron configuration is crucial (see Section 3.3.1): for this kind of problem it is relatively unimportant whether Cr^{3+} replaces Mg^{2+} in MgO, Al^{3+} in Al_2O_3, or Ti^{4+} in TiO_2. But when we are interested in semiconduction it is not the actual ionic charge of a defect that matters, but the charge *imbalance* that it introduces. Cr^{3+} can replace Al^{3+} without the need for any compensating defects or electronic carriers. When the same ion replaces Mg^{2+}, however, there must be a compensating negative charge somewhere, either as another defect or as a conduction-band electron. In a similar way, Cr^{3+} replacing Ti^{4+} must be compensated by a hole or a suitable defect. The defect notation introduced by Kröger and Vink[27] deals with this problem by specifying the charge on a defect *not* as the normal ionic one, but as the *difference* between this and the charge present at the same site in the perfect lattice. Thus Cr^{3+} is defined as a *neutral* defect when it replaces Al^{3+}, but as a *positive* defect replacing Mg^{2+} and a *negative* one replacing Ti^{4+}. The advantage of this convention is that it makes it easier to see how the charges on defects should be balanced, and the types of electronic carrier that are introduced.

Three features of a defect must be specified in the Kröger–Vink notation:

(1) the atom actually present at the defect site, using its normal chemical symbol. In the standard notation V is used to represent a vacancy. However, this is potentially confusing if vanadium might be present and the alternative vacancy symbol □ is preferable;

(2) the lattice site, given as a subscript; either this is the symbol for the element normally present at a regular site, or i in the case of an interstitial site;

(3) the relative charge of the defect. To make clear the difference between this and the normal ionic charge, the notation used is $^{\bullet}$ for a positive charge, or $^{2\bullet}$ for two, $'$ etc. for negative charges, and (optionally) ^{x} for a neutral defect.

Some examples will help to clarify the notation.

$Li_i^{\cdot}$ Interstitial Li^+: clearly +1 charge because the interstitial site normally has no charge present.

Li'_{Mn} Li^+ replacing Mn^{2+}, giving a net −1 charge.

$Mn_{Mn}^{\cdot}$ Mn^{3+} in place of normal Mn^{2+}, equivalent to a hole in the occupied Mn^{2+} 3*d* levels.

Cr_{Al}^{x} Cr^{3+} replacing Al^{3+}: no net charge.

$\square_O^{\cdot\cdot}$ O^{2-} vacancy with no electron present, so net +2 charge.

$\square_O^{\cdot}$ O^{2-} vacancy with one electron trapped, net +1 charge. (In some other notations this is referred to as an F^+ centre.)

$\square_O^{x}$ O^{2-} vacancy with two electrons trapped and no net charge ('F centre').

The notation is completed by the obvious e' and $h^{\cdot}$, for the electron and hole carriers giving *n*- and *p*-type semiconduction.

The notation may be used for writing equations for chemical reactions involving the production or elemination of defects in oxides. Examples are:

$$Li = Li_i^{\cdot} + e' \quad (4.31)$$

representing the insertion of lithium into an oxide, to give interstitial Li^+ and a conduction band electron;

$$Li_2O = 2Li'_{Mn} + O_O^x + \square_O^{\cdot\cdot} \quad (4.32)$$

showing the (compensated) doping of Li into MnO, with the charge being compensated with an oxygen vacancy; and

$$\tfrac{1}{2}O_2 + \square_O^{\cdot\cdot} = O_O^x + 2h^{\cdot} \quad (4.33)$$

the filling of an oxygen vacancy by reaction with oxygen gas: two holes are generated because neutral oxygen takes up electrons to make O^{2-} in the lattice. In all these examples, elements or compounds not written in the Kröger–Vink notation represent other (solid or gaseous) phases.

These equations also illustrate the balancing rules which have to be obeyed. The elements actually present, and the defect charges, must be conserved. The balancing for lattice sites is a little more flexible, as extra sites may be created or destroyed, but only in combinations corresponding to the ideal stoichiometry of the oxide: for example, in eqn. (4.32) two Mn and two O sites are created in the right-hand side. In TiO_2, two oxygen sites would have to accompany each titanium site added or subtracted. Sometimes one writes 0 (zero) for the perfect lattice. Thus:

$$0 = \square_{O}^{\cdot\cdot} + \square_{M}'' \tag{4.34a}$$

could represent the formation of a pair of vacancies (a Schottky defect), and

$$0 = e' + h^{\cdot} \tag{4.34b}$$

the generation of intrinsic carriers.

4.2.2 *The law of mass action*

The real power of the Kröger–Vink notation becomes apparent when it is used to write down equilibrium relations between defect concentrations. Consider the reaction

$$\tfrac{1}{2}O_2 = O_{O}^{x} + \square_{M}'' + 2h^{\cdot} \tag{4.35}$$

representing the oxidation of an oxide MO to give $M_{1-x}O$. It is assumed here that each metal vacancy liberates two free holes; i.e. that we are in the *exhaustion regime* (see Section 4.1.1) when the binding of carriers to the defect may be neglected. According to the **ideal law of mass action** the chemical **equilibrium constant**[19] for this reaction is

$$K_1 = [\square_{M}'']\, p^2/p_{O_2}^{1/2}. \tag{4.36}$$

Here [x] represents the concentration of defect x, p the hole ($h^{\cdot}$) concentration and p_{O_2} the oxygen pressure. Note that normal lattice ions such as O_O are ignored in this; as for water molecules in dilute aqueous solution they can be considered to be present in constant concentration. Strictly speaking, concentrations should be replaced by thermodynamic **activities** in such equilibria. By using concentrations we are effectively assuming **ideal** thermodynamic behaviour for point defects; the errors inherent in this approximation will be discussed below.

If oxygen uptake is the *only* source of defects in MO, then according to eqn (4.35),

$$p = 2[\square_{M}''] \tag{4.37}$$

and eqn (4.36) gives

$$p = (2K_1)^{1/3} p_{O_2}^{1/6}. \tag{4.38}$$

The equilibrium constant K_1 will depend on temperature, but if this is fixed we have a prediction that p-type conductivity under these conductions should be proportional to the $\frac{1}{6}$ power of the oxygen partial pressure.

An alternative possibility is that one hole is strongly bound to the metal vacancy, giving the reaction

$$\tfrac{1}{2}O_2 = O_O^x + \square_M' + h^{\cdot}. \tag{4.39}$$

Then the equilibrium constant is

$$K_2 = [\square_M']p/p_{O_2}^{1/2}. \tag{4.40}$$

With

$$p = [\square_M'] \tag{4.41}$$

this gives

$$p = (K_2)^{1/2}p_{O_2}^{1/4}. \tag{4.42}$$

We can see that the alternative models for the vacancies give different predictions of how the carrier (hole) concentration should vary with the oxygen pressure. This variation should be reflected in the conductivity in a *p*-type semiconductor where only holes are involved. Figure 4.4 (a) shows some data for NiO, where conduction is assumed to be of this type.[28] Logarithmic plots of conductivity against oxygen pressure are shown for different temperatures, and compared with the power-law predictions. No simple power-law is obeyed precisely over a wide pressure range, but the conductivity variation is close to $p_{O_2}^{1/4}$ at lower temperatures and higher pressures, and something nearer $p_{O_2}^{1/6}$ at higher temperatures and lower pressures. This suggests that the predominant defect type may change according to the conditions. In other words, the equilibrium

$$\square_M' = \square_M'' + h^{\cdot} \tag{4.43}$$

is expected to lead to a mixture of singly- and doubly-ionized vacancies, in a proportion that changes with temperature and defect concentration.

Similar equilibria can be written for *n*-type oxides, where electronic carriers are produced by loss of oxygen. Again the p_{O_2} dependence is a function of the charge on the defects. For example if oxygen vacancies are formed, one possibility is a doubly ionized model:

$$O_O^x = \square_O^{\cdot\cdot} + \tfrac{1}{2}O_2 + 2e'. \tag{4.44}$$

With the assumption

$$n = 2[\square_O^{\cdot\cdot}] \tag{4.45}$$

this leads to an electron concentration

$$n = (2K_3)^{1/3}p_{O_2}^{-1/6}. \tag{4.46}$$

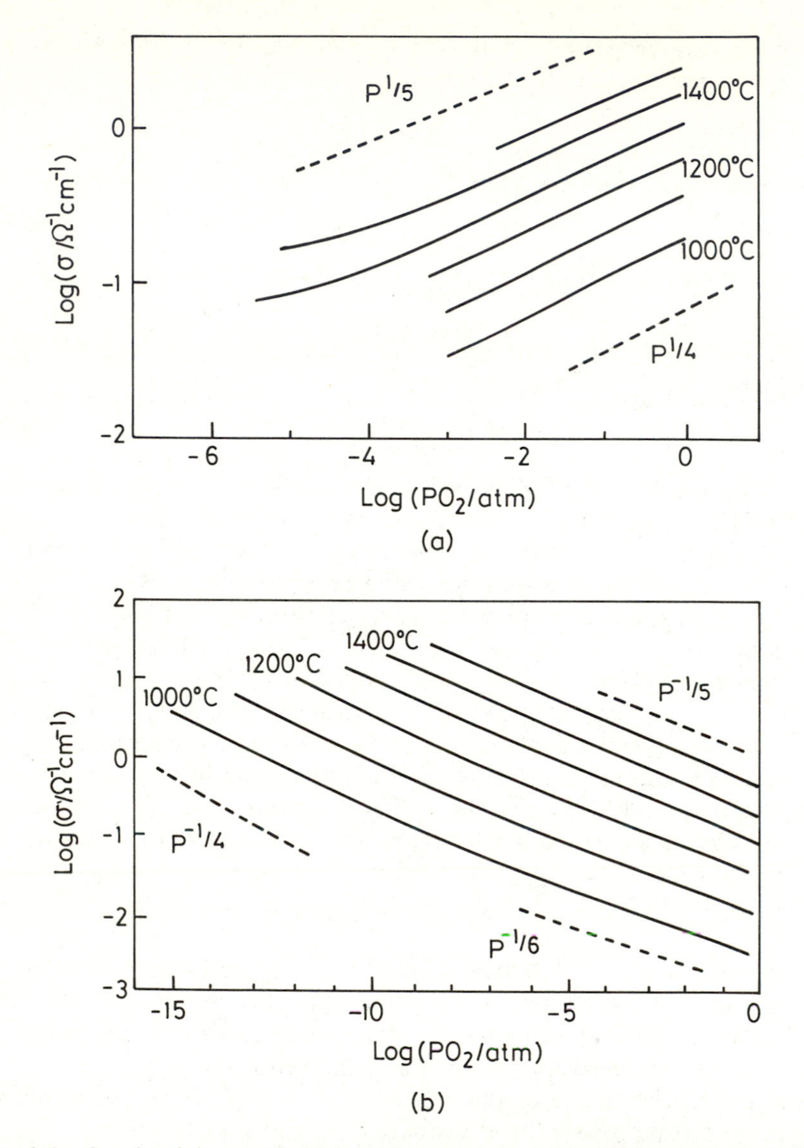

FIG. 4.4 Conductivity as a function of oxygen partial pressure. (a) *p*-type NiO; (b) *n*-type TiO_2. The broken lines show the slopes predicted by different *n* exponents in the power law $p_{O_2}{}^{1/n}$. (Based on refs 28, 29.)

Alternatively, we may have singly ionized defects with one electron strongly bound to the vacancy:

$$O_O^x = \square_O^{\cdot} + \tfrac{1}{2}O_2 + e' \tag{4.47}$$

giving the different exponent

$$n = (K_4)^{1/2} p_{O_2}^{-1/4}. \tag{4.48}$$

Thus the *inverse* dependence of carrier concentration on oxygen pressure is a characteristic of *n*-type behaviour, and is illustrated by the data for rutile, TiO_2 in Fig. 4.4 (b).[29] Again there is some ambiguity about the precise power-law obeyed, with the appearance of a change-over at different pressures. However, even if the simple power-law works well, this does not constitute a proof of any particular defect model. It is also possible that reduced oxides may contain *metal interstitials*. An equilibrium such as

$$0 = M_i^{\cdot\cdot} + \tfrac{1}{2}O_2 + 2e' \tag{4.49}$$

together with the balancing condition

$$n = 2[M_i^{\cdot\cdot}] \tag{4.50}$$

also leads to a $p_{O_2}^{-1/6}$ dependence. The variation of conductivity with oxygen pressure *cannot* therefore be used to distinguish between vacancy and interstitial models. Further information about the defect type can sometimes be obtained from diffusion measurements (see Section 4.2.3).

Table 4.2 summarizes the predictions of the mass-action law for defect

Table 4.2 *Defect equilibria and oxygen power-law exponents: showing how carrier concentration is predicted to vary with $p_{O_2}{}^{1/n}$ according to the ideal law of mass action*

Defect reaction	Predicted *n*
$\tfrac{1}{2}O_2 = O_O^x + \square_M' + h^\cdot$	4
$\tfrac{1}{2}O_2 = O_O^x + \square_M'' + 2h^\cdot$	6
$O_O^x = \square_O^\cdot + \tfrac{1}{2}O_2 + e'$	−4
$O_O^x = \square_O^{\cdot\cdot} + \tfrac{1}{2}O_2 + 2e'$	−6
$\square_O^\cdot + \tfrac{1}{2}O_2 = O_O^x + h^\cdot$	2[a]
$\square_O^{\cdot\cdot} + \tfrac{1}{2}O_2 = O_O^x + 2h^\cdot$	4[a]
$M_M^x + O_O^x = M_i^\cdot + \tfrac{1}{2}O_2 + e'$	−4[b]
$M_M^x + O_O^x = M_i^{\cdot\cdot} + \tfrac{1}{2}O_2 + 2e'$	−6[b]
$M_M^x + 2O_O^x = M_i^{3\cdot} + O_2 + 3e'$	−4[c]
$M_M^x + 2O_O^x = M_i^{4\cdot} + O_2 + 4e'$	−5[c]

[a] Under conditions where $p \ll [\square_O^\cdot]$.
[b] For compounds of formula MO.
[c] For compounds MO_2.

Table 4.3 *Oxygen exponents and probable defect type for some semiconducting oxides*[a]

Compound	Temperature range (K)	p_{O_2} range ($\log_{10}$ atm)	n	Defect type
MnO	1500–1800	> -9	6	$\square''_{Mn}$
		< -11	-6	Intrinsic
CoO	1300–1700	< 0	4	$\square'_{Co}$
NiO	1300–1700	-5 to 0	4	$\square'_{Ni}$
	1500–1700	< -5	6	$\square''_{Ni}$
TiO_2	1300–1800	-16 to 0	-5	$Ti_i^{4\cdot}$
$SrTiO_3$	1300	-18 to -12	-6	$\square_O^{\cdot\cdot}$
$BaTiO_3$	1300	-18 to -12	-6	$\square_O^{\cdot\cdot}$
$LiNbO_3$	1300	-20 to -10	-4	$\square_O^{\cdot}$

[a] From data in refs 6, 13.

reactions where variable oxygen content is the only source of non-stoichiometry. The experimental behaviour of some semiconducting oxides is shown in Table 4.3. As it is necessary to allow chemical reactions to come into equilibrium, which requires the diffusion of ions in the lattice, quite high temperatures are needed for meaningful measurements. Sometimes, also, the oxygen pressures required are much outside the range that can be controlled in pure oxygen atmosphere. Even when this is not the case, it is desirable to 'buffer' the oxygen pressure against changes produced in the reactions. Both these problems can be overcome by making use of gas-phase equilibria involving oxygen. Thus CO/CO_2 mixtures make use of the equilibrium

$$CO + \tfrac{1}{2}O_2 = CO_2. \tag{4.51}$$

To generate lower p_{O_2} values, the H_2/H_2O reaction can be used:

$$H_2 + \tfrac{1}{2}O_2 = H_2O. \tag{4.52}$$

According to these equations, the effective oxygen pressure is controlled by the ratios p_{CO_2}/p_{CO} or p_{H_2O}/p_{H_2}, the equilibrium constants for the gas-phase reactions being well known as a function of temperature.[30]

As we have illustrated for the cases of NiO and TiO_2, the oxygen pressure exponents given in Table 4.3 are often quite approximate, and may change with temperature and pressure. The principal defects shown are sometimes rather uncertain. We have already emphasized that the defect type cannot be found unambiguously from the semiconducting behaviour. As we shall see later, both the point-defect model and the assumption of ideal mass-action behaviour are also open to criticism.

The Kröger–Vink notation can also be used to show the effect of impurities. Thus p-type behaviour is induced in monoxides such as NiO by lithium doping according to the equation

$$Li_2O + \tfrac{1}{2}O_2 = 2Li'_{Ni} + 2O^{x}_{O} + 2h^{\cdot}. \qquad (4.53)$$

If this reaction goes to completion, and no other source of holes is present, we expect conductivity to be proportional to Li content. This is found in MnO, CoO, and NiO over a wide range of conditions, although at very low Li levels and high oxygen pressures, reactions such as (4.35) and (4.39) involving metal vacancies come into play.[31]

Some compounds show a transition from n- to p-type semiconduction as the oxygen pressure is changed. An example given in Table 4.3 is MnO, where a change in carrier type with conditions is also indicated by thermopower measurements, as discussed earlier.[16] In principle, n-type behaviour at low p_{O_2} values might come from the presence of ionized Mn interstitials or oxygen vacancies. A more likely alternative already noted is the possibility of extrinsic carrier generation. At a given temperature in a semiconductor we expect from eqn (4.8) that

$$n \propto p^{-1}. \qquad (4.54)$$

Then if doubly ionized Mn vacancies dominate, we expect

$$n \propto p_{O_2}^{-1/6}. \qquad (4.55)$$

As suggested previously, n-type conduction may predominate even if $p \gg n$, because electrons have much the greater mobility (see Table 4.1).

The occurrence of p-type conduction in some d^0 oxides under high oxygen pressures is in many ways more surprising, because there is no possibility of further oxidizing the metallic element. That such behaviour does occur is shown both by the oxygen-pressure dependence and Seebeck measurements.[32] It seems to be associated with the presence of impurities that replace the d^0 element by one with a lower oxidation state. Thus Al may be introduced into TiO_2 according to the reaction

$$Al_2O_3 = 2Al'_{Ti} + 3O^{x}_{O} + \square^{\cdot\cdot}_{O}. \qquad (4.56)$$

At high oxygen pressure the vacancies fill:

$$\tfrac{1}{2}O_2 + \square^{\cdot\cdot}_{O} = O^{x}_{O} + 2h^{\cdot} \qquad (4.57)$$

giving rise to holes in the valence band. This would lead to a $p_{O_2}^{1/4}$ dependence for conduction if the vacancy concentration is sufficiently large.

4.2.3 Defect mobility and diffusion

Defects in a solid may not only provide electronic carriers; they are themselves potentially mobile species, and are involved in processes such as diffusion and ionic conduction.[33,34] These have considerable importance in their own right, but diffusion measurements can also play a part in studying the relation between defect chemistry and semiconducting properties.

Typical defect mobilities are several orders of magnitude lower than those for electrons or holes. Electronic carriers therefore dominate the conductivity of a solid if they are present in significant concentrations. True ionic conduction is a feature of insulating solids with relatively high band gaps; the best-known example among transition metal oxides is zirconia, ZrO_2. The oxide ion conductivity in this compound is much enhanced by doping with lower valency oxides such as CaO or Y_2O_3; then oxygen vacancies are produced by a reaction similar to (4.56) above. As an oxide conductor, this material is used as a solid electrolyte in electrode oxygen sensors which depend on the electrochemical cell:

$$O_2(p_1)\,|\,\mathrm{ZrO_2.CaO}\,|\,\mathrm{O_2}(p_2). \qquad (4.58)$$

The platinum-catalysed electrode reaction at each side is

$$\tfrac{1}{2}\mathrm{O_2} + 2\mathrm{e}' + \square_{\mathrm{O}}^{\cdot\cdot} = \mathrm{O_O^x}. \qquad (4.59)$$

The diffusion of oxide ions through the zirconia lattice is essential to complete an electrical circuit. The potential difference between the two sides is essentially a measure of the free-energy change in the overall cell reaction, which consists of transfering O_2 between gases with pressure p_1 and p_2. If the electrons in the external circuit are all provided by the electrode reactions (4.59), this gives a cell potential equal to

$$E = kT/(4e)\ln(p_2/p_1). \qquad (4.60)$$

Zirconia-based oxygen sensors are used to monitor car exhaust systems in conjunction with catalytic converter systems for reducing the pollution content, and for many other purposes.[35]

Even in zirconia a small amount of electronic conduction takes place. In solids with an appreciable number of electronic carriers, it is generally these that dominate the conductivity. Only occasionally, when carrier concentrations are very low, is ionic conduction also significant in this respect. But although ionic motion often makes a negligible contribution to the *conductivity*, it is important in other ways in the chemistry of oxides. Ionic diffusion is an essential process in most solid state syntheses, and may be especially important with methods of the *chimie douce* kind, depending on low-temperature processes such as intercala-

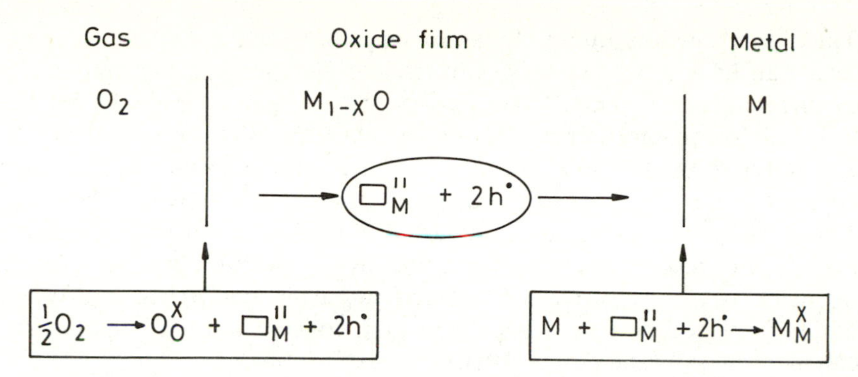

FIG. 4.5 Reactions and transport processes involved in the growth of an oxide film on a metal, assuming that metal vacancies are the principal source of non-stoichiometry.

tion or ion exchange. Many batteries depend on solid-state oxidation/reduction reactions that involve atomic transport through a solid. Corrosion is another such process, and a typical situation involving the growth of an oxide film is illustrated in Fig. 4.5.[35] In a non-stoichiometric oxide $M_{1-x}O$ containing metal vacancies, the reactions taking place are

(a) at the oxide/air interface:

$$\tfrac{1}{2}O_2 = O_O^x + \square_M'' + 2h^\cdot \tag{4.61}$$

and (b) at the metal/oxide boundary:

$$M + \square_M'' + 2h^\cdot = M_M^x. \tag{4.62}$$

Growth of the film depends on the flux of both lattice defects (vacancies in this case) and electronic carriers (holes) through the oxide. If diffusion is the rate-limiting process in the film growth, the film thickness (d) increases with time (t) according to the **parabolic law**:

$$d^2 = kt. \tag{4.63}$$

Processes other than bulk diffusion may be important, but one can nevertheless understand why metals are often more susceptible to corrosion if they have oxides which are prone to non-stoichiometry and hence high defect concentrations (compare iron with aluminium).

The **diffusion coefficient** (D) is normally defined in terms of **Fick's Law**,[19] which relates the flux of material j to a concentration gradient ∇c:

$$j = -D\nabla c. \tag{4.64}$$

There are, however, many different ways in which a diffusion experiment can be set up, and these can lead to a rather bewildering variety of different coefficients.[36] The two most commonly measured are the **tracer diffusion coefficient** D^*, and the **chemical diffusion coeffient** $\tilde{D}$. The latter refers to diffusion in a chemical composition gradient, for example under the conditions of oxide film growth shown in Fig. 4.5. Tracer diffusion, as its name suggests, is measured with a radioactive isotope, normally diffusing into a chemically uniform sample. Tracer diffusion is closely related to the **self diffusion coefficient** D_s. When one type of mobile defect is present at a concentration x, this in turn can give the intrinsic **defect diffusion coefficient** D_d:

$$D_s = xD_d. \tag{4.65}$$

Figure 4.6 (a), for example, shows how the tracer diffusion coefficient for $Mn_{1-x}O$ is proportional to stoichiometry deficit x at different temperatures. The mobile species in this case are almost certainly metal vacancies.

Diffusion in a chemical gradient is unfortunately more difficult to interpret.[37] This is because the true driving force for diffusion is not really a concentration gradient, as implied by Fick's Law, but a gradient of thermodynamic *chemical potential*. For this reason, chemical diffusion coefficients depend not only on the intrinsic mobility of defects, but also on the variation of their thermodynamic activities with concentration. With suitable approximations a very useful equation can be obtained. If the defects obey ideal thermodynamics (as assumed in the equilibria discussed above), if only one kind of defect is mobile, and if the motion of ions rather than electronic carriers is rate limiting in chemical diffusion, it can be shown that [38]

$$\tilde{D} = (1 + |z_d|)D_d. \tag{4.66}$$

As above, D_d refers to the intrinsic diffusion of the mobile defect, and z_d is the number of separate electronic carriers that must diffuse with it. This is the same value of defect charge that is used in the Kröger–Vink notation. It is remarkable that $\tilde{D}$, unlike D^* or D_s, should not depend on the deviation from stoichiometry in this approximation.

Experimental results show that chemical diffusion in many oxides is indeed much less sensitive to oxygen pressure (and hence to defect concentration) than in tracer diffusion. However, as the data for CoO in Fig. 4.6 (b) show, small variations *do* occur, and they suggest that the assumption of ideal behaviour is not correct.[38, 40] Nevertheless, eqn (4.66) is sometimes used to make a comparison between different types of diffusion coefficient and thus to estimate an effective defect

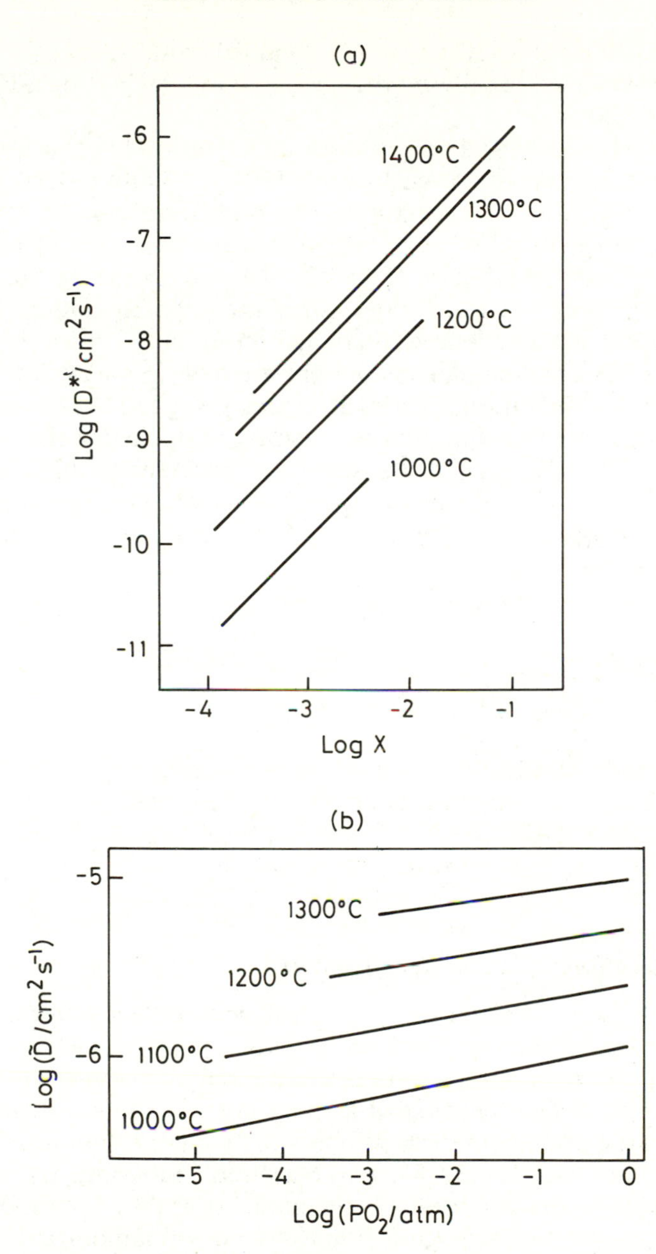

Fig. 4.6 (a) Variation of the Mn tracer diffusion coefficient with the stoichiometry deficit x in $Mn_{1-x}O$. (b) Chemical diffusion coefficient in CoO as a function of oxygen pressure. (Based on refs 39, 40.)

charge z_d. For CoO a value $z_d = 1$ is obtained, which is consistent with the $p_{O_2}^{1/4}$ dependence for the p-type conductivity, suggesting $\square_{Co}'$ as the principal defect.

As an example of the application of different kinds of measurement we may cite the case of reduced rutile, TiO_{2-x}.[41] Point defects in rutile probably only exist to a maximum concentration $x = 10^{-4}$, making their study quite difficult. The oxygen-pressure dependence of the conductivity, shown previously (Fig. 4.2), has an exponent somewhere between $-\frac{1}{4}$ and $-\frac{1}{6}$, which does not allow unambiguous assignment of the defect charge; indeed there has been much dispute even as to whether the principal defects present are oxygen vacancies or interstitial titanium. Diffusion experiments seem to suggest the latter. Tracer experiments show that titanium is much more diffusible than oxygen: in fact $D_{Ti}^*/D_O^* = 30$. As with conductivity, the titanium diffusion coefficient shows a p_{O_2} exponent of close to $-\frac{1}{5}$, which is consistent with interstitial titanium liberating four electronic carriers according to the following scheme:

$$TiO_2 = Ti_i^{4\cdot} + 4e' + O_2. \tag{4.67}$$

Then we find

$$n \propto [Ti_i^{4\cdot}] \propto p_{O_2}^{-1/5}. \tag{4.68}$$

On the other hand, the effective defect charge z_d deduced from chemical diffusion experiments appears to be closer to 3 than 4. It is possible therefore that both $Ti_i^{4\cdot}$ and the interstitial with one electron trapped, $Ti_i^{3\cdot}$, are both present in equilibrium.

4.2.4 Limitations of point-defect theories

In spite of the wide application of ideal point-defect models, and of their considerable success from an empirical point of view, a number of quite severe reservations can been expressed about their validity. At a microscopic level, the models assume not only that isolated point defects are the principal source of non-stoichiometry, but also that they do not interact significantly. There is reason to doubt both assumptions.

As we have noted in Chapter 1, the model of isolated point defects is often contradicted by structural studies on non-stoichiometric oxides.[34] It was in $Fe_{1-x}O$ that diffraction results first suggested the presence of a proportion of *interstitials* as well as the expected vacancies. Various types of defect cluster have been proposed for this compound, some of which were illustrated in Fig. 1.10. Although the deviations from stoichiometry attainable in FeO are much larger than in many oxides, there is no reason to think that this compound is unique in showing pronounced defect interactions.

Another type of behaviour is the formation of crystallographic shear planes in some reduced d^0 systems (see Fig. 1.11). Structural studies are usually performed on compounds with significant deviations from stoichiometry, higher than is often the case for conductivity investigations. There is reason to think, however, that shear planes in many oxides may form at very low deviations from perfect stoichiometry. In TiO_2 for example, the point-defect concentration may never be higher than about 10^{-4}.

Whatever the type of defect aggregation, one would expect some kind of equilibrium to be set up, with isolated point defects predominating in nearly stoichiometric materials. It was mentioned in Section 3.3.2 how such equilibria may occasionally be studied by spectroscopic methods. ESR can be sometimes be used in this way. In TiO_{2-x} for example, an ESR signal attributed to point defects (probably interstitial Ti^{3+}, i.e. $Ti_i^{3\cdot}$ using the Kröger–Vink notation) is found to reach maximum intensity when x is about 10^{-4}, after which it declines with a further increase in x. This has been taken as evidence for point defects, present only at very low levels of reduction.[42] Similar experiments can rarely be done with doped magnetic solids, and in general there is little quantitative information available about equilibria involving defect clusters.

Although models of isolated point defects must be suspect except close to perfect stoichiometry, it is quite hard to assess the implications of this problem for semiconduction, especially as there are no serious models of the electronic structure of defect aggregates. But another type of criticism can be made of the ideal mass-action law, even under conditions where isolated point defects predominate. Since the important defects in semiconductors all carry a net charge, they must interact through long-range Coulomb forces. The effect of such long-range interactions can be treated using the **Debye–Hückel** theory, which is widely applied to aqueous ionic solutions.[19] Approximations valid under sufficiently dilute conditions allow one to calculate **activity coefficients,** correction factors that should be applied to the concentrations appearing in chemical equilibria. This theory shows that deviations from ideal thermodynamics can be important even under very dilute conditions. As well as altering equilibria, Coulomb forces also modify the mobilities of charged species, so that conductivities are no longer strictly proportional to their concentration.

The Debye–Hückel theory has been applied to charged defects in solids, and often predicts an appreiable deviation from ideal behaviour. For example, in p-type CoO under conditions where $p_{O_2}^{1/4}$ behaviour is found experimentally, the predicted activity coefficients for Co vacancies may be as low as 0.1.[43] It has even been suggested that when this non-ideal behaviour is taken into account, the electronic and ionic transport properties of CoO may be better described with *doubly* charged vacancies $\square_{Co}''$, rather than the singly charged $\square_{Co}'$ suggested by the ideal law.

Unfortunately the Debye–Hückel theory itself depends on approximations that are really only applicable when the defect concentrations are small, with stoichiometry deficits less than around 10^{-4}.

All these criticisms of the ideal point-defect model are seriously damaging. Although it continues to be used as a helpful *empirical* scheme for interpreting defect and transport behaviour in non-stoichiometric oxides, many of the detailed conclusions drawn from these interpretations must be suspect.

4.3 Carrier binding energies and spectroscopy

So far this chapter has concentrated on transport properties. Further information on the electronic properties of defects and their associated carriers is available from spectroscopic techniques of various kinds. After discussing these in the next section, we shall return to the subject of carrier ionization energies, and look briefly at the interpretation of these in terms of various theoretical models.

4.3.1 Spectroscopic studies of free and bound carriers

The introduction of carriers on doping an oxide generally gives rise to strong absorption bands. These are always broad, and when they extend through the visible spectrum they may even make the solid black, and so obscure the intrinsic absorptions such as crystal field transitions. In other cases, absorption may be strongest in the IR, and its tail into the red end of the visible spectrum leads to a blue coloration frequently found with reduced or lightly doped d^0 oxides. The assignment of these optical transitions is not always easy, as there are various possibilities, for example free-carrier absorption, Franck–Condon transitions of localized polarons, and excitations within various types of defect level.[26]

Truly 'free' electrons cannot absorb a photon, because the requirements of energy and momentum conservation cannot be met simultaneously. For band electrons in a solid, absorption is possible either as a direct band-to-band transition allowed electronically, or as a $\boldsymbol{k}$-non-conserving indirect process in which phonon modes or other scattering centres are involved. **Free-carrier absorption** is of the latter kind, and implies the absorption of photons by carriers within a band, mediated by some kind of scattering process.[44] It is expected in doped semiconductors with a significant number of ionized carriers, when these carriers do not form small polarons. Theory suggests that absorption should fall with increasing energy according to λ^n, where λ is the wavelength, and the exponent n depends on the dominant scattering mechanism: values of 2, 2.5, and 3 are predicted respectively for acoustic-mode scattering, optic-mode scattering, and ionized-impurity scattering. At sufficiently high free-carrier concentrations, the optical response changes, as we shall see later (Section 4.4.1), and the solids develop a reflectivity edge and a plasma frequency characteristic of a metal.

Optical absorption in lightly reduced WO_3, $SrTiO_3$, and $KTaO_3$ has been attributed to free-carrier absorption.[45,10,11] No simple power law is obeyed precisely, but an approximate λ^2 behaviour is found in WO_3, and something between $\lambda^{2.5}$ and λ^3 in the ternary compounds. It is hard to draw definite conclusions from these wavelength dependencies, especially as other types of absorption process may be contributing; for example in $SrTiO_3$ it seems that both defect transitions (probably involving oxygen vacancies such as $\square_O^{\bullet}$) and specific transitions within the conduction band are involved.[10]

In the low temperature ε phase of WO_3 a different type of optical transition is found, as illustrated in Fig. 4.7.[46] The very broad band

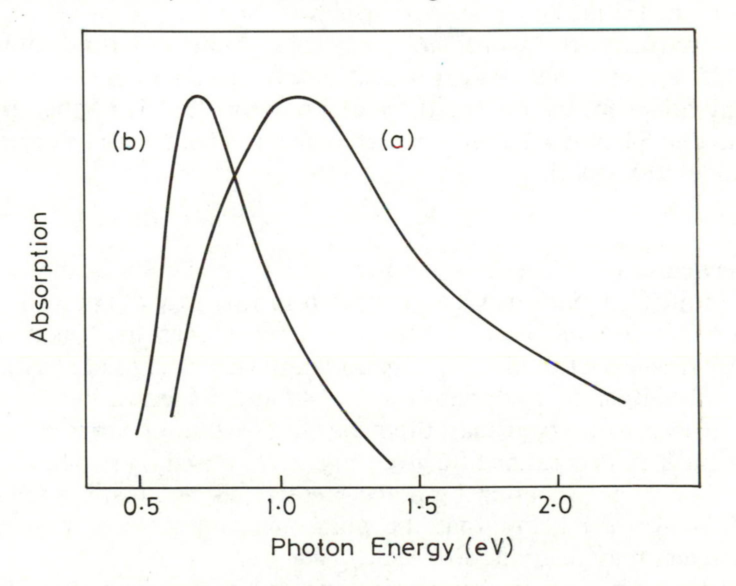

FIG. 4.7 Optical absorption spectra for ε-WO_3. (a) Spectrum of ESR-silent defects, attributed to bipolarons. (b) Spectrum of paramagnetic defects appearing after illumination.[46]

peaking at 1.1 eV appears under dark conditions, but after strong illumination this shifts to lower energy and is replaced by a peak at 0.7 eV. The broad nature of these transitions suggests that vibrational modes are strongly excited; this is expected from to the Franck–Condon principle, when a large change of equilibrium geometry occurs in the excited state. The transitions have been assigned to *polarons*, with carrier trapping being accompanied by strong local lattice relaxations. As explained in Section 2.4.5 a small polaron is expected to show optical behaviour analogous to the **intervalence transitions** found in mixed-valence compounds. Thus in reduced WO_3 one can imagine an electron undergoing a transition from W^{5+} to a neighbouring W^{6+}, with at least part of the energy coming from the difference of equilibrium geometry surrounding these two oxidation states. However, the polarons in WO_3 may be less localized than this simple chemical picture suggests. A very interesting observation is that the normal state of ε-WO_3, which gives the higher-energy absorption in Fig. 4.7, does not show the ESR signal expected from unpaired electrons trapped at defect sites. ESR appears under illumination, and is associated with the modified optical absorption found then. A likely interpretation is that electrons in the ground state are trapped not singly but in *pairs*, probably at neighbouring W^{5+} sites. This situation is known as a **bipolaron**, and may be associated with some metal–metal bonding occurring between edge-sharing WO_6 octahedra in the reduced compound.

The configuration coordinate model for small polarons shown in Fig. 2.25 suggests that optical transition energies should be related to the activation energy for the thermal hopping: that is to the term E_μ discussed in Section 4.1.2. If the electronic band width can be neglected, the model predicts that

$$E_{\text{opt}} = 4E_\mu . \tag{4.69}$$

In a few cases of polaronic absorption, it has been claimed that optical and conductivity data may be correlated in this way.[47] The validity of eqn (4.69) is dubious, however, for a number of reasons. It is doubtful whether the neglect of electronic band width is ever a good approximation, especially in d^0 compounds of the 4*d* and 5*d* series. Furthermore, the transitions observed may often be due to *bound* carriers, so that the simple symmetrical picture given Fig. 2.25, based on two lattice sites far from any perturbation by defect centres, is not really applicable. We shall also see below that the polaronic character of bound and free carriers may be quite different.

A distinction between free-carrier and polaronic absorption is hard to make experimentally, and indeed is probably not justifiable theoretically. The low mobilities in most semiconducting oxides suggest that carriers interact strongly with vibrations, and that electronic excitations will have a large vibrational component even when small polarons are not present. This is equally true for transitions of bound carriers into the conduction or valence bands. Many optical transitions observed are probably of this kind, as the experiments are performed under conditions where a majority of carriers is bound rather than free. But one should *not* expect to see a transition occurring simply at the defect ionization energy, because of the vibrational excitations also present. In Section 3.3.2 we considered the problem of relating optical and thermal transition energies of transition metal impurities; similar difficulties must always arise in interpreting defect transitions in oxides.

The intense absorption associated with defect transitions can make them hard to study by conventional optical techniques unless their concentration is very low. A useful alternative to optical absorption is electron energy-loss spectroscopy (EELS) in the low-energy regime. The application of this to crystal-field excitations of magnetic insulators was mentioned previously (Section 3.4.1). Figure 4.8 shows EELS spectra of some sodium–tungsten bronzes, Na_xWO_3, with varying x.[48] Above $x = 0.3$ the compounds are metallic and the loss features correspond to plasmon excitations, but at low x values a broad loss peak is found with a maximum around 1 eV, and thus similar to the optical absorptions in reduced WO_3.

The polaronic character of optical transitions in $Na_{0.1}WO_3$ has been confirmed by measurements using the resonance Raman effect.[49] When a vibrational Raman spectrum is excited with radiation at a frequency where electronic absorption occurs, a resonant enhancement in the intensity of some bands is found. The vibrational modes most affected are those which reflect the change in equilibrium geometry of the appropriate excited electronic state. In $Na_{0.1}WO_3$ an enhancement of the W–O stretching vibrations is found, showing that the impurity absorption is indeed strongly coupled to this mode.

Also presented in Fig. 4.8 are photoelectron spectra of the tungsten-bronze series. The low-binding-energy peak shown in detail in the spectra comes from electrons in orbitals of mostly W 5*d* character. In the metallic samples the shape of this band changes with x, and can be interpreted in terms of electrons in a partially filled conduction band. But below $x \sim 0.3$ the *shape* of the band does not change with composition, although its intensity relative to the stronger valence band ionization is roughly proportional to the donor concentration. If the small-polaron picture is valid, we can think of ionization from a localized W^{5+} state with its associated lattice distortion, giving W^{6+} with a regular equilibrium geometry. The width of the 'conduction band' signal is therefore controlled by vibrational excitations in accordance

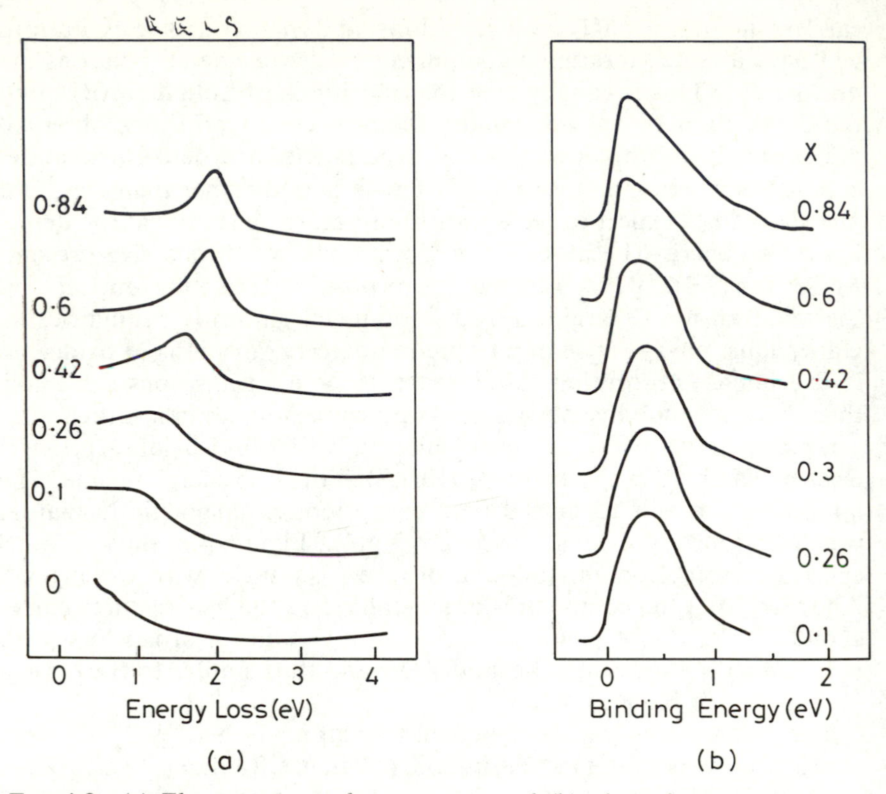

FIG. 4.8 (a) Electron energy-loss spectra, and (b) photoelectron spectra, of Na_xWO_3 for different x values.[48] For $x > 0.3$ these compounds are metallic (see Sections 4.4.1 and 5.1.2). Peaks in non-metallic samples show (a) excitations and (b) ionizations of defect levels, with Franck–Condon broadening due to vibrational interactions.

with the Franck–Condon principle, and not by the electronic band width.

Small polarons have another type of spectroscopic signature, in a very different energy range from the techniques discussed above. This is the phenomenon of **dielectric loss** in measurements in the radio-frequency region of the spectrum. Carriers give such losses when they hop from site to site at a frequency comparable to that of the radiation employed. By studying the temperature dependence of the dielectric properties it is possible to estimate the activation energy for hopping.[33] This behaviour has been observed in Li-doped CoO and NiO, where activation energies attributed to hole hopping near the Li'_M impurities are around 0.2 and 0.1 eV respectively.[50] The observation that *bound* holes apparently form small polarons in these oxides is especially interesting, as the evidence from transport measurements (discussed in Section 4.1.2) suggests that this is not the case with free carriers. As we have seen earlier, much evidence suggests that the energy balance between small and large polarons is often fairly even. Thus the behaviour found in CoO and NiO, where a transition to small polaron behaviour is induced by carrier binding, may not be uncommon.

Most spectroscopic measurements do not give a very direct idea of the *atomic* nature of the orbitals holding the carriers. Electrons in doped d^0 oxides must be in perturbed conduction-band levels, of predominantly metal d character. For doped magnetic insulators, the nature of the top-occupied and bottom-empty levels is frequently less certain, as was explained in Section 3.4.2. It is often assumed that electrons and holes occupy respectively the lower and upper Hubbard sub-bands formed from the metal d orbitals, but as we have seen there are other possibilities. The comparatively high mobility of electrons in MnO suggests a wider band of metal s type. This would be consistent with the particularly large Hubbard splitting expected for the high-spin d^5 ion, which could make the upper Hubbard $3d$ levels higher in energy than the Mn $4s$ band, as in Fig. 3.15 (b) or (d). An alternative possibility

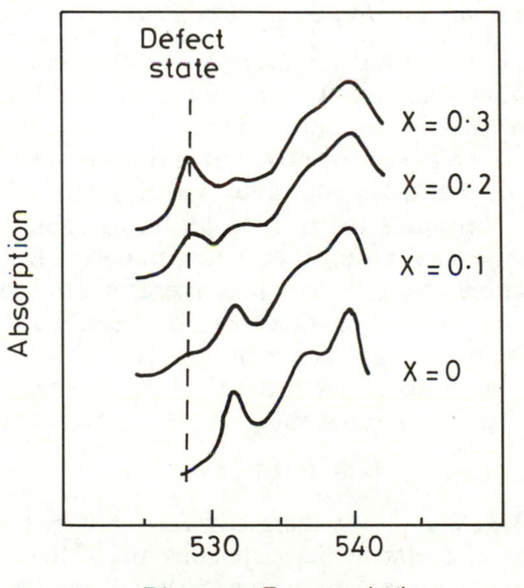

FIG. 4.9 X-ray absorption spectra showing the oxygen $K(1s)$ edge in Li-doped NiO. The defect peak marked corresponds to transitions into hole states which are shown to have appreciable oxygen $2p$ character.[52]

for the top-occupied level is the oxygen $2p$ band, as suggested by Fig. 3.15 (c) or (d). Doping in this situation would give rise to *oxygen holes*, as opposed to the more conventional idea of *metal holes*: i.e. in an ionic picture, O^- rather than an oxidized metal ion.

Evidence for oxygen holes has now been claimed for a number of p-type oxides of the later transition elements.[51] For example, Fig. 4.9 shows the oxygen K X-ray absorption spectrum (XAS) of $Li_xNi_{1-x}O$ for a range of doping levels.[52] The defect peak marked in the figure shows transitions into the hole levels introduced by doping. The appearance of these transitions in the oxygen XAS indicates that the holes must have considerable oxygen $2p$ character. The weakness of similar transitions in the metal-based X-ray spectra suggests that the O^- interpretation may be the best ionic picture here, with rather little Ni^{3+} contribution. But some care is needed in the interpretation of these spectra, because the core hole present in the final state may strongly perturb the electronic structure, so that it is not really the ground-state electron distribution that is being observed.[53] This type of problem will be discussed further in Section 5.4.2 in connection with the charge carriers present in superconducting copper oxides.

4.3.2 Interpretation of defect ionization energies

There are many difficulties involved in the theoretical interpretation of carrier binding energies. As we have seen, a direct spectroscopic measurement of these is generally impossible, and even with transport measurements there are frequently ambiguities of interpretation. The precise nature of the defect sites and their geometrical configurations is often quite uncertain. Even if these problems could be overcome, it is doubtful whether any existing theoretical model could give numerically accurate predictions. Nevertheless, it is interesting to look briefly at the application of the two defect-state theories described in Section 2.5.3; these are the *hydrogenic* and the *ionic* models.

The hydrogenic model predicts that the binding energy of a carrier to a single-ionized defect is equal to

$$E_c = R_H(m^*/m_0)/\varepsilon_r^2 \tag{4.70}$$

where $R_H = 13.6$ eV is the Rydberg constant, m^* is the carrier's effective mass, and ε_r the relative dielectric constant of the undoped solid.[1,2] There are several questions concerning the use of this formula to oxides: for example, what effective mass to use, and how will subtleties in the band structure affect it; also, should one use the static or the high-frequency value for ε_r? It can be argued that when the effective mass is strongly enhanced by polaron formation, as is very often the case, the carrier will interact with the defect via the static dielectric constant.[54] However, the hydrogenic model should really only be used when the predicted impurity orbital radius a_0, related to the hydrogenic Bohr radius $a_H(= 53\,\text{pm})$ by

$$a_0 = a_H \varepsilon_r / (m^*/m_0) \tag{4.71}$$

is considerably larger than the lattice spacing. If $(m^*/m_0) = 10$, this requires a dielectric constant of the order of 100 or more. Such very large values are found only in a few d^0 oxides. This is the case for example with $SrTiO_3$ and $KTaO_3$, where they are associated with soft modes that almost, but not quite, induce a ferroelectric transition at low temperatures (see Section 3.1.3). For $SrTiO_3$ the values $m^* = 12m_0$ and $\varepsilon_r = 220$ (applicable at room temperature) lead to a predicted binding energy of 3×10^{-3} eV, and with $KTaO_3$, less than 10^{-4} eV is predicted. One would not expect to be able to detect these ionization energies except at very low temperatures. In fact most samples of these reduced compounds show *metallic* behaviour, since (as discussed in Section 4.4) the highly extended impurity orbitals overlap strongly even at low doping levels. The very weak interaction with ionized impurities is also apparent in the carrier mobilities found in these compounds, which are large at low temperatures when phonon scattering is unimportant.[10,11]

For most transition metal oxides, the hydrogenic model predicts impurity state radii comparable to the metal–metal distance. Bound carriers are likely to be highly localized, either on the impurity or at a neighbouring transition metal size. Such a highly localized charge will also produce a significant lattice relaxation, and any electronic calculation which neglects this is unlikely to be satisfactory. The only theory currently capable of dealing with this situation is the ionic model with polarization effects included, as described in Section 2.1.2.

Calculations have been performed on the Li-doped monoxides MnO, FeO, CoO, and MnO, using the shell-model formalism for calculating the polarization terms.[55] As explained earlier, it is possible with this method to separate the ionic and electronic contributions to polarization, and thus to simulate small-polaron motion, where the slow ionic motions give rise to the activation energies for hopping. It is impossible to *predict* directly whether small or large polarons are formed, because this also requires a knowledge of the band width, which of course the ionic model itself cannot provide. Nevertheless, the ionic calculations predict that small-polaron formation is more likely for holes in MnO than with CoO and NiO. This is consistent with the observations. The activation energy for mobility of small polarons in MnO is calculated as 0.34 eV, close to the measured value. Binding energies for small-polaron holes to the Li'_M impurity are estimated at 0.48, 0.52, and 0.54 eV for MnO, CoO, and NiO respectively, which are not far different from the observed carrier activation energies of 0.4, 0.3, and 0.3 eV. A more precise comparison between theory and experiment is difficult. For one thing the degree of compensation in the measured samples is uncertain, so that it is not clear whether observed carrier activation energies are really equal to the hole-binding energies. Furthermore, free holes in CoO and NiO, unlike ones bound to the impurities, are probably *not* small polarons. Thus an unknown correction should be applied to the calculation, equal to the energy difference between small and large polarons. Another worrying feature of ionic model calculations is that they invariably predict holes to be in *metal* orbitals, in conflict with the evidence from X-ray absorption spectroscopy discussed above.

4.4 Transition to the metallic state

It is common in semiconductors to see a transition to metallic behaviour at sufficiently high doping levels.[54,56] Doped silicon, for example, shows metallic conductivity when the donor concentration is above 10^{18}–10^{19} cm^{-3}. Many such transitions are known with oxides, both in doped d^0 compounds and with semiconductors based on magnetic insulators. The following sections discuss some features of the transition itself; the properties of compounds well on the metallic side of the borderline will be described in more detail in Chapter 5. Some oxides show transitions as a function of temperature or pressure which are not associated with changing the doping level, and are also discussed in Chapter 5.

4.4.1 Properties of heavily doped oxides

As the doping level increases in a semiconducting oxide, the activation energy for conduction tends to decrease. A typical sequence is shown in Fig. 4.10, where the variation of conductivity with reciprocal temperature is plotted for $La_{1-x}Sr_xVO_3$ with different x values in (a), and the apparent activation energy for conduction against concentration in (b). The activation energy falls to zero at a critical doping level around $x = 0.2$, above which metallic conduction is found.[12,57] A selection of compounds showing similar behaviour is given in Table 4.4; values for both the critical concentration and the roughly temperature-independent conductivities at the transition are approximate, since there are difficulties in defining the precise transition point.

A drop in activation energy with increased doping is found even when

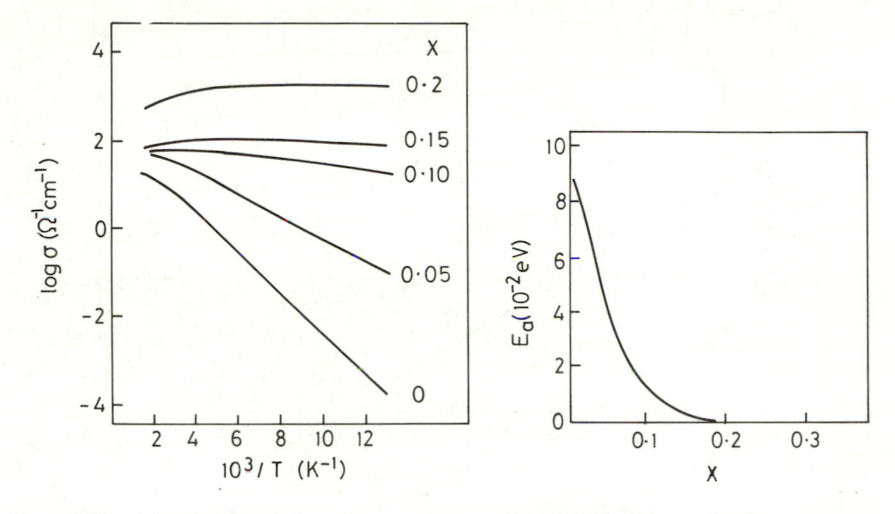

FIG. 4.10 (a) Conductivity–temperature plots, and (b) activation energy deduced from them, for $La_{1-x}Sr_xVO_3$. The disappearance of the activation energy at $x \sim 0.2$ indicates the onset of metallic behaviour.[56]

no transition takes place, for example in Li-doped NiO.[6] In some cases such as this, the change may be associated with an increase of dielectric constant due to the extra polarizability introduced by impurity states, which weakens the carrier binding to the impurity. As one approaches the transition region in the compounds in Table 4.4, however, an additional factor must be involved: impurity states overlap to form a narrow *band*, thus also lessening the gap required for carrier ionization. A more detailed model is put forward in the next section. One interesting point is that the Arrhenius law for conduction (eqn (4.19)) is often *not* well obeyed for heavily doped compounds. Thus the notion of a well defined activation energy associated with carrier ionization or mobility is no longer appropriate. Although the precise conductivity–temperature relationship is often difficult to establish experimentally, it has been found that some compounds obey a law of the form

$$\sigma = A \exp[-(T_0/T)^{1/4}] \tag{4.72}$$

where A and T_0 are constants. This is illustrated by the plot shown in Fig. 4.11 for a series of oxides of general formula $Nb_{18-n}W_{8+n}O_{69}$.[58] The carrier concentration increases with the W/Nb ratio determined by n, and the virtually temperature-independent conductivity for the highest concentration shows that the metallic point has been reached. The '$T^{1/4}$' dependence is thought to be associated with the *variable-range hopping* of carriers between states localized by disorder.[54,56] It

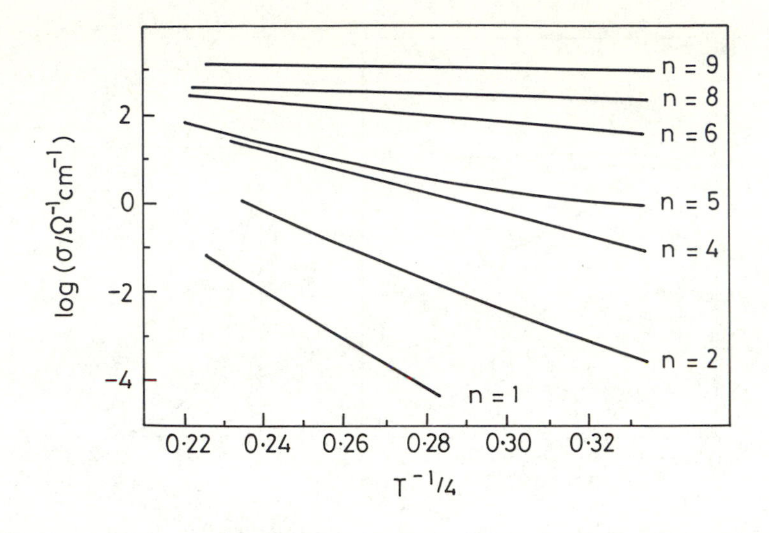

FIG. 4.11 Plot of log (conductivity) against $1/T^{1/4}$ for some reduced oxides of formula $Nb_{18-n}W_{8+n}O_{69}$. The linear plots are suggestive of a *variable-range hopping* mechanism. The horizontal plot for the compound with the highest *n* value indicates metallic conduction.[58]

is generally found only at low temperatures, and no simple law applies over a wide temperature range.

Other transport properties have been used to show signs of the transition to metallic behaviour. The Seebeck coefficient (thermopower) of a metal is generally much smaller than that of a semiconductor, and indeed large drops in thermopower have been shown to occur around the transition point in many of the compounds in Table 4.4.

In parallel with the transport properties, we might expect the optical and spectroscopic characteristics of compounds to change as the metallic point is reached. Some typical trends can be seen in the PES and EELS of the tungsten-bronze compounds Na_xWO_3 (Fig 4.8). Beyond the critical value of $x \sim 0.3$, an edge starts to appear at the Fermi level in PES, characteristic of the Fermi surface of a metal, where the occupation number of states drops to zero within a partially filled band. Similar effects have been seen in other systems, such as $La_{1-x}Sr_xVO_3$.[64] The EELS loss increases in energy, and the optical properties change in character, from the absorption spectra of localized carriers in the semiconducting material, to the reflecting behaviour of a metal.[65] At high x values the EELS shows a **plasma excitation** which is associated with the optical reflectivity edge; we shall see in Chapter 5 how a measurement

Table 4.4 *Transitions to the metallic state in some doped oxides*[a]

Compound	Critical carrier concentration per TM atom	per cm^3	Conductivity at transition $(\Omega\ \mathrm{cm})^{-1}$
$SrTiO_3$	–	3×10^{18}	10
$KTaO_3$	–	$< 10^{18}$	–
Na_xWO_3	0.3	4×10^{21}	1000
$Na_xW_{1-y}Ta_yO_3$	0.3	4×10^{21}	500
Nb–W–O	0.4	6×10^{21}	1000
$La_{1-x}Sr_xVO_3$	0.2	4×10^{21}	500
$La_{1-x}Sr_xMnO_3$	0.2	4×10^{21}	100
$Li_{1+x}Ti_{2-x}O_4$	0.3	–	–
$Li_xMg_{1-x}V_2O_4$	0.3	–	–

[a] Data from refs 10–12, 57–63.

of the plasma frequency can give important information about the carriers in a metal.

The change in electronic properties within the tungsten-bronze series is associated with an alteration in the appearance of these compounds. At low x values they are blue, because of the carrier absorption which peaks in the infrared, and tails into the red end of the visible spectrum. As x increases and metallic behaviour is found, the solids become first red and then golden bronze. This change is associated with strong reflectivity below a plasma edge, which increases in energy with x, thus shifting through the visible spectrum with increasing carrier concentration. It is after this series that the general class of *oxide bronzes* has been named: like Na_xWO_3 they are d^0 oxides doped with an electron donor such as an alkali or other electropositive metal, or hydrogen.[66] Most of them are *not* bronze in colour, however, and many are not even metallic. For example in the $M_xV_2O_5$ series a transition to metallic behaviour does not occur even at the highest attainable x values.[20] 4*d* and 5*d* bronzes frequently do show a transition like the tungsten series, although the electronic properties may be influenced by unusual band-structure effects: we shall look at some of these, such as occur with the 'blue bronze' $K_{0.3}MoO_3$, in Section 5.3.1.

One of the features shown by both transport and spectroscopic measurements is that the metal–insulator transition in doped compounds takes place in some sense *gradually*; thus activation energies fall continuously to zero, and optical transitions *evolve* from localized to metallic excitations. It is only the conductivity at absolute zero that shows an actual discontinuity at the transition point in most cases. Sometimes this gradual transition may be complicated, however, by structural changes that also occur as the doping level is increased. In Na_xWO_3 the cubic perovskite phase is only stable for $x > 0.4$; below this value there is a series of slightly distorted structures. The proximity of the structural and the metal/non-metal transition might lead to a suspicion that they could be related. However, by making various substitutions it is possible to show that this is probably not the case.[63] For example, in $Na_xTa_yW_{1-y}O_3$, where the carrier concentration is $x - y$, it is found that the electronic transition occurs at a similar carrier concentration to that in the unsubstituted bronze, but in a cubic phase and without the complication of a structural change.

One interesting observation about both the sets of values in Table 4.4 is the near universality in these numbers (within a factor of two for concentrations, and ten for conductivities) over a wide range of oxides. Critical concentrations are in the range 0.2–0.3 carriers per transition metal atom, with a conductivity around 10^2–10^3 $(\Omega\,\mathrm{cm})^{-1}$. Two compounds shown are exceptional: with reduced $SrTiO_3$ and $KTaO_3$ the critical carrier concentration is three or four orders of magnitude less than is the general rule. We shall see in the next section how this behaviour can be related to very weak carrier binding in these compounds, as discussed above (Section 4.3.2).

4.4.2 Models of the transition

In order to understand the behaviour of electrons in strongly doped semiconductors, it is necessary to take account of many types of interaction: the band structure; the lattice behaviour giving rise to polarons; binding to defect sites; lattice disorder; and Coulomb forces between electrons.[54,56] Because all these effects occur *simultaneously*, it is clearly hard to give a theoretically satisfactory model for the metal–insulator transition. Different models have been proposed, most of which emphasize a limited subset of the interactions at the expense of others. The picture suggested in Fig. 4.12 is certainly not immune to this criticism, but seems to provide a reasonably satisfactory account of many features. The following stages are shown for an *n*-type material, with a defect that provides one ionizable electron (for a *p*-type compound the energy scale could be simply reversed):

(a) Very dilute impurities give a sharp donor level E_d below the conduction band. In the ground state this level has one electron per impurity.

(b) As the impurity concentration increases, the overlap of donor orbitals leads to the formation of an **impurity band**, split off from the

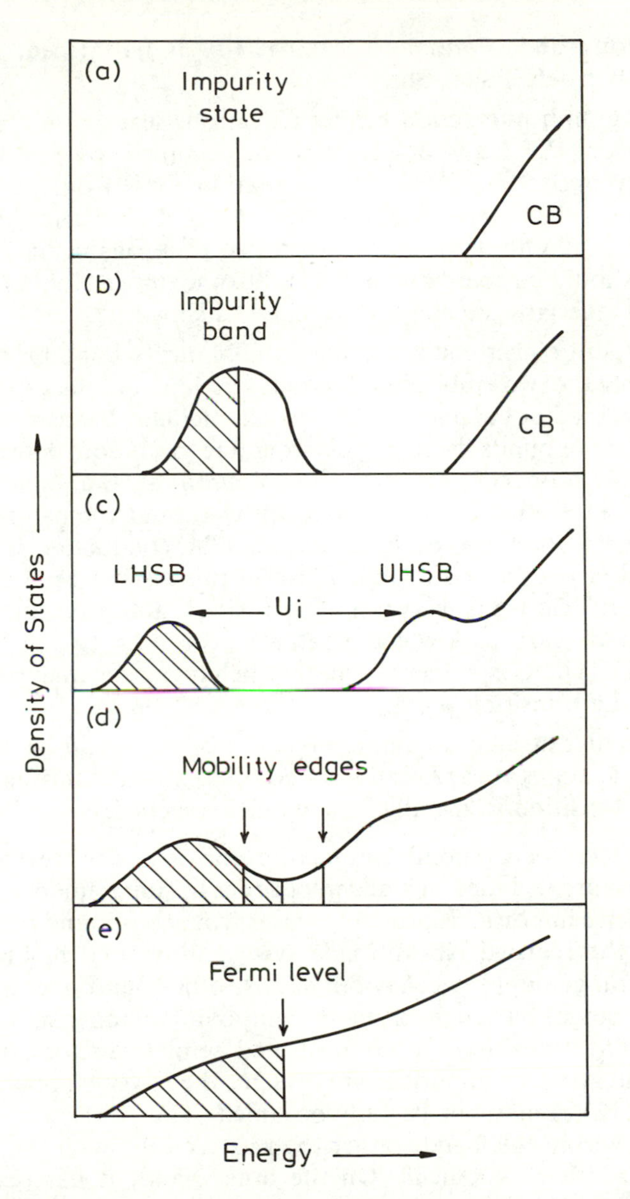

FIG. 4.12 Model of the transition to metallic behaviour in a doped oxide. Stages (a)–(e) are described in the text.

conduction band. Formally speaking, this is half filled, having one electron per defect site. But:

(c) Coulomb interaction between electrons leads to a Hubbard gap (see Sections 2.4.1 and 3.4.2), with an occupied lower sub-band and an empty upper one. A simple calculation, based on the hydrogenic model for defect states, suggests that the Hubbard gap (given by the screened Coulomb repulsion between two electrons in the same defect orbital) should be comparable to E_d: thus the upper Hubbard sub-band probably overlaps the conduction band as shown.

(d) At still higher concentrations, the impurity band broadens sufficiently that the two Hubbard sub-bands overlap, and the gap disappears. In an ordered solid one would predict metallic behaviour now, but in doped compounds the strong disorder due to random defects is important. This produces *Anderson localization* at the band edges, as described in Section 2.4.4. For a small overlap of the bands, the states at the Fermi level are localized and metallic conduction is prevented. Conduction is expected to take place in this regime by *variable-range hopping* of electrons between defect sites, and the conductivity is predicted to vary at low temperatures according to the $T^{1/4}$ law of eqn (4.72). There is evidence for this behaviour in many compounds close to the transition point.

(e) Finally, at high doping levels the impurity band broadens sufficiently that states at the Fermi level become extended through the solid, and the transition to metallic behaviour is reached.

The effects of Coulomb interaction and disorder treated in stages (c) and (d) are included in a different order in some models, and indeed their relative importance probably varies from compound to compound. Thus in the reduced Nb–W oxide system illustrated in Fig. 4.11, the variable-range hopping behaviour seems to be found over a wide composition range, although in many compounds it only manifests itself close to the transition. In compensated semiconductors, where both donor and acceptor impurities are present, the lower Hubbard sub-band in Fig. 4.12 (c) may not be fully occupied. The Anderson localization of states within this band, rather than its crossing with the upper sub-band, may then be crucial. On the other hand, it has been claimed that *long-range* Coulomb repulsions between carriers are important in the Anderson transition regime.[67] These may reduce the density of states close to the Fermi level, and lead to a stronger temperature dependence for conductivity than the normal variable-range hopping formula. One theory[68] predicts that

$$\sigma = A' \exp[-(T_0'/T)^{1/2}] \qquad (4.73)$$

but there seems to be no good evidence for this in oxides.

The critical concentration for metallic conduction must depend on how much overlap of impurity orbitals is required to overcome the localizing effects of Coulomb repulsion and disorder. The **Mott criterion**, based essentially on the Hubbard-model condition $W > U$ for hydrogenic orbitals,[54] is

$$n_c^{1/3} a_0 \sim 0.25 \tag{4.74}$$

where n_c is the critical carrier concentration and a_0 the radius of the impurity orbital, given by eqn (4.71). As we have noted, very large values for the dielectric constant ε_r are required for the hydrogenic model to be applicable, although the Mott criterion has been found to apply quite well, outside the oxide field, to transitions occurring over an enormous range of n_c.[69] Among transition metal oxides, this model is probably most applicable to the $SrTiO_3$ and $KTaO_3$ systems: the extremely high dielectric constants lead to very weak carrier binding and large donor radii. With $m^* = 12$ and $\varepsilon_r = 220$ appropriate to $SrTiO_3$, one predicts $a_0 = 1$ nm and $n_c \sim 5 \times 10^{18}$ carriers per cm^3, not far from the observed 3×10^{18}.

These compounds are exceptional because of their extremely high dielectric constants at low temperatures. For most transition metal oxides carrier binding is stronger, and a picture of impurity states localized on the sites neighbouring the impurity atom is more appropriate: as suggested in Section 4.3 there is likely to be strong polaron formation, which further assists the carrier localization. Thus it is generally necessary for a significant fraction of lattice sites to have impurity electrons or holes present before their overlap is sufficient to reach the transition point. A rather different model for the transition has been put forward, in which it is imagined that *clusters* of impurity states start to form when neighbouring sites become occupied. At low concentrations these remain isolated, but at a certain concentration known as the **percolation threshold**, they start to form a connected network through the solid.[70] Estimates of the percolation threshold depend on the structure and connectivity of the solid, but for a simple cubic lattice the value $x = 0.31$ is found, which is close the observed critical concentration. One attractive feature of this theory is that it explains the near-universality of the critical x values for many doped oxides, as shown in Table 4.4.

The percolation model suggests that a doped solid consists in some sense of two phases, one non-metallic and the other metallic, which interpenetrate each other in intimate contact. Variable-range hopping is predicted below the percolation threshold. Some evidence for this 'two-phase' nature has been claimed from a number of studies. In the superconducting spinel $Li_{1+x}Ti_{2-x}O_4$ the fraction of superconducting material apparently varies with x as expected for a percolating network of Ti^{3+} sites occupied by carriers.[61] The optical properties of some reduced Nb and W oxides, where localized (polaron) and free-carrier absorptions are found to *coexist* some way into the metallic regime, seem to suggest a similar picture.[58] Limited electron delocalization over clusters has also been suggested as an interpretation of thermopower measurements in the vanadium-spinel system $Li_xMg_{1-x}V_2O_4$.[62]

A rather different picture of the metal/non-metal transition in a disordered solid is obtained by starting from the metallic side. It can be argued that carriers should become localized when the scattering by disorder reduces their mean-free path to about the lattice parameter (or intersite metal–metal distance) a. Based on a model of Anderson localization, Mott predicted that the **minimum metallic conductivity**, found at the transition point, should be[54]

$$\sigma_{min} = 0.026\, e^2/\hbar a. \tag{4.75}$$

For a typical a value of 0.3 nm, this gives 300 $(\Omega\,cm)^{-1}$, although the detailed numerical constant in eqn (4.75) is rather sensitive to the directional properties of the impurity orbitals, and for impurity bands based on d orbitals a rather larger value (around 1000 $(\Omega\,cm)^{-1}$) is appropriate. Although the concept of minimum metallic conductivity may not have entirely general validity, it is notable that conductivities of this order are found at the transition point for many of the compounds listed in Table 4.4.

References

1. Kittel, C. (1976). *Introduction to solid state physics*, (5th edn). Wiley, New York.
2. Cox, P. A. (1987). *The electronic structure and chemistry of solids*. Oxford University Press.
3. Hamnett, A. (1987). Transport properties. In *Solid state chemistry: techniques* (ed. A.K. Cheetham and P. Day). Oxford University Press.
4. Kofstad, P. (1972). *Non-stoichiometry, diffusion and electrical conductivity in binary metal oxides*. Wiley, New York.
5. Dominguez-Rodriguez, A., Castaing, J., and Marquez R. (eds) (1984) *Basic properties of binary oxides*. University of Seville.
6. Hamnett, A. and Goodenough, J. B. (1984). Binary transition metal oxides. In *Physics of non-tetrahedrally bonded compounds*, (ed. O. Madelung), Landolt-Bornstein New Series III, Vol. 17g. Springer, Berlin.
7. See for example Kleinpennig, T. G. M. (1976). *J. Phys. Chem. Solids*, **37**, 925.

8. Karim, D.P. and Aldred, A.T. (1979). *Phys. Rev.* B, **20**, 2255.
9. Poumellec, B., Marucco, J.F., and Lagnel, F. (1986). *J. Phys. Chem. Solids*, **47**, 381.
10. Lee, C., Destry, D. and Brebner, J.L. (1975). *Phys. Rev.* B, **11**, 2299.
11. Whemple, S.H. (1965) *Phys. Rev.* A, **137**, 1575.
12. Webb, J.B. and Sayer, M. (1980). *J. Phys. C: Solid State Phys.*, **9**, 1415.
13. Arizmendi, L., Cabrera, J.M., and Agullo-Lopez, F. (1984). Point defects in double-oxide crystals. In ref. 5.
14. Heikes, R.H. and Ure, R.W. (1961). *Thermoelectricity*. Interscience, New York.
15. Jonker, G.H. (1959). *J. Phys. Chem. Solids*, **9**, 165.
16. Hed, A.Z. and Tannhauser, D.S. (1967). *J. Chem. Phys.*, **47**, 2090.
17. Bransky, I. and Tallen, N.M. (1971). In *Conductivity in low-mobility materials*, (ed. N. Klein). Taylor and Francis, London.
18. Holstein, T. (1959) *Ann. Phys. NY*, **8**, 343.
19. Atkins, P.W. (1990). *Physical chemistry*, (4th edn). Oxford University Press.
20. Galy, J., Darriet, J., Casalot, A., and Goodenough, J.B. (1970). *J. Solid State Chem.*, **1**, 339.
21. Cannon, R.D. (1980). *Electron transfer reactions*. Butterworths, London.
22. Purcell, K.F. and Kotz, J.C. (1977). *Inorganic chemistry*. Saunders, W.B. Philadelphia.
23. Arket, G.A. and Volger, J. (1964). *Physica*, **30**, 1667.
24. Friedman, F., Weichmann, F.L., and Tannhauser, D.S. (1975). *Phys. Status Solidi* (a), **27**, 273.
25. Bevan, D.J.M. (1973). Non-stoichiometric compounds: an introductory essay. In *Comprehensive inorganic chemistry*, (ed. A.F. Trotman-Dickenson). Pergamon, Oxford.
26. Hayes, W. and Stoneham, A.M. (1985). *Defects and defect processes in non-metallic solids*. Wiley, New York.
27. Kröger, F.A. and Vink, H.J. (1956). *Solid State Phys.*, **3**, 310.
28. Bransky, I. and Tallen, N.M. (1968). *J. Chem. Phys.*, **49**, 1243.
29. Blumenthal, R.N., Kirk, J.C., and Hirthe, W.M. (1966). *J. Phys. Chem. Solids*, **28**, 1077.
30. Navrotsky, A. (1987). Thermodynamic aspects of inorganic solid-state chemistry. In *Solid state chemistry: techniques*, (ed. A.K. Cheetham and P. Day). Oxford University Press.
31. Bosman, A.J. and Van Dach, H.J. (1970). *Adv. Phys.*, **19**, 1.
32. Smyth, D.M. (1984). *Prog. Solid State Chem.*, **15**, 145.
33. West, A.R. (1984). *Solid state chemistry and its applications*. Wiley, Chichester.
34. Tilley, R.J.D. (1987). *Defect crystal chemistry*. Blackie, Glasgow.
35. West, J.M. (1986). *Basic corrosion and oxidation*, (2nd edn). Ellis Horwood, Chichester.
36. Barr, L.W. and Lidiard, A.B. (1970). In *Physical chemistry: an advanced treatise, Vol. 10: solid state*. Academic, New York.

37. Wagner, J.B. (1968). In *Mass transport in oxides*, NBS Special Publication 296, Washington; also (1975). *Prog. Solid State Chem.*, **10**, 3.
38. Morin, F. (1984). Transport of matter in transition metal oxides. In ref. 5, p. 195; also (1979). *J. Electrochem. Soc.*, **126**, 760.
39. Dieckmann, R. (1984). *Solid State Ionics*, **12**, 1.
40. Petot-Ervas, G. (1984). Electrical properties of oxides at high and intermediate temperatures. In ref. 5, p. 131.
41. Millot, F., Blanchin, M.G., Tetot, R., Marucco, J.F., Poumellec, B., Picard, C., and Touzelin, B. (1987). *Prog. Solid State Chem.*, **17**, 263.
42. Hasiguti, R.R. (1972). *Ann. Rev. Mater. Sci.*, **2**, 69.
43. Picard, C. and Gerdanian, P. (1989). *J. Phys. Chem. Solids*, **50**, 385.
44. See discussion and references in refs. 10, 11, 45.
45. Berak, J.M. and Sienko, M.J. (1970). *J. Solid State Chem.*, **2**, 109.
46. Schirmer, O. and Salje, E. (1980). *J. Phys. C: Solid State Phys.*, **13**, L1067; Salje, E. and Gutler, B. (1984). *Phil. Mag.* B, **50**, 707.
47. Kudinov, R.K., Mirlin, D.N., and Firsov, Yu. A. (1970). *Sov. Phys. Solid State*, **11**, 2257.
48. Egdell, R.G. and Hill, M.D. (1983). *J. Phys. C: Solid State Phys.*, **16**, 6205.
49. Egdell, R.G. and Jones, G.B. (1989). *J. Solid State Chem.*, **81**, 17.
50. Austin, I.G. and Mott, N.F. (1969). *Adv. Phys.*, **18**, 41.
51. Kuzmany, K., Mehring, M., and Fink, J. (eds) (1990). *Electronic properties of high-*T_c *superconductors and related compounds*, Springer Series in Solid State Science, Vol. 99. Springer, Berlin.
52. Kuiper, P., Kruizinga, G., Ghijsen, J., Sawatzky, G.A., and Werweij, H. (1989). *Phys. Rev. Lett.*, **62**, 221.
53. Kanamori, J. and Kotani, A. (eds), (1988). *Core-level spectroscopy in condensed systems*, Springer Series in Solid State Sciences, Vol. 81. Springer, Berlin.
54. Mott, N.F. (1974). *Metal–insulator transitions*. Taylor and Francis, London.
55. Catlow, C.R.A., Mackrodt, W.C., and Norgett, M.J. (1977) *Phil. Mag.*, **35**, 177.
56. Friedman, L.R. and Tunstall, D.P. (eds) (1974). *The metal non-metal transition in disordered systems*. Scottish Universities Summer School in Physics Publications, Edinburgh.
57. Dougier, P. and Castalot, A. (1970). *J. Solid State Chem.*, **2**, 396.
58. Ruscher, C., Salje, E., and Hussain, A. (1988). *J. Phys. C: Solid State Phys.*, **21**, 4465.
59. Jonker, G.H. and Van Santen, J.H. (1950). *Physica*, **16**, 337; 599.
60. Harrison, M.R., Edwards, P.P., and Goodenough, J.B. (1985). *Phil. Mag.* B, **52**, 679.
61. Ueda, Y., Tanaka, T., Kosagi, K., Ishikawa, M., and Ysnoka, H. (1988). *J. Solid State Chem.*, **77**, 401.
62. Goodenough, J.B. (1970). *Matter Res. Bull.*, **5**, 621.
63. Doumerc, J.P. (1974). Tungsten bronzes: fluorine or tantalum substitution. In ref. 56, p. 313.

64. Egdell, R. G., Harrison, M. R., Hill, M. D., Porte, L., and Wall, G. (1984). *J. Phys. C: Solid State Phys.*, **17**, 2889.
65. Cox, P. A. (1983). *Solid State Commun.*, **45**, 91.
66. Dickens, P. G. and Wiseman, P. J. (1975). Oxide bronzes. In *Solid state chemistry*, MTP International Review of Science, Inorganic Chemistry Series 2. Butterworths, London.
67. Efros, A. L. and Pollak, M. (eds), (1986). *Electron–electron interactions in disordered systems*. Elsevier, Amsterdam.
68. Efros, A. L. and Schlovksii, B. I. (1975). *J. Phys. C: Solid State Phys.*, **8**, L49.
69. Edwards, P. P. and Sienko, M. (1982). *Int. Rev. Phys. Chem.*, **3**, 83.
70. Pollak, M. (1974). Percolatron and hopping transport. In ref. 56, p. 95.

5

METALLIC OXIDES

A true metal is characterized by a resistivity that increases with temperature. Thus metals remain conducting down to the lowest attainable temperatures. In *superconductors*, of which there are a number of oxide examples, resistance vanishes below a critical temperature T_c. The term *simple metal* is normally understood to refer to solids where the unmodified free-electron theory provides a good description of the band structure and transport properties.[1,2] Probably no transition metal oxides fall within this category, as the conduction bands based on the metal *d* orbitals do not show the simple isotropic and parabolic behaviour of the free-electron model. Nevertheless, for a limited group of oxides band theory seems to provide a good description of most of their properties. These were classified as 'simple metals' in Chapter 1, and are discussed in the first part of the present chapter.

The factors that upset simple metallic behaviour in oxides are numerous, and the division of this chapter reflects two important effects. Electron–electron interaction, or *correlation*, is important when bands are narrow. It gives rise to anomalies in magnetic and other properties, and in the limit may cause the solid to behave as a magnetic insulator rather than a metal. Section 5.2 discusses this type of effect, and includes some compounds that are close to the borderline of metallic behaviour. Some chemical trends apparent in looking at this borderline are also described. Section 5.3, on the other hand, discusses compounds where metallic behaviour is disturbed by the interaction of electrons with the crystal *lattice*. This can happen in various ways, and is frequently associated with a transition from metallic properties at higher temperature to a low-temperature non-metallic form. The final part of the chapter, Section 5.4, discusses superconductivity, especially the 'high-T_c' copper oxide compounds.

5.1 Simple metals

As noted above, band-structure effects are important in all oxides. However, the number of transition metal oxides where band theory alone can provide an adequate account of their properties is quite limited, and probably confined to some compounds of the 4*d* and 5*d* series.[3] Most of this section is concerned with some of the most 'typical' compounds of this type: ReO_3, and the widely studied sodium-tungsten bronze series, Na_xWO_3.

5.1.1 Band structure and Fermi surfaces

The band structure of RuO_2 was discussed in Section 2.3.1. That of ReO_3, shown in Fig. 5.1, is in many ways simpler.[4] The simple cubic structure of ReO_3, with only one formula unit per cell (see Fig. 1.7), makes this simple to interpret using the ideas discussed in Chapter 2. There are bands from the oxygen 2*s* and 2*p* levels, with some bonding admixture of rhenium orbitals. Above these come the conduction bands, derived largely from Re 5*d*, but with antibonding contributions from oxygen orbitals. An analysis of this band structure confirms the effects expected from the earlier discussion (see Section 2.3.1). There is a 'crystal field' splitting of the Re 5*d* band, into a lower one of t_{2g} type, and an upper e_g, which comes from antibonding combinations (of π and σ type respectively) of rhenium 5*d* and oxygen 2*p* orbitals. The widths of these bands, which are all-important for the metallic properties of ReO_3, are also largely caused by the same bonding interactions, and *not* by direct Re–Re overlap. Thus the dispersion of the t_{2g} band, in which the Fermi level lies in this d^1 compound, can be understood from the simple model presented in Fig. 2.12, showing how the π-type interaction changes with wavevector. As in the one-dimensional model system, the

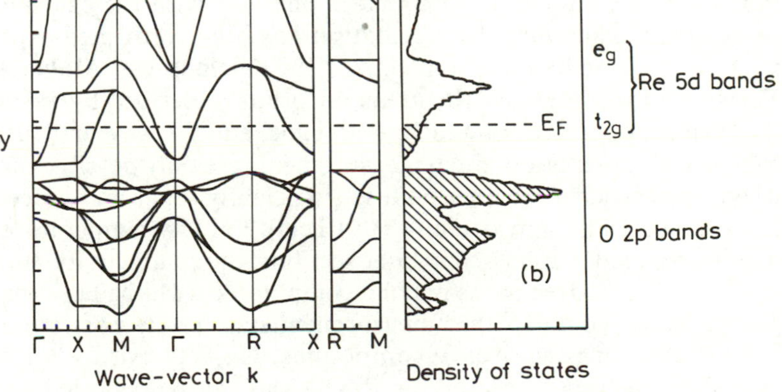

FIG. 5.1 Band structure and density of states calculated for ReO_3 (from ref. 4).

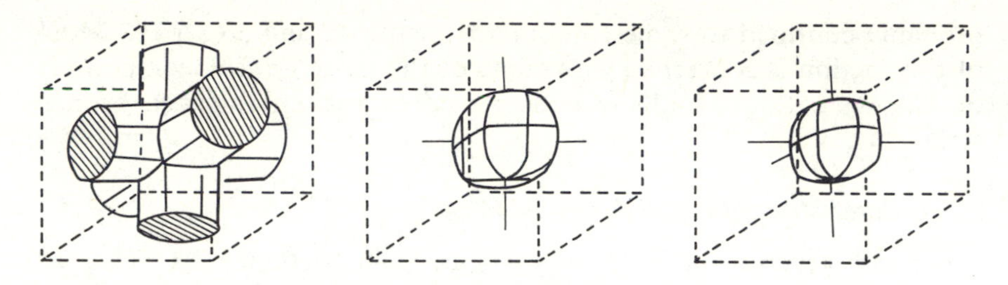

Fig. 5.2 Fermi surface calculated for ReO_3; the three 'sheets' arise from bands which cross the Fermi level at different wavevectors (from ref. 4).

cubic symmetry of ReO_3 guarantees that oxygen $2p$ and rhenium $5d$ orbitals have no mixing at the Γ point of the Brillouin zone, where $\boldsymbol{k} = (0,0,0)$. The structure of the bands is, however, more complex in this three-dimensional compound. The three t_{2g} orbitals making the lower part of the conduction band are degenerate at Γ, but this degeneracy is split at other parts of the zone, due to the different bonding interactions of orbitals with different orientations with respect to $\boldsymbol{k}$. It is particularly interesting that one 'branch' of this band remains nearly flat in the ΓX direction, with $\boldsymbol{k} = (x,0,0)$. This is due to the d_{yz} orbitals oriented *perpendicular* to this direction, which cannot achieve any overlap with orbitals on oxygen.

Figure 5.2 shows the Fermi surface (see Section 2.3.2) calculated for ReO_3. The three separate 'sheets' arise from different parts of the t_{2g} conduction band, which cross the Fermi level at different $\boldsymbol{k}$ points. Spin–orbit coupling in the Re $5d$ orbitals has been included in this calculation, and is found to have some effect on the detailed shape of the Fermi surface, although not a major one.[4] As with the calculation on RuO_2 described earlier, the calculation has been confirmed experimentally: in this case by measurements of the **de Haas–van Alphen effect.**[5] This experiment measure the areas of 'orbits' described by electrons at the Fermi surface in strong magnetic fields, and the ways in which these change with orientation. Such measurements are only possible on metals where the carrier mean-free path is sufficiently long, and indeed ReO_3 has a conductivity approaching 10^7 $(\Omega\,\mathrm{cm})^{-1}$ at low temperatures.[6] For most of the oxides discussed in later sections, the conductivities are much lower, and mean-free paths may be comparable with the lattice spacing. Thus detailed Fermi-surface measurements are rarely possible.

The cubic tungsten-bronze compounds, Na_xWO_3 with $x > 0.4$, have essentially the same structure as ReO_3, and can be regarded as 'defect perovskites' with Na^+ occupying a fraction of the large 'interstitial' sites of the ReO_3 structure.[7] As might be expected, calculations show the band structure to be extremely similar to that of ReO_3.[8] The conduction bands have W–O antibonding character, with the width coming again from the metal–oxygen bonding interaction. Additional bands are expected from the valence orbitals on sodium, but these are predicted to be at much higher energy, and hardly mixed with the occupied bands. That sodium orbitals are not involved in the main conduction bands is confirmed by NMR measurements on ^{23}Na. The **Knight shift** in NMR of metals is due to the paramagnetic effects of electrons at the Fermi surface, coupling with the nuclear spin through the *Fermi-contact interaction*. Involvement of Na 3*s* orbitals at the Fermi surface would give a significant Knight shift in the ^{23}Na resonance; this is not seen, however, showing that any such involvement is very small.[9] The photoelectron spectra shown earlier (Fig. 4.8), together with measurements of the change in relative peak areas with exciting photon energy, are also consistent with a majority W 5*d* character for the conduction band.[10,11]

A number of measurements on the tungsten bronzes have been aimed at determining the density of states at the Fermi level, $N(E_F)$, and especially how this varies with conduction electron concentration. The results of two types of measurement are shown in Fig. 5.3 (a). Most direct of all measurements of $N(E_F)$ comes from the electronic specific heat. At low temperatures the specific heat of a metal can be fitted to the equation

$$C = aT^3 + \gamma T. \tag{5.1}$$

The T^3 term is the low-temperature contribution from lattice vibrations, as predicted by the Debye model. It is γ that represents the electronic contribution, from which $N(E_F)$ can be calculated according to eqn (2.24).[12]

The other set of estimates shown in Fig. 5.3 (a) comes from photoelectron spectroscopy. The height of the Fermi edge in the spectra of Fig. 4.8, compared with the total area of the conduction band signal, enables $N(E_F)$ to be calculated easily, if the total conduction-electron concentration is known. In the case of Na_xWO_3 the intensity of conduction band ionization is found to be proportional to x, and the simple assumption that each sodium provides one conduction electron is confirmed.[10]

As can be seen from Fig. 5.3 (a) there is quite good agreement between the specific heat and the PES estimates. Both measurements, however, suggest an almost *linear* variation of $N(E_F)$ with the electron carrier concentration x. This is in contrast with the predictions of a free-electron model:

$$N(E_F) \propto m^* x^{1/3}. \tag{5.2}$$

The appropriate curve for $m^*/m_0 = 1$ is shown by the broken curve in

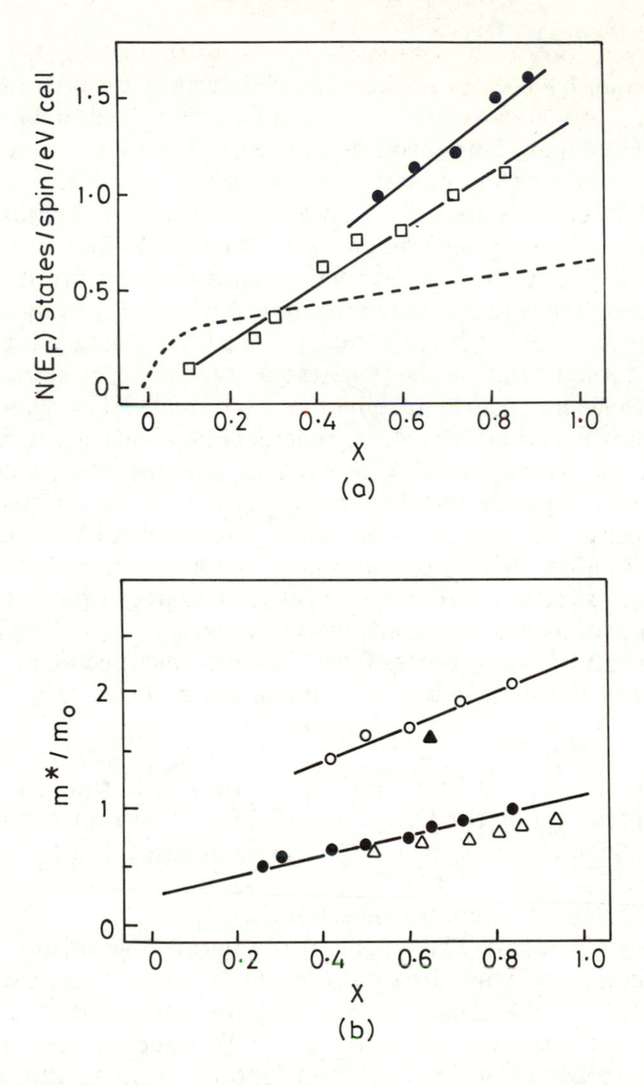

Fig. 5.3 (a) Density of states at the Fermi level in the tungsten-bronze series, Na_xWO_3: □, estimated from photoelectron spectroscopy;[10] ●, from specific heats.[12] The broken curve is the prediction of a rigid-band free-electron model, with $m^*/m_0 = 1$. (b) Variation of effective mass ratio in Na_xWO_3 from photoelectron spectroscopy (○) and magnetic susceptibilities[13] (▲), with values estimated from the plasma frequency either optically (△) or from electron energy loss (●).[10]

Fig. 5.3 (a), and is obviously very different (even allowing for an adjustment in m^*) from the experimental values. In Fig. 5.3 (b) this discrepancy has been illustrated in a different way, by plotting the variation in effective mass with composition required to make the density of states fit the free-electron model.[10]

Another way of measuring the density of states at the Fermi level is from the Pauli susceptibility (eqn (2.25)). Measurements show that as predicted for a metal, the susceptibilities are almost temperature independent.[13] To obtain the conduction-electron contribution, however, it is necessary to make a correction for the diamagnetic susceptibilities of closed-shell ions. In Na_xWO_3 this can be done by using the measured susceptibility of the parent d^0 compound WO_3, and the known value for Na^+. There is also a diamagnetic *Landau term* from the conduction electrons, which does not depend directly on $N(E_F)$, and is hard to estimate except for free-electron metals.[14] Most interpretations of the magnetic susceptibility of metallic solids use a modified free-electron model, and assume that band-structure effects can be taken account of by an isotropic effective mass, m^*. Then the full conduction-electron contribution is given as

$$\chi_e = \chi_P [1 - m_0^2/3m^{*2}] \tag{5.3}$$

where χ_P is the Pauli contribution of eqn (2.25). As we have seen with ReO_3, the complex Fermi surface is far from being spherical as the modified free-electron assumption implies, and the use of this formula is only approximate. Nevertheless, a reasonable agreement is found with the tungsten bronzes between the measured susceptibilities and those estimated on the basis of specific-heat measurements.[13] Figure 5.3 (b) shows how the effective masses estimated from susceptibility data also agree with those from the photoelectron spectra.

The linear variation of $N(E_F)$ with the sodium content x shown in Fig. 5.3 has given rise to some controversy. The essential point at issue is whether the electronic structures can be interpreted by a *rigid-band* model, where the band structure is assumed to remain unchanged as the electron concentration is increased.[8,15] For $x < 0.3$, the tungsten bronzes are semiconducting, and should be discounted here. The metal/non-metal transition was discussed in Section 4.4: the disordered occupation of Na^+ sites, polaron formation, and Coulomb interactions may all be involved, and it is likely that these influences persist some way into the metallic regime, and may distort the density of states. But we have also seen for ReO_3 that the conduction-band structure is far from free-electron-like, so that there is no good reason to expect eqn (5.2) to work well, even if the perturbations important at low x can be neglected. It has been suggested that a linear variation of $N(E_F)$ with x can be explained with a rigid-band density of states which varies *exponentially* with energy close to the band edge,[15] although the shapes of the photoelectron spectra do not support this. An argument against the rigid band assumption is based on the structural observation that the W–W distance increases somewhat with x.[3,7] This is expected as electrons are introduced into antibonding orbitals. It should affect the band structure by decreasing the orbital overlap, thus narrowing the bands and increasing the effective mass m^*. One should probably conclude that the linear variation of $N(E_F)$ is largely fortuitous and results from a combination of effects.

Effective-mass values from optical and EELS measurements are also shown in Fig. 3.3 (b), and will be discussed later. These do *not* represent direct measurements of the density of states, and perfect agreement with other measurements is not expected. The good agreement between the three measurements just discussed is important as it confirms the validity of the band model, without important correlation effects.

5.1.2 Transport properties

As already mentioned, the conductivity of ReO_3 is exceptionally high for an oxide. Values approaching 10^7 $(\Omega\,\text{cm})^{-1}$ are reported at low temperatures, and the room-temperature conductivity is an order of magnitude higher than the most conducting tungsten bronze.[6] The comparison between these compounds probably reflects the more ordered structure of ReO_3, and indeed a detailed study of the Na_xWO_3 series has highlighted the importance of disorder in the partially occupied Na^+ sites.[16]

Figure 5.4 (a) shows how the conductivity varies with composition and temperature in the metallic Na_xWO_3 compounds. Not only the conductivities themselves but their temperature variations are much less than in ReO_3. According to **Matthiessen's rule** one can (approximately) separate two contributions to the resistivity:[1]

$$1/\sigma = \rho = \rho_{dis} + \rho_{lat} \tag{5.4}$$

ρ_{dis} is the residual resistivity at low temperatures due to scattering by static disorder, and ρ_{lat} the 'lattice' contribution from scattering by thermally excited vibrations. At 4 K the latter term should be negligible, whereas the disorder contribution is roughly independent of temperature: thus the two contributions may be separated from the data in Fig. 5.4 (a). Figure 5.4 (b) shows values for ρ_{lat} at 300 K; they decline with x, probably due to the greater efficiency of screening as the conduction-electron concentration increases.

The disorder scattering term remaining at 4 K is exceptionally large

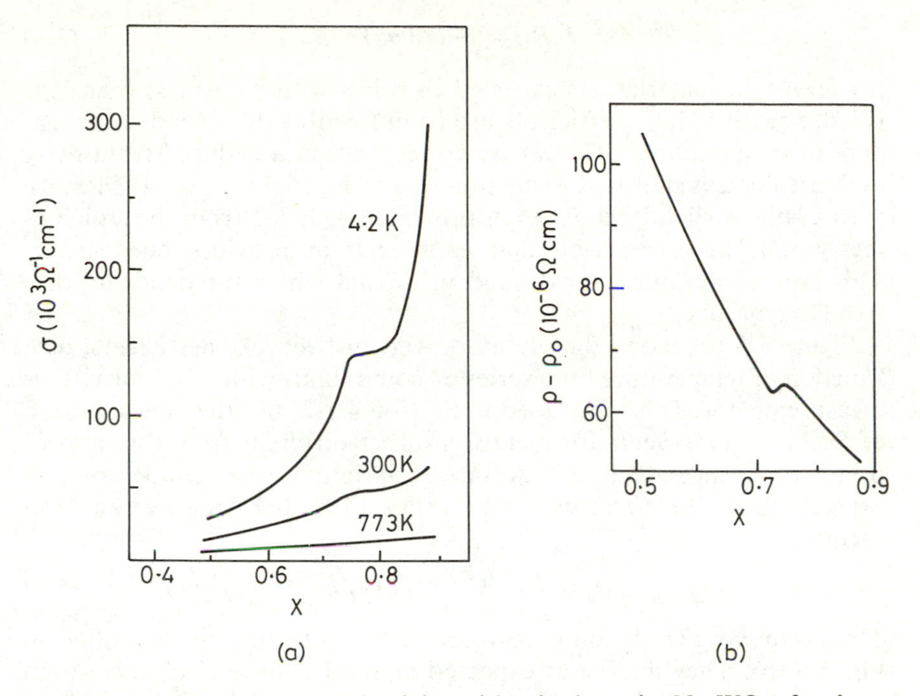

FIG. 5.4 (a) Variation of conductivity with x in the series Na_xWO_3, for three different temperatures. (b) Lattice (vibrational) contribution to the room-temperature resistivity in the series.[16]

in comparison with values expected for 'pure' metals: for example at $x = 0.5$ it contributes over half of the room-temperature resistivity. It almost certainly has a major contribution from disorder in the occupation of Na^+ sites. The strong variation with x of conductivity at 4 K, by a factor of 10 between $x = 0.5$ and $x = 0.9$, shows how the importance of such scattering decreases, partly due to the nearly complete filling of the Na^+ sites, and partly because of the more efficient screening as the carrier concentration increases. This variation seems to be not entirely regular, however, and shows some indication of a 'plateau' around $x = 0.75$. Neutron diffraction studies suggest that there is some partial ordering of the Na^+ at this concentration;[17] this would be expected to influence the disorder scattering, and the rapid increase of conductivity with x just below 0.75 may be a consequence of reduced scattering.

The results of some other transport measurements on the tungsten bronzes are shown in Fig. 5.5. The Hall effect, as mentioned in connection with semiconductors (Section 4.1.2), is sensitive to the carrier type and concentration. Free-electron theory predicts:[1,2]

$$R_{\mathrm{H}} = -1/(ne) \tag{5.5}$$

for an n-type material. As expected in solids with bands less than half filled, negative Hall coefficients are found, with values nearly independent of temperature.[16] The carrier concentration n deduced from these measurements is shown as a function of x in Fig. 5.5 (a). It does increase with x, but is slightly larger than the value deduced from the stoichiometry, on the assumption that each sodium provides one carrier. This type of deviation is expected in a solid where the bands are not free-electron-like.

Figure 5.5 (b) shows the thermopower, or Seebeck coefficients, as a function of temperature for a series of compounds with different x. This measurement was also discussed in Section 4.1.2, but the interpretation of Seebeck coefficients for metals is rather different from that appropriate to semiconductors.[14] A linear variation with temperature is expected from the *diffusion contribution*, predicted from free-electron theory as

$$\alpha_{\mathrm{d}} = (\pi^2/3)(k^2/e)T/E_{\mathrm{F}}. \tag{5.6}$$

This seems to give the dominant contribution to the values plotted in Fig. 5.5 (b). They decline as expected as the Fermi level E_{F} (measured from the bottom of the conduction band) increases with x; but the values

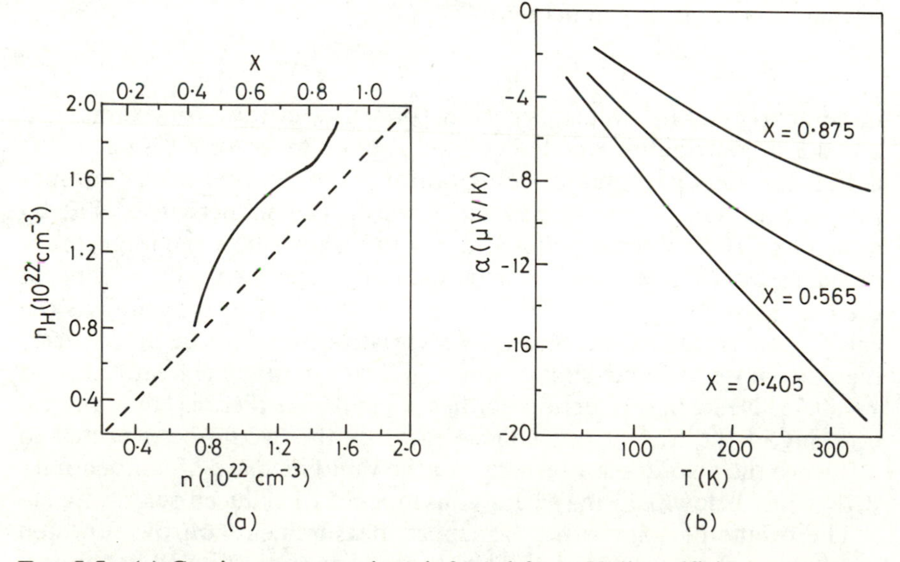

FIG. 5.5 (a) Carrier concentrations inferred from Hall coefficients as a function of x in Na_xWO_3. (b) Seebeck coefficients as a function of temperature for different x values.[17]

once again do not fit a free-electron model which would predict E_F to be proportional to $x^{2/3}$. There is another contribution to the thermopower, known as *phonon drag*, which varies roughly as $1/T$. It is due to scattering processes which depend on the detailed shape of the Fermi surface and is hard to calculate.

5.1.3 Optical and spectroscopic properties

A characteristic feature of the optical properties of a 'good' metal is high reflectivity at low frequencies, up to an edge at the *plasma frequency*. The metallic oxides under discussion in this section show this expected behaviour, although their optical properties may be complicated by a variety of intra- and interband transitions.

Figure 5.6 shows the reflectivity of ReO_3, and an analysis in terms of the dielectric function,[18] obtained from the complex refractive index n by:

$$n^2 = \varepsilon_1 + i\varepsilon_2. \tag{5.7}$$

The negative sign of the real part ε_1 at low frequencies is normal for a metal, and predicted by the *Drude Theory* (see Section 2.3.2). The plasma frequency can be defined by the change in sign of ε_1; i.e. by the condition

$$\varepsilon_1 = 0. \tag{5.8}$$

According to the Drude prediction, based on a free-electron model,[1,2]

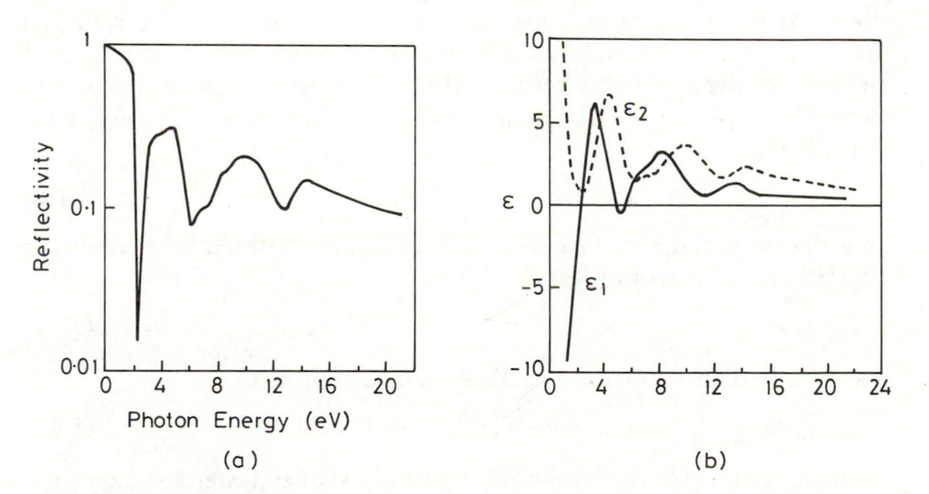

FIG. 5.6 (a) Reflectivity spectrum of ReO_3, and (b) real (ε_1) and imaginary (ε_2) parts of the dielectric function deduced from it.[18]

$$\omega_p^2 = ne^2/(\varepsilon_0 \varepsilon_{opt} m^*) \tag{5.9}$$

where ε_{opt} is the contribution to ε_1 coming from the electronic polarizability of closed shells and n is the carrier concentration. Although this equation is often used on an empirical basis for discussing the plasma frequency of metallic compounds, it is approximate. As we have seen, band structures are not free-electron-like; also, as Fig. 5.6 shows, the dielectric constant varies considerably above the plasma frequency due to interband transitions.

The plasma frequency of 2.3 eV can be used to derive an *optical effective mass* of $0.86m_0$ for ReO_3. Unlike the situation with genuinely simple metals such as sodium, the presence of the 'screening term' ε_{opt} in eqn 5.9 is very important: without this term an unscreened plasma frequency of 5.5 eV would be predicted.[18]

The lowest energy interband transition in ReO_3 occurs at 4.2 eV, and may be attributed to excitations from the valence band into the conduction band. This is similar to the band-gap excitation in a d^0 oxide (see Section 3.1.1) except that the lower part of the conduction band is now occupied. Transitions can only occur to states above the Fermi levels, and so the minimum excitation energy is larger than the 'band gap'.

As with the Fermi surface and transport measurements, there has been extensive work on the tungsten bronzes. It was mentioned in Section 4.4.1 that the reflectivity edge is responsible for the colour of the metallic compounds, and the way in which this changes with sodium content. Plasma frequencies have been measured from optical reflectance, and also by *electron energy-loss spectroscopy* (EELS) as described in Section 4.4.1. At low incident energies (100 eV or less), electrons are reflected from the surface and undergo energy-loss processes due to the dielectric response of the *surface* rather than the bulk. If the sample is uniform in behaviour right up to the surface, the probability of loss at energy $h\omega$ is predicted as[19]

$$P(\omega) \propto -Im\, 1/[\varepsilon(\omega) + 1]. \tag{5.10}$$

In consequence, the *surface-plasma frequency* is defined by a condition slightly different from (5.9)

$$\varepsilon_1 = -1. \tag{5.11}$$

The Drude theory gives the surface-plasma frequency as

$$\omega_{sp}^2 = ne^2/[\varepsilon_0(\varepsilon_{opt} + 1)m^*]. \tag{5.12}$$

Values obtained for the 'optical effective mass' of tungsten bronzes were shown in Fig. 5.3 (b), with a comparison of values from other sources. Optical and EELS data agree well with one another, but yield m^* values considerably smaller than from measurements of $N(E_F)$, interpreted by the free-electron model.[10,20] The variation with x reflects the limitations of the free-electron-like rigid-band approach discussed above. The discrepancy between density-of-states and optical values should not be surprising in view of the failure of the free-electron model to describe the band structure. Although the plasma frequency is, in a sense, a Fermi-surface property, it does not depend just on the density of states, but also on the velocity of Fermi-surface electrons, according to eqn (2.17). Thus deviations from free-electron theory will affect it differently from $N(E_F)$.

Photoelectron spectra for some metallic oxides have been shown already (Figs 2.16 and 4.8). Their principal features can be interpreted readily with a band approach, as the main ionizations reflect the energies of occupied bands. Densities of states must, however, be weighted by the relative *ionization cross-sections* for different orbitals, in order to understand the intensities of the different bands. A striking illustration of this is shown in the spectrum of ReO_3 measured at X-ray photon energies, in Fig. 5.7.[21] The two bands are the occupied part of the conduction band and the valence bands. The relative occupancy of these two bands in ReO_3 is 1: 18, however, which is very different from the ratio

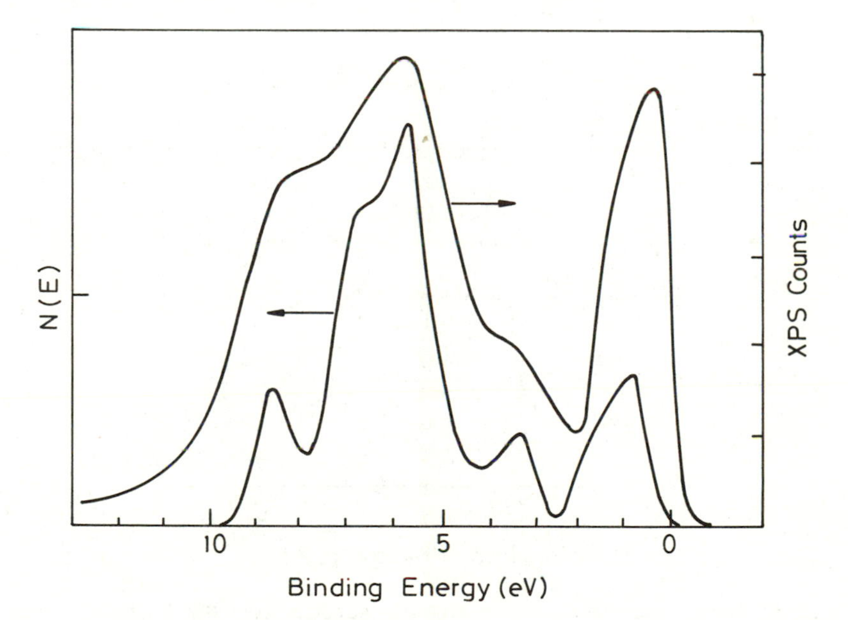

FIG. 5.7 Photoelectron spectrum of ReO_3, measured with X-ray photons. This spectrum is unlike the total density of states, but reflects the Re 5*d* contribution shown (from ref. 21).

of intensities seen. With high-energy photons the ionization cross-section for rhenium 5*d* orbitals is very much larger than for oxygen 2*p*. Thus to a good approximation the spectrum in Fig. 5.7 reflects the Re 5*d* partial density of states, and not the total density of states. The strong mixing of Re 5*d* orbitals into the valence band can be seen, and is in good agreement with the band calculation described above.

Although simple 'independent-electron' interpretations of photoelectron spectra seem to work well for the valence levels in these compounds, the same is not true for core-level spectra. Figure 5.8 shows the metal core-level PES for $Na_{0.7}WO_3$ and RuO_2.[22,23] The spectra are expected to show splitting due to the strong spin–orbit coupling in the core levels: W

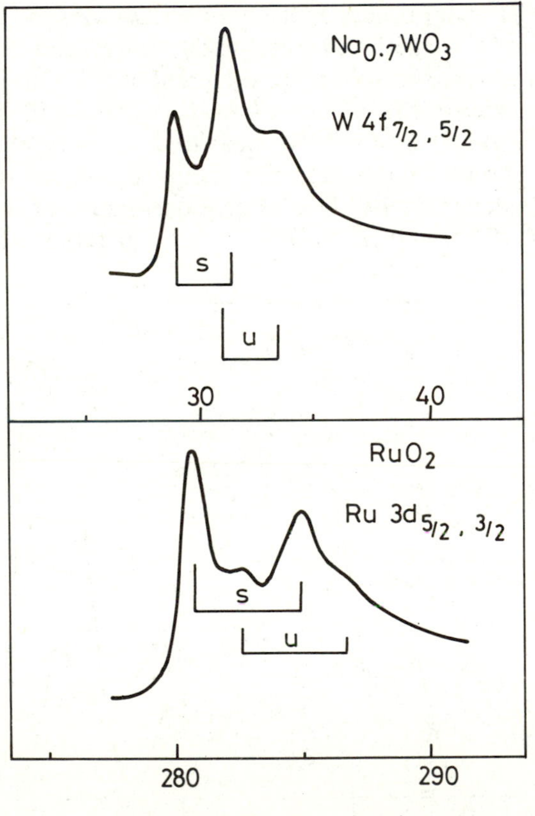

FIG. 5.8 Metal core-level photoelectron spectra for the metallic oxides $Na_{0.7}WO_3$[22] and RuO_2.[23] In each case a spin–orbit doublet is expected. The spectra show additional structure, however, with overlapping doublets from screened (s) and unscreened (u) final core-hole states.

$4f$ gives components with $j = \frac{7}{2}$ and $\frac{5}{2}$ and Ru $3d\, j = \frac{5}{2}$ and $\frac{3}{2}$. But each expected spin–orbit component can be resolved into at least two signals. Similar features seem to be present in many other metallic oxides, and have given rise to some controversy.[24]

The 'simplest' explanation that has been advanced to explain the structure of core peaks is that one is seeing two different oxidation states for the transition element. Thus in Na_xWO_3 it can be argued that W^{5+} and W^{6+} are simultaneously present, which might be expected to have different core ionization energies.[25] The metallic nature of these compounds on the other hand suggests a uniform distribution of electrons, with all tungsten present in the same, fractional oxidation state. This might be countered by considering the very rapid time-scale of the PES technique, which is able to 'catch' the different oxidation states present in a rapidly fluctuating system. We shall return to this argument below. The main disadvantage of a naïve mixed-valency interpretation, however, is that it does not predict the correct relative intensities of the different signals, and most importantly, that it cannot explain why a similar structure is seen in spectra from compounds such as RuO_2, where no mixed valency is present. In RuO_2, it has been argued that there might be a different *surface* composition, for example corresponding to RuO_3.[26] It is very hard to rule out this kind of effect in general, but in the case of RuO_2 both valence-region PES and EELS studies are entirely compatible with the expected *bulk* electronic structure.[23]

These 'mixed-valency' interpretations are motivated by the belief of many surface chemists, that core-level photoelectron spectra can provide a straightforward guide to the oxidation states of different elements present. They are probably over-naïve in that they neglect the important *perturbation* of the electronic structure that is produced in core ionization. Strong 'satellite' structure appears in both core and valence ionizations of many other compounds; in the case of magnetic insulators the spectra can be interpreted satisfactorily in terms of different accessible final states of the ionized solid, using the cluster and impurity-state models described earlier (see Sections 2.2.2 and 3.4.2). A similar final-state picture, although of a more qualitative nature, has been proposed for core ionizations in metallic oxides.[27,28] The basic idea is expressed in Fig. 5.9. There is a strong Coulomb interaction between a core orbital and the d orbital involved in the conduction band. Ionizing a core electron may thus lower the energy of a conduction-band orbital sufficiently to pull it below the band, forming an impurity state. Two different final states are possible: a *screened* state in which the impurity orbital is occupied, and an *unscreened* state at higher binding energy with no electron in the impurity orbital.

It appears from the spectra of simple metals such as aluminium that

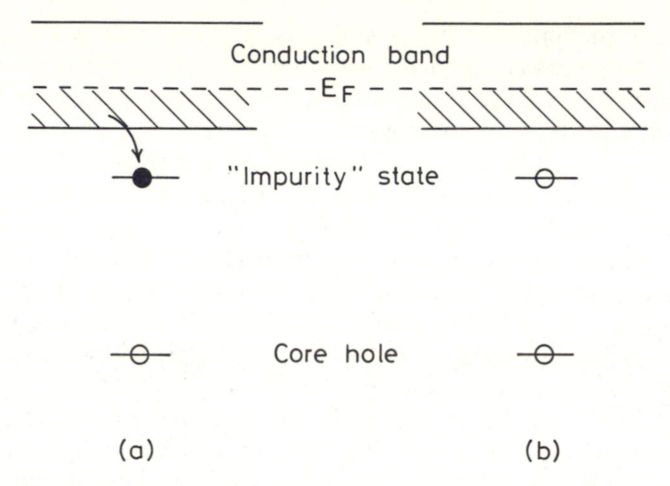

FIG. 5.9 Model for final-state structure appearing in core ionizations. The presence of a core hole pulls one valence orbital down in energy out of the conduction band, forming an impurity state. 'Screened' and 'unscreened' final states result according to whether this impurity level is occupied or empty, as in (a) and (b) respectively.

the Coulomb effect of the core hole is always important in ionization, as well as in X-ray absorption and emission spectra.[29] In PES, core levels of metals often have a strong asymmetry due to the response of conduction electrons. In these cases the bands are broader than in oxides, and indeed a 'free-electron' conduction band extends to infinity. The different behaviour of metal oxides arises because the width of the conduction band is comparable to the strength of the core–hole perturbation. Final-state structure appears less often in ionization from core levels of oxygen or of ternary elements present, probably because the Coulomb interaction of these core levels with conduction-band orbitals is weaker. A two-peak structure has been seen in the oxygen 1*s* level in some cases, and has been suggested as evidence for 'oxygen holes': that is, for some oxygen present as O^- instead of O^{2-}. Once again, this simple mixed-valency argument is probably too naïve, and a final-state interpretation is preferable.[30] It is probably true nevertheless that a strong final-state effect in oxygen ionization is evidence for appreciable oxygen character in the conduction band.

Although the final-state interpretation of core structure is probably more satisfactory than a mixed-valency picture, they need not always be incompatible.[31] In some metallic oxides it is possible to picture conduction as taking place by the rapid transfer of electrons between two oxidation states, such as W^{5+} and W^{6+} in the tungsten bronzes. The rate of transfer ν is related to the *hopping integral* t that determines the conduction-band width, according to

$$\nu \sim t/\hbar. \tag{5.13}$$

With $t \sim 1$ eV, $\nu \sim 10^{16}$ s^{-1}, and so most measurements show only an average electron distribution, corresponding to some fractional oxidation state. Core ionization, however, gives rise to a large energy difference between the two oxidation states: if this difference is V, then the ionization process can be regarded as a probe of the oxidation state, on a time scale τ given by:

$$\tau \sim \hbar/V. \tag{5.14}$$

It can be seen that if $V \gg t$ the time-scale of the measurement is much faster than that of electron transfer. Thus we expect to see two core peaks from core ionization, with relative intensities given by the ratio of oxidation states present in the ground state. The condition $V \gg t$ is certainly valid in the mixed-valence compounds discussed later in Section 5.2.2. These are barely metallic, however; with the much larger band widths appropriate to the compounds of interest here, we probably have $V \sim t$. The simple mixed-valence picture is no longer valid here, and the relative intensity of the different final states is a sensitive function of the different parameters.[31]

In metallic compounds without mixed valency, the detailed structure of core ionizations also depends on the relative magnitude of a number of parameters, including the band width and the Coulomb repulsion energy between conduction-band orbitals.[32] A simple qualitative prediction is that the relative intensity of the lower binding energy (screened) final state increases with the conduction band width. This is borne out in the series of ruthenium compounds discussed in section 5.2.3.

5.2 Electron correlation and magnetic anomalies

The band model is inadequate for describing the properties of many metallic oxides. Conductivities are frequently much lower than those of the compounds discussed in Section 5.1, and measurements of the density of states at the Fermi surface may give very different results according the method employed, none agreeing well with expectations of band theory. One reason for the discrepancies is that conduction bands are often too narrow for the effects of electron–electron interaction — that is, *correlation* — to be ignored. Magnetic properties are especially sensitive to electron correlation, and so a discussion of anomolous magnetic behaviour forms a major part of this section. In the limit when electron correlation comes to dominate, we expect the Fermi surface to disappear completely and a magnetic insulator to be formed. This situation, which can be described as a *Mott transition*, will also be illustrated, with some examples that are apparently borderline between the metallic and magnetic insulator states.

5.2.1 Band magnetism

The Pauli paramagnetism of a metal provides a useful way of estimating the density of states at the Fermi level, and we have seen above that in some cases the results agree well with values found from other techniques. As discussed in Section 2.3.3, however, the exchange interaction (K) between electrons in a band enhances the magnetic susceptibility, by an amount depending on the value of $KN(E_F)$.[14] Many oxides do in fact show significant **Stoner enhancements** of this kind.[3] Examples are $LaNiO_3$ and $LaCuO_3$, where the magnetic susceptibilities are at least 10 times that expected for a reasonable value of $N(E_F)$.[33] In $LaNiO_3$ the photoelectron spectrum shows an anomalously *low* value for $N(E_F)$, which also seems to be a typical sign of electron correlation.[34] In these compounds the conduction band is made up of e_g-type metal $3d$ orbitals, undoubtedly containing a strong admixture of oxygen $2p$, which may provide a sufficient band width to overcome the Hubbard U; an alternative model is that it is the gap between the upper Hubbard sub-band and the *oxygen* 2p *band* that has become zero with the abnormally high oxidation states Ni^{3+} and Cu^{3+} (see Section 5.2.4). In any case, it is clear from the magnetic properties that the bands are narrow enough for electron–electron interaction to be important.

If

$$KN(E_F) > 2 \tag{5.15}$$

the Stoner model predicts that the paramagnetic state of a metal will be unstable. The simplest consequence is *ferromagnetism* where the ground state has an excess of electrons of one spin; however, the simple theory does not prove that this state is necessarily more stable than an *antiferromagnetic* one with *spin-density waves* where the dominant spin direction changes periodically from site to site. A number of ferromagnetic and antiferromagnetic metallic oxides are known, although it is not clear in all cases that magnetic properties have been very well characterized and interpreted; for example spin–orbit coupling can also give rise to temperature-dependent magnetic susceptibilities that might sometimes be mistaken for magnetic-ordering effects.[35]

The complex factors involved in the magnetic ordering of a narrow-band metal have been discussed by several authors, particularly Goodenough.[3] Magnetic ordering temperatures are expected to increase as the band width decreases, reaching a maximum around the critical width where itinerant metallic behaviour breaks down, and a transition to a magnetic insulator takes place. The nature of the ordering depends in a sensitive way on the degree of band filling and its width. With a half-filled band, antiferromagnetism is generally expected; but with other fractional occupancies, ferromagnetism may be found with broader bands, and a spiral antiferromagnetic ordering as bands become narrow. The metallic compounds $CaRuO_3$ and $SrRuO_3$ illustrate the subtlety of this problem, as $SrRuO_3$ is ferromagnetic with a Curie temperature of 160 K, whereas the calcium compound is not magnetically ordered at 4.2 K.[36] The electronic structure of these low-spin $4d^4$ metals must be very similar, and analagous to RuO_2 which shows no magnetic anomalies.

One of the best-known ferromagnetic metals is CrO_2, which is used in magnetic recording tape. It has the rutile structure, and the Fermi level in the d^2 compound is expected to lie in the t_{2g} part of the Cr $3d$ band. Qualitatively, one can understand the ferromagnetism on the basis of the rather contracted $3d$ orbitals, which provide both a fairly narrow band and strong exchange interactions between electrons. Both band-structure and spectroscopic studies show that the exchange splitting between spin-up and spin-down electrons is actually comparable to the t_{2g} band width (see Fig. 5.10).[37] This means that there is a 100 per cent spin polarization of the d electrons, which agrees with the magnetic observation of a saturation moment of $2\mu_B$ per chromium.[38] It also has the consequence that the Fermi level lies in a *gap* for the minority spin electrons, a situation described as a *half-metallic ferromagnet*. The Curie temperature of CrO_2 is 177°C. In a band model the disappearance of a static moment would be caused by thermal excitation of electrons into the minority spin band, causing an eventual disappearance of the exchange splitting. This is not a satisfactory description of the paramagnetic state, however, as in ferromagnetic solids disordered *local moments* persist above the transition.

5.2.2 Mixed valency and double exchange

We have seen in Section 4.4 that some magnetic insulators can be made metallic by high levels of doping. The function of such doping is to provide extra electrons or holes, and thus in a chemical sense to induce mixed valency where an element is present in different oxidation states. In discussing the Hubbard model in Section 2.4.1, it was noted that electrons can then move from site to site without incurring the penalty of extra electron repulsion. Mixed valency does not always give a metal: as discussed in Chapter 4 the extra carriers may be localized by the potential field of defects and by polaron formation; and we shall see in Section 5.3.3 below that *charge ordering* of carriers can take place even when defects are not present. Nevertheless, a number of oxides are known where mixed valency is associated with high conductivity, but where the same transition element in a single valency gives a magnetic insulator. The examples we shall discuss here are Fe_3O_4 and the Mn^{3+}/Mn^{4+} perovskite systems, where the conductivity associated with mixed valency gives rise to ferromagnetic interactions by a mechanism known as **double exchange.**[39]

The double-exchange mechanism was first formulated to explain the ferrimagnetic properties of *magnetite* Fe_3O_4. This is an unusual example of a magnetic mineral, and its properties were exploited for centuries as crude compass needles. Magnetite particles are also found in some bacteria, where the magnetic properties are apparently used for orientation with respect to the *vertical* component of the Earth's magnetic field.[40] In the inverse spinel structure, equal proportions of Fe^{2+} and Fe^{3+} reside in octahedral sites, and the conductivity (around $200(\Omega\ cm)^{-1}$ at room temperature) can best be interpreted in terms of facile electron transfer between these two states. One should probably

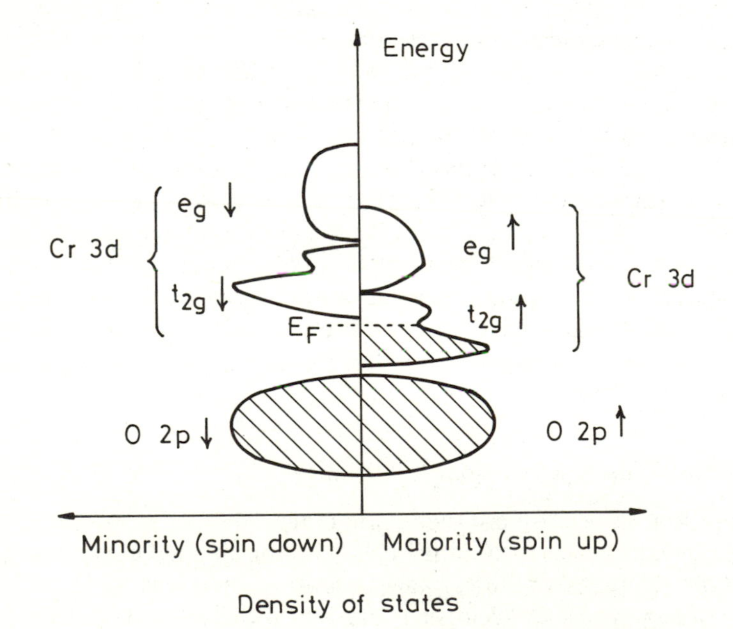

FIG. 5.10 Schematic density of states calculated for ferromagnetic CrO_2. Note that the Fermi level lies in the band gap for minority spin electrons. (Based on ref. 37.)

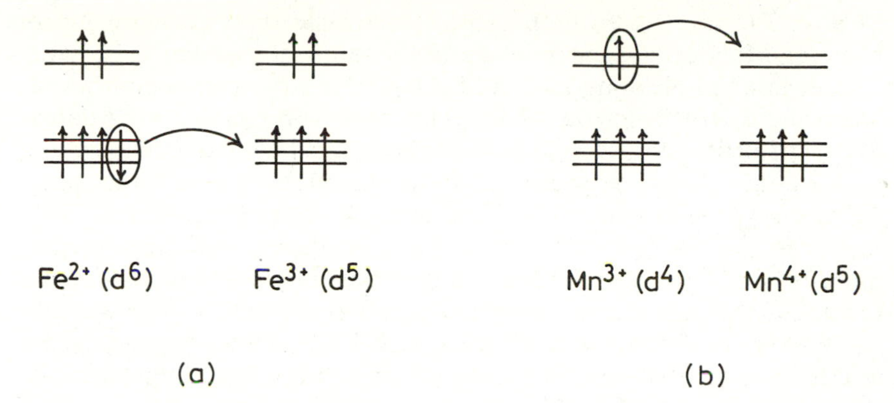

FIG. 5.11 The double-exchange mechanism giving ferromagnetic coupling between ions participating in electron transfer: (a) Fe^{2+}/Fe^{3+}, and (b) Mn^{3+}/Mn^{4+} transfer. In each case the ferromagnetic alignment of neighbouring ions is required to maintain the high-spin arrangement of electrons on both the donating and the receiving ion.

reject a *band* interpretation here for two reasons: oxides containing Fe^{2+} or Fe^{3+} alone are invariably magnetic insulators; also the relatively low conductivity of Fe_3O_4 indicates electron mobilities only around 1 $cm^2V^{-1}s^{-1}$, and mean-free paths of the order of the Fe–Fe distance.

The magnetic consequences of electron transfer between Fe^{2+} and Fe^{3+} are shown in Fig. 5.11 (a). In order to leave the final Fe^{3+} in a high-spin state, it is the minority spin electron that must move; thus the receiving Fe^{3+} ion must have its spins aligned parallel to the initial ion, or transfer would be inhibited by the Pauli exclusion principle. A more detailed theoretical treatment shows that this simple pictorial argument is slightly misleading, although the conclusion is qualitatively correct: electron transfer is strongly favoured by a ferromagnetic alignment of neighbouring ions.[41,42] This agrees with the magnetic structure of Fe_3O_4 deduced by neutron diffraction: Fe ions in octahedral sites are ferromagnetically aligned. The overall ferrimagnetic properties arise because the Fe^{3+} ions in tetrahedral sites have spins aligned antiparallel to the majority spin direction. The tetrahedral ions do not appear to participate in the conduction process, and antiferromagnetic coupling with the octahedral sites is provided by a superexchange mechanism as discussed for magnetic insulators (Section 3.4.4). The cancellation of Fe^{3+} spins gives a net saturation moment close to 4 μ_B per Fe_3O_4 formula unit, the value expected for Fe^{2+}.[43]

The $(La_{1-x}A_x)MnO_3$ perovskites, where A is a divalent element such as Sr or Pb, are interesting because the end members containing only Mn^{3+} or Mn^{4+} are magnetic insulators with antiferromagnetic order.[44] As discussed in Sections 3.4.3 and 3.4.4, the Mn^{3+} compound has an unusual form of magnetic order, associated with a cooperative Jahn–Teller distortion of this high-spin d^4 ion. Metallic conductivity occurs with a doping level x greater than about 0.2; simultaneously the Jahn–Teller distortion vanishes, and it becomes *ferromagnetic*, with a Curie temperature in the range 300–400 K. The disappearance of the distortion indicates that the e_g electron is no longer localized, and it is presumably the transfer of this electron from site to site that gives the high conductivity. A picture rather similar to that in Fe_3O_4 shows how again the electron transfer between two high-spin ions is favoured by their ferromagnetic alignment (Fig. 5.11(b)).

The interaction between the electronic conduction and magnetic ordering also gives rise to an unusual variation in conductivity with temperature, shown for three $La_{1-x}Pb_xMnO_3$ compositions in Fig. 5.12.[45] As the temperature increases the resistivity rises sharply close to T_C. From the double exchange argument above, this behaviour is not surprising: sites with the *wrong* spin orientation will act as strong scattering centres for conduction electrons. Thus as local spin disorder increases (an effect which begins below the disppearance of long-range order at T_C), scattering and resistivity increases. Above T_C the electron mean-free path is

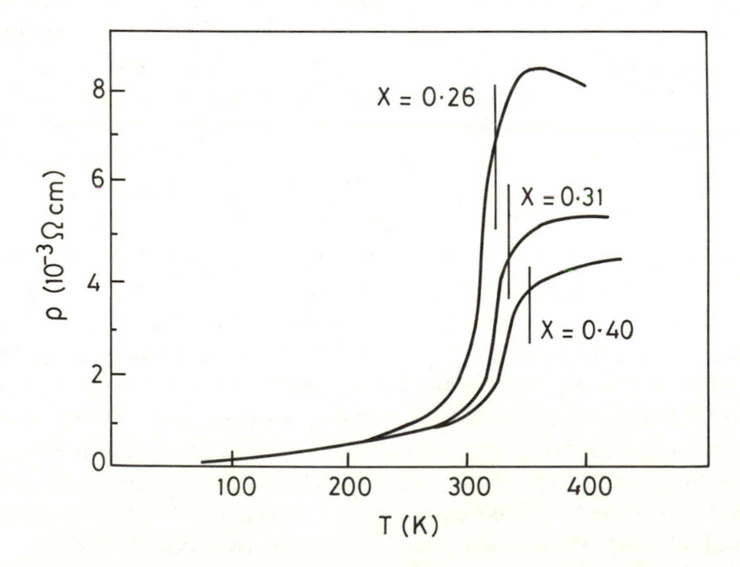

FIG. 5.12 Resistivity of three $La_{1-x}Pb_xMnO_3$ compounds as a function of temperature.[45] The Curie temperature of each composition is shown by a vertical line.

of the same order of magnitude as the lattice parameter, as in Fe_3O_4, although in the latter compound it is also probable that lattice scattering is also important; as we shall see later Fe_3O_4 undergoes a transition associated with *valence ordering* at 120 K.

The double-exchange mechanism is so called because it involves the single-centre (Hund's rule) exchange interaction on the two ions involved in the transfer process. Its importance probably extends beyond the mixed-valency compounds with high conductivity as just described. As we have seen, the motion of carriers is strongly perturbed in a magnetic lattice unless the local moments are ferromagnetically aligned. Magnetic scattering from this type of effect may make an important contribution to the low carrier mobilities found in many oxide semiconductors. It may also be favourable for an isolated carrier in an antiferromagnetic compound to produce a *local* ferromagnetic ordering of surrounding ions, so that its energy can be lowered by delocalization. This situation is known as a **spin polaron**, and has been postulated, although without definite evidence, for a number of mixed-valency oxides.[46]

Interactions between electronic carriers and spins are important in some mechanisms proposed for superconductivity in copper oxides.[47] Doping La_2CuO_4 with carriers does indeed rapidly destroy the antiferromagnetic order of the pure compound, but there is an important difference here from the cases we have considered in this section. Putting holes in the Cu^{2+} compound probably gives rise to *diamagnetic* centres, so that the double-exchange mechanism does not operate. The interaction between carriers and spins in doped copper oxides is therefore a little more subtle. It will be discussed in Section 5.4.3.

5.2.3 The 'Mott transition'

The term *Mott transition* has been applied rather loosely to many metal/non-metal (MNM) transitions in solids; probably it should be reserved for the transition between a metal and a magnetic insulator which occurs as the band width (or more specifically the ratio W/U) decreases.[46] In several series of compounds one can see this transition taking place as the transition element is changed, and some of the chemical trends observed here will be discussed below (Section 5.2.4). One remarkable compound (V_2O_3) seems to be poised right on the borderline between the two states, and undergoes a Mott transition as a function of temperature and pressure. After describing V_2O_3, the present section will discuss a series of Ru^{4+} oxides which show a Mott transition as the ternary element is varied.

Figure 5.13 (a) shows how the resistivities of V_2O_3, and some solid

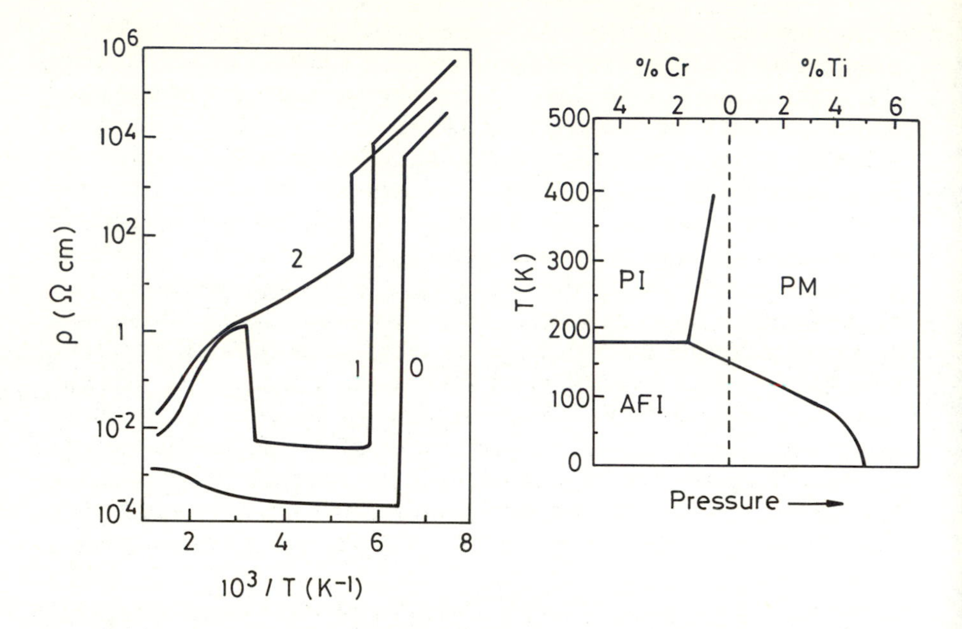

FIG. 5.13 (a) Resistivity of V_2O_3 and two $V_{2-x}Cr_xO_3$ compositions (labelled with percentage Cr content) against inverse temperature. (b) Phase diagram showing the behaviour of doped V_2O_3 against composition and pressure (combined horizontal axis) and temperature. The phases marked are: PM, paramagnetic metal; AFI, antiferromagnetic insulator; PI, paramagnetic insulator. (From ref. 48.)

solutions containing a few per cent of Cr_2O_3, vary with temperature.[48] Pure V_2O_3 shows a sharp MNM transition at 150 K; the corundum lattice undergoes a small structural change involving alterations in the near-neighbour V–V distances, although this is much less marked than that occurring with the transition in Ti_2O_3 described later (Section 5.3.2). Adding less than one per cent of Cr_2O_3 raises the transition temperature, and also induces a remarkable 're-entrant' transition back to non-metallic behaviour at higher temperatures. Adding Ti_2O_3 instead of Cr_2O_3 has the reverse effect, lowering the MNM transition temperature. The same effects as substitution can also be produced by pressure: increasing pressure has the same effect as Ti substitution, whereas the changes induced by Cr substitution are reversed by high pressures.

The temperature-dependent behaviour of V_2O_3 and the way this is altered by substitution and pressure are summarized in the *generalized phase diagram* of Fig. 5.13(b). The three phases shown are: (PM) a metal, which is paramagnetic although with a considerably enhanced susceptibility showing that correlation is important; (AFI) a magnetic insulator with the local moments of V^{3+} antiferromagnetically ordered; and (PI) a paramagnetic insulator with the same local moments without long-range order. The horizontal boundary between the AFI and PI phases at 180 K thus represents the Néel temperature associated with a conventional magnetic-ordering transition of the type described in Section 3.4.3. The other boundaries show transitions from a metal to a magnetic insulator: it is this that characterizes the Mott transition.

The effect of pressure is fairly easy to understand, as compression decreases the interatomic distances, enhancing orbital overlap and broadening the bands. W/U is therefore increased and eventually the non-metallic phase disappears. Titanium doping seems to work in the same way, probably because the Ti 3*d* orbitals are more extended than those of vanadium, and so increase the overall band width by their stronger overlap. Chromium has the reverse effect; its more contracted orbitals may form localized states which do not contribute to the band, and by removing some states, cause it to become narrower.

The temperature dependence of the MNM transition is more subtle. On general thermodynamic grounds, one expects that phases stable at higher temperatures are those with more entropy. It can be argued that the extra entropy of metallic electrons relative to that of a magnetically ordered insulator is what drives the transition at 150 K in V_2O_3 itself.[46] The entropy of conduction electrons depends on their effective mass (through the dependence of the heat capacity on $N(E_F)$ given in eqn (2.24), and it is probably only because of a considerable enhancement due to correlation that the metallic entropy in V_2O_3 is large enough to induce the transition observed. On the other hand, the magnetically *disordered* insulator has more electronic entropy than the metal, thus accounting for the 'backward' slope of the PI/PM boundary which shows the 're-entrant' transition for Cr-doped samples.

In spite of the *qualitative* understanding possible for the V_2O_3 system, there seems to be no *quantitatively* satisfactory theory of the Mott transition. Even the idealized Hubbard model cannot be solved to a reasonable degree of approximation when $U \sim W$, and as we have noted in Section 2.4.1 this model is itself rather inadequate to describe the complex interactions taking place in metal oxides.

V_2O_3 is a rare example of a genuinely borderline material; in most other cases the Mott transition can only be studied in series of compounds where the band width varies systematically. Sometimes quite minor substitutions are required. For example, changing the size of the lanthanide element Ln in the series $LnTiO_3$ seems sufficient to induce a Mott transition.[49] Another example of the transition from metallic to magnetic insulator behaviour is found in the ternary oxides of Ru^{4+}.[36]

$Bi_2Ru_2O_7$ is a Pauli paramagnetic metal, whereas $Ln_2Ru_2O_7$ with Ln = Y or a later lanthanide such as Gd is a magnetic insulator. This example is especially interesting as it contains the $4d^4$ ion Ru^{4+} in a *low-spin* configuration, which is relatively unusual for magnetic insulators. These compounds have the pyrochlore structure which is quite complex, but which has the same type of three-dimensional network of corner-shared metal–oxygen octahedra found in perovskites. The main difference is that the M–O–M angle is considerably smaller than the 180° of ideal perovskites. The Ru^{4+} perovskites $CaRuO_3$ and $SrRuO_3$ are metallic, but both have temperature-dependent susceptibilities and the strontium compound undergoes ferromagnetic ordering.

The properties of these compounds suggest that the Ru 4*d* conduction band must get narrower between $Bi_2Ru_2O_7$ and $Y_2Ru_2O_7$, with $SrRuO_3$ and $CaRuO_3$ being intermediate. A series of spectroscopic investigations on these compounds have confirmed these trends, while at the same time highlighting the difficulties of interpreting results from compounds close to the MNM borderline.[36]

Figure 5.14 shows (a) photolectron spectra of the Ru 4*d* conduction band region, and (*b*) electron energy-loss spectra of the series, including the metallic $Pb_2Ru_2O_{7-x}$ which has an oxygen-deficient pyrochlore structure. The Pb and Bi compounds show the features expected of 'good' metals, as illustrated above in Section 5.1. The PES shows a reasonable conduction-band shape and density of states at the Fermi level. The plasmon at ~1.5 eV in the EELS spectrum suggests effective masses $m^* \sim 2m_0$, consistent with $N(E_F)$ deduced from the PES. At the other end of the series, with non-metallic $Y_2Ru_2O_7$, both Fermi edge and plasmon have disappeared. The broad Ru 4*d* feature in the PES *cannot* now be interpreted simply in terms of ionization from a band, because it is necessary to take account of interactions occurring in the localized electron configuration. Crystal field theory is the most appropriate model for this. Thus ionization from the $^3T_{1g}$ ground state of (t_{2g}^4) (see Table 2.1) gives states of the (t_{2g}^3) configuration. Using theoretical estimates of the relative intensity and energy of these final states allows a reasonable fit to the spectrum, as shown in the Figure. The Gaussian broadening of each state is due to lattice vibrations excited in accordance with the Franck–Condon principle, as the equilibrium geometry of the localized ionized state will be different from that of the ground electronic state.

The transition from metallic $Bi_2Ru_2O_7$ to the magnetic insulator $Y_2Ru_2O_7$ must be caused by a narrowing of the conduction band, and indeed the PES of the insulating compound has been interpreted without any electronic band width at all. But the appearance of final-state electronic and vibrational interactions in PES means that we cannot expect

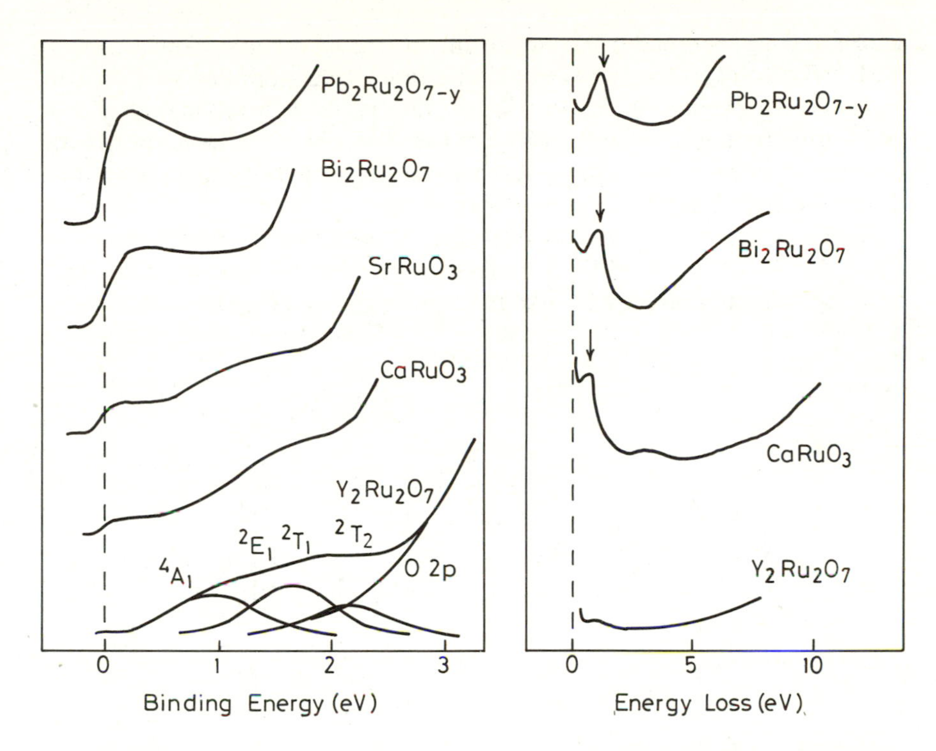

FIG. 5.14 (a) Photoelectron, and (b) electron energy-loss (EELS) spectra of some ternary Ru^{4+} oxides. In (a) the signal from Ru 4*d* ionization is shown in detail; for insulating $Y_2Ru_2O_7$ the bottom curves show a fit to the different final states expected for the ionized $4d^3$ configuration. In (b) the losses corresponding to the plasma frequency are marked by arrows. (Based on ref. 36.)

to see such band narrowing directly in this type of measurement. The spectra of the correlated 'borderline' metals $CaRuO_3$ and $SrRuO_3$ are somewhat intermediate in appearance. A plasmon is still visible in the EELS, with a reduced energy indicating a larger effective mass than in $Bi_2Ru_2O_7$. This is consistent with a narrower conduction band, but correlation may further enhance the effective mass. For example one theory of correlation in narrow-band metals suggests the formula

$$m^{**} = m^* [1 - (U/U_0)^2]^{-1} \tag{5.16}$$

where m^{**} is the enhanced mass, U is the Hubbard parameter, and U_0 the critical value at the Mott transition.[50] In a simple theory a narrower band would also be expected to increase $N(E_F)$, but the photoelectron spectra show a *smaller* Fermi edge than in $Bi_2Ru_2O_7$. The increased width and appearance of structure in the Ru 4*d* signals from $CaRuO_3$ and $SrRuO_3$ suggest that some final-state effects similar to those in $Y_2Ru_2O_7$ are beginning to contribute; this type of broad signal with a small apparent $N(E_F)$ seems to be typical of photoelectron spectra of correlated metals, and shows again that electron–electron interactions dominate over band-structure effects.

Some core-level photoelectron spectra of the compounds are shown in Fig. 5.15. Ruthenium 3*d* PES signals have the same type of final-state structure already described for metallic oxides; in Section 5.1.3 the

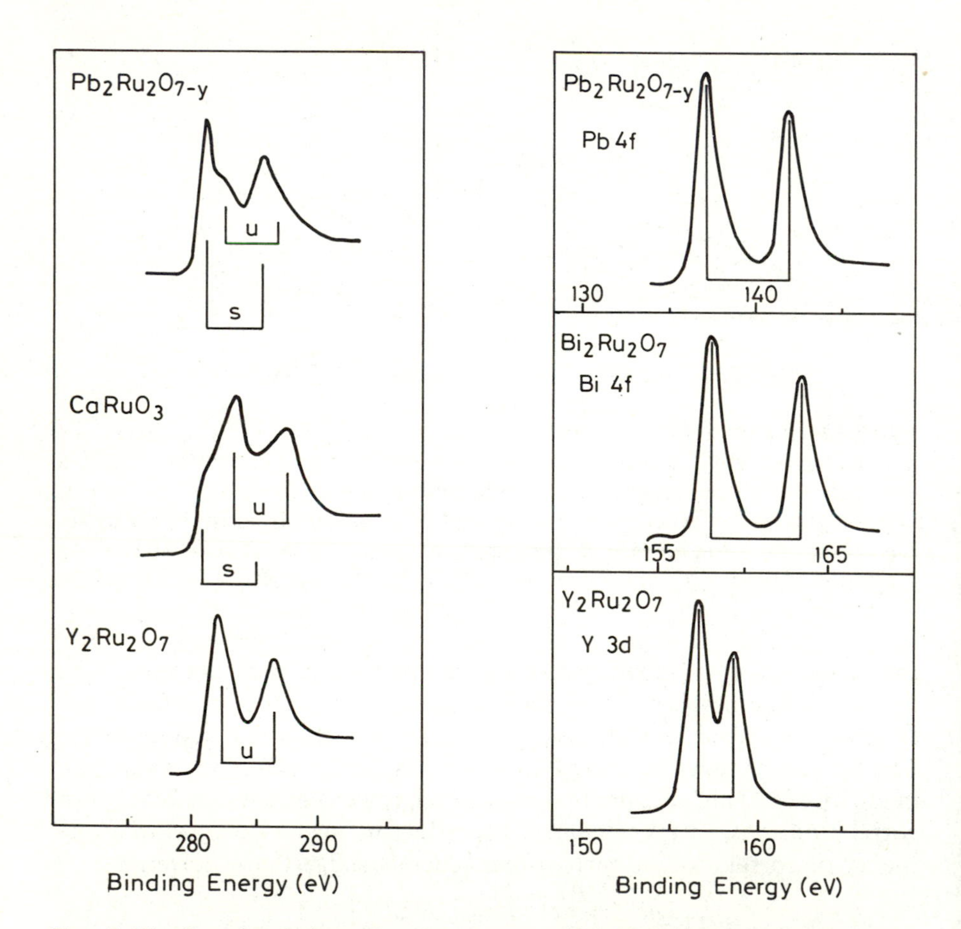

FIG. 5.15 Core-level photoelectron spectra of (a) Ru 3*d* levels and (b) ternary elements in some Ru^{4+} oxides.[36] All lines show spin–orbit splitting; note the final-state structure present on Ru 3*d* ionizations in the metallic compounds, and the Pb 4*f* and Bi 4*f* asymmetry. Neither feature is present in non-metallic $Y_2Ru_2O_7$.

lower-binding-energy component of these peaks was attributed to the *screening* effect of metallic electrons responding to the production of a core hole. Between the 'good' metals and the poor ones in the Ru^{4+} series the relative intensity of this screened peak diminishes, consistent with a narrowing of the band which decreases the screening ability of conduction electrons. In non-metallic $Y_2Ru_2O_7$ the screened peak has disappeared completely.

The Pb, Bi, and Y core spectra are also interesting. Pb and Bi $4f$ levels show a pronounced asymmetry that is lacking in the Y $3d$ spectrum. This type of asymmetry is found in core PES of many metals,[20] and is also a consequence of screening by conduction electrons, when the perturbation produced by the core hole is not strong enough to give the bound state shown in Fig. 5.9. The asymmetry of the Pb and Bi core levels suggests that orbitals from these atoms make some contribution to the density of states at the Fermi level. Pb^{2+} and Bi^{3+} have the electron configuration $6s^2$, the $6s$ electrons being relatively weakly bound and forming states above the oxygen $2p$ band in energy. Interaction between the $6s$ level and the ruthenium $4d$ orbitals is probably the main factor leading to a conduction band wider than in the compounds with pre-transition ternary elements.

The difference between $CaRuO_3$, $SrRuO_3$, and $Y_2Ru_2O_7$ must be more subtle. On the one hand, the difference between the perovskite and pyrochlore structures may be significant, with the smaller Ru–O–Ru angle of the latter weakening in the indirect π overlap on which the conduction band width depends. On the other hand, some degree of covalent interaction between the ternary element and oxygen $2p$ orbitals may compete with the Ru–O bonding and also cause the band to narrow. The higher charged Y^{3+} would be expected to compete more in this way, and there could also be a difference between Sr^{2+} and the smaller Ca^{2+}; there are some indications, both from the magnetic properties and the photoelectron spectra, that the conduction band is slightly narrower in $CaRuO_3$ than in $SrRuO_3$.

Changes which seem to occur along the series $LnTiO_3$ must presumably also be related to the smaller radii of the heavier lanthanide Ln^{3+} ions, more localized behaviour being found for smaller lanthanides.[49] As with Ca^{2+} *versus* Sr^{2+}, increased Ln–O bonding may be involved, but it is possible that small structural distortions, giving a decrease of the Ti–O–Ti angle with the smaller ternary ion, are also important. The fact that rather small structural and electronic changes can cause a large alteration of electronic properties demonstrates how close to the Mott transition borderline many transition metal oxides are.

Another parameter that has been found to be important is the dimensionality. Layer-perovskite analogues of metallic $LaNiO_3$ and the mixed-valency Mn perovskites discussed earlier are found *not* to be metallic.[51,52] The main structural difference is that metal–oxygen octahedra form a two-dimensional planar network in the layer perovskite, as opposed to the three-dimensional network of the normal perovskite (see Fig. 1.9). Reducing the number of near-neighbour interactions in this way would decrease the band width somewhat, and might be sufficient to tip the balance across the metal/non-metal borderline. But it is also possible that there may be some more subtle effects connected with the dimensionality itself. Both Fermi-surface nesting and the effect of disorder are more likely to cause a breakdown in the normal properties of a metal in two dimensions rather than three. The possibility of making *intergrowths* between cubic and layer perovskites offers an opportunity to explore the influence of band width and dimensionality in oxides, and may lead to a better understanding of these effects.

5.2.4 Chemical trends in the Mott transition

We have just seen how changes in the crystal structure or in the ternary element present in an oxide may sometimes cause a sufficient change in band width to tip the balance between a metallic state and a magnetic insulator. Far greater changes can take place as the transition element itself is varied. Metallic compounds are commoner in the 4*d* and 5*d* series than for 3*d* oxides. This difference is consistent with many other trends in electronic properties, such as the greater prevalence of metal–metal bonding and of low- rather than high-spin configurations in the lower series; all these difference are a reflection of the greater ability of 4*d* and 5*d* orbitals to overlap with their surroundings. Band widths are therefore larger than in comparable 3*d* oxides. At the same time electron-repulsion terms are smaller with the larger orbitals, so that W/U is larger.

Changes within the 3*d* series are a little harder to interpret, and seem to depend on a number of trends.[3,53] Increasing nuclear charge leads to a contraction of the orbitals, and the degree of direct metal–metal overlap decreases. In binary compounds of earlier elements (such as TiO_x and VO_x) such direct overlap is sufficient to give metallic behaviour, although VO_x seems to be very close the borderline in behaviour. In some compounds such as Ti_2O_3 which are clearly metallic at higher temperatures, metal–metal bonding intervenes (see Section 5.3.2) to give non-metallic forms at low temperatures. But the later binary oxides are magnetic insulators, the boundary coming before MnO, Cr_2O_3, and MnO_2 in oxides with these three different compositions.

Unlike the binary series, some ternary series such as the perovskites show a transition from metal to magnetic insulator early in the series, and a transition back to metallic behaviour for later elements. For the

M^{3+} compounds $LaMO_3$, Ti is apparently already on the borderline, and V, Cr, Mn, and Fe form magnetic insulators. Co^{3+} is exceptional, as the diamagnetic low-spin $3d^6$ configuration can be understood equally in localized or band-theory terms, as explained in Section 3.2. However, $LaNiO_3$ and $LaCuO_3$ are metals, although with enhanced susceptibilities showing the importance of electron correlation. There may be several trends at work here. Band width in these perovskites comes principally from metal–oxygen covalent interaction, and with increasing nuclear charge there is an effect which tends to counteract the orbital contraction. The stabilization of the $3d$ orbitals brings their energies closer to that of the oxygen $2p$ orbitals, thus increasing the tendency to covalent mixing. Without such mixing it is possible that the Fermi level in $LaNiO_3$ and $LaCuO_3$ would fall within the oxygen $2p$ valence band, so that they would be metals with 'oxygen hole', rather than metal d, carriers.[54] But probably it is better to imagine substantial mixing between metal and oxygen orbitals, giving a band broad enough to overcome the Hubbard repulsion.

Another factor that complicates the treads is that for early transition elements (up to d^3 in high-spin, and d^6 in low-spin compounds) only the t_{2g} part of the d band is occupied, whereas for later elements the Fermi level lies in the e_g band. The σ-bonding effects that determine the dispersion of the e_g band are expected to be stronger than with the π-bonding t_{2g} levels.

Increasing the oxidation state from M^{3+} to M^{4+} in the perovskites also increases the covalent interaction and broadens the bands. Thus in the series $SrMO_3$, metals are formed with V, Cr, and Fe, and a magnetic insulator only with Mn; the later elements from Co onwards do not form stable compounds of this type.

It would be wrong, however, to consider changes in band width as the only factor, as the Hubbard U parameter must also change. We have seen some of the difficulties involved in finding reliable estimates in U, but some qualitative trends can be suggested. Due to exchange-energy changes at the half-filled shell, free-ion values are considerably larger for the d^5 configuration than elsewhere (see Section 2.4.1, Fig. 2.22 (a)). There is no reason why solid-state screening effects should remove this difference. Indeed, all $3d^5$ oxides are high-spin magnetic insulators, and among compounds with less electronegative non-metals, the behaviour of Mn^{2+} is often exceptional.[53] In a low-spin compound such as IrO_2 ($5d^5$), the exchange effect in the half-filled shell is of course absent. But as suggested in Section 2.4.1, the trends shown by *ground states* of the free ions may not be quite appropriate for discussing the metal/insulator borderline. The values given in Fig. 2.22 (b) incorporate some of the effects expected from spin-pairing in the metallic state, and show a more even trend towards higher U values at d^5. One might also expect the free-ion trends to be complicated by crystal field effects. Thus the t_{2g}^3 configuration of an octahedral ion has a half-filled sub-shell, which could give an extra contribution to U as found at d^5 in the free ion. It may be significant that the first unequivocal examples of magnetic insulators in M_2O_3 and MO_2 compounds of the $3d$ series occur with the d^3 configuration.

5.3 Lattice interactions

The interaction of conduction electrons with the crystal lattice can give rise to distortions of the regular structure. Sometimes these introduce temperature-dependent energy gaps and cause metal/non-metal transitions which are quite different from the Mott transition discussed above. The following sections look at some different types of distortion found in transition metal oxides, and the associated MNM transitions.

5.3.1 Fermi-surface instabilities

Section 2.3.4 discussed the concept of **Fermi-surface nesting**, and the possibility of instabilities leading to charge- or spin-density waves. Charge-density wave (CDW) instabilities are a prominent feature of low-dimensional solids, where electronic interactions are strongly anisotropic and predominate in only one, or sometimes two, dimensions. Familiar examples are the 'one-dimensional' metal chain compounds such as $K_2Pt(CN)_4Br_{0.3}.3H_2O$, and the layered chalcogenides like TaS_2. The low dimensionality of the electronic band structure in these compounds is a consequence of their particular crystal structures. Oxides rarely have layer- or chain-type structures, and so low-dimensional electronic phenomena are less common in this field. One exception to this general rule, however, is with some of the high-oxidation-state oxides of molybdenum. The d^0 insulator MoO_3 has an unusual structure based on layers of edge-sharing octahedra as illustrated in Fig. 5.16 (a). The octahedra are actually strongly distorted, as was shown in Fig. 1.6; it is probably the instability of Mo^{6+} in a regular octahedral site and its tendency to form one or two short Mo–O bonds that give rise to the structure. Layers rather similar to that of MoO_3 survive in a number of reduced molybdenum oxide phases, and in bronzes formed by insertion of metal ions into sites between the layers. These structural features give rise to effects which can be associated with either one- or two-dimensional band structures, depending on the connectivity of groups within the layers.[55]

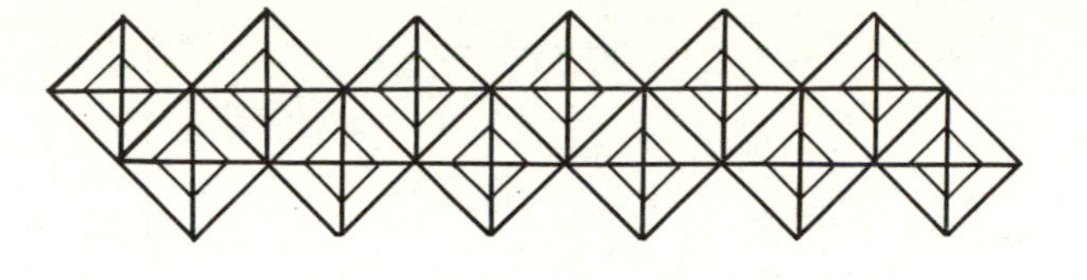

(a)

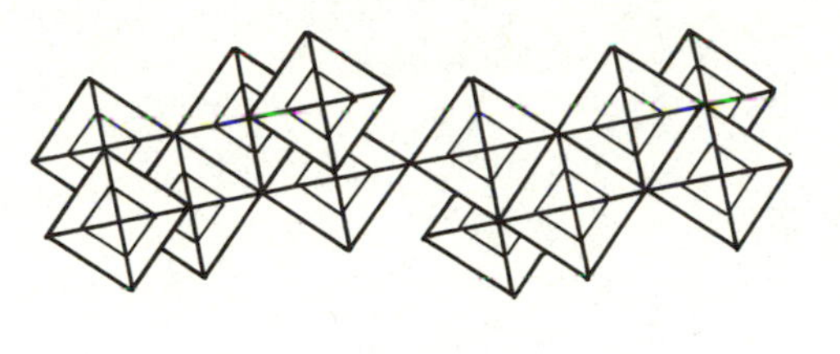

(b)

FIG. 5.16 Structures of (a) MoO_3 and (b) $A_{0.33}MoO_3$. In each case a projection of one layer is shown, with edge- and corner-sharing contacts between (idealized) MoO_6 octahedra.

The most extensively studied 'low-dimensional' oxides are the so-called *blue bronzes* $A_{0.3}MoO_3$, where A can be K, Rb, or Tl. Figure 5.17 shows how the conductivity of the potassium compound varies with temperature.[56] Above 180 K it behaves as a metal, but below this temperature the resistivity starts to rise, and at low temperatures there appears to be an intrinsic band gap of about 0.15 eV. A study of the infrared reflectivity of this compound shows highly anisotropic features. Perpendicular to the crystal *b* axis can be seen the vibrational modes expected of an insulator. But with radiation polarized along the *b* axis a high metallic reflectivity is found at 300 K; below the transition at 180 K this decreases somewhat and vibrational features start to appear.[57]

Both conductivity and reflectivity results on $K_{0.3}MoO_3$ are characteristic of a one-dimensional metal, and can be understood from the structure shown in Fig. 5.16 (b). The view down the *b* axis shows one layer composed of corner- and edge-sharing octahedra. The arrangement of these leads to closer Mo–Mo connections along the *b* axis than in other crystal directions. There is probably no direct metal bonding, but the shorter distances in one direction increase the indirect Mo–O–Mo overlap which is the principal cause of conduction band width. Thus the

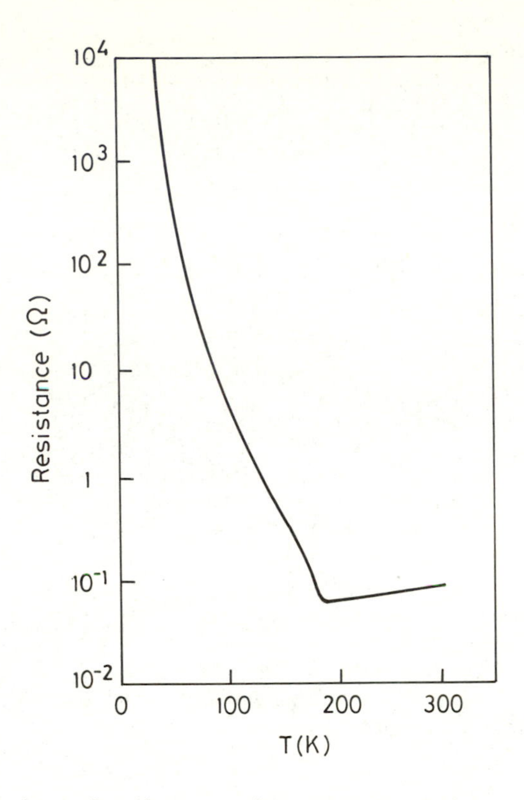

FIG. 5.17 Variation of resistance with temperature for the 'blue bronze' $K_{0.3}MoO_3$.[56]

partially occupied Mo 4*d* band is strongly anisotropic in nature, having considerably more dispersion along the *b* direction than in perpendicular ones.

The idea that the blue bronzes have a Peierls instability driven by Fermi-surface nesting is supported by diffraction studies. *Superlattice* reflections with non-integral indices are observed, indicating a periodic lattice distortion along *b*.[58] As in other solids with Peierls instabilities, diffuse scattering can be seen well above the transition temperature, showing the onset of distortions, probably dynamic in nature, without long-range order. Below 180 K the superlattice reflections sharpen as the periodic distortion 'locks in' and becomes ordered in three dimensions.

The electrical properties of $K_{0.3}MoO_3$ below the Peierls transition temperature show some very interesting non-linear features.[59] As the electric field on a sample is increased above a critical value (which depends on the temperature and the sample history) the activation energy for conduction suddenly decreases. It is believed that in high fields, conduction occurs by a collective motion of *sliding charge-density waves,* rather than by individual thermally excited electrons and holes. Below the critical electric field, the charge density waves are probably *pinned* by defects. This interpretation is supported by measurements on samples where some molybdenum has been replaced by tungsten. The impurities decrease the activation energy for normal conduction at low fields, but they also increase the critical field required to 'de-pin' the charge density waves, presumably because of the disordered potential produced by random substitution. Close to the critical field itself, remarkable fluctuations and 'spikes' occur in the current, associated with momentary de-pinning and subsequent re-pinning of charge-density waves. This *chaotic* behaviour is characteristic of highly non-linear systems.

Given the structural features of MoO_3, it might be expected that two-dimensional electronic behaviour could be found in some reduced oxides and bronzes of molybdenum. There are indeed a number of examples known, including Mo_4O_{11}, and the 'purple bronze' $K_{0.9}Mo_6O_{17}$.[55] Both compounds show conductivity anomalies at around 100 K, although of a much less marked kind than in $K_{0.3}MoO_3$. These seem to be associated with charge-density wave instabilities; as explained in Section 2.3.4, Fermi-surface nesting in two (and occasionally three) dimensions can lead to periodic distortions which may reduce the area of the Fermi surface without destroying it completely.

5.3.2 Metal–metal bonding

In contrast to the rather subtle physics of charge-density waves, some oxides show a form of structural distortion that can easily be interpreted in chemical terms. We noted in Section 1.3.1 that early transition elements in low oxidation states have a tendency to involve their *d* electrons in direct bonding between the metal atoms. In some cases such metal–metal bonding can dominate the structure, but other cases occur in compounds where the basic structure is that expected for an ionic compound dominated by metal–oxygen interactions. Interesting electronic behaviour often occurs, as the *d* electrons may occupy a metal–metal bonding band that is split off from the rest of the *d* band at low temperatures, preventing metallic behaviour. At higher temperatures lattice expansion leads to weakening or even the disappearance of the M–M bonds, and an MNM transition may occur.

Ti_2O_3 is semiconducting with a small gap (~ 0.2 eV) below 390 K, but becomes metallic at higher temperatures. Associated with this transition is an anomalous change in the lattice constants, and particularly the c/a

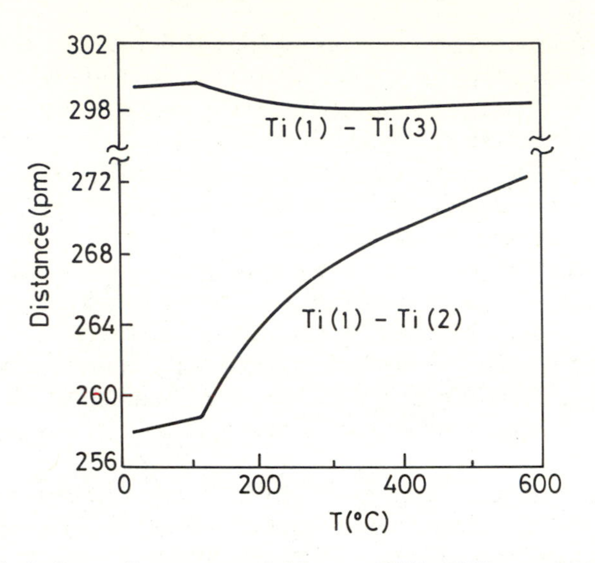

FIG. 5.18 Variation of nearest-neighbour Ti(1)–Ti(2), and next-nearest-neighbour Ti(1)–Ti(3) distances in Ti_2O_3 as a function of temperature.[61] The change in slope at 390 K is associated with the transition to metallic conductivity.

ratio of the corundum structure.[3,60] The main features of this structure were illustrated in Fig. 1.7. The closest Ti–Ti distance is that which occurs along the *c* axis between pairs of face-sharing octahedra; next-nearest neighbours are found in the perpendicular directions, between octahedra sharing only corners. Figure 5.18 shows how these two distances change with temperature in Ti_2O_3.[61] Above the transition the close Ti(1)–Ti(2) pairs move apart considerably, whereas the basal-plane Ti(1)–Ti(3) distances decrease slightly. The low-temperature Ti(1)–Ti(2) distance is close to the value (250 pm) found in the element, and the structural features therefore suggest appreciable Ti–Ti bonding, which weakens at high temperatures.

A number of empirical and band-structure schemes have been proposed for the electronic levels in Ti_2O_3, which mostly agree on the essential features shown in Fig. 5.19.[3,62] The Ti 3*d* 't_{2g}' set of orbitals is split by the trigonal symmetry into a lower a_1 orbital and a doubly-degenerate *e* pair. The spatial orientation of these is such that the a_1 orbitals overlap strongly within the near-neighbour pairs, forming a low-energy bonding combination and a corresponding antibonding one at higher energy. Thus the M–M bonding in the d^1 compound is provided by two electrons per Ti pair, in the bonding a_1 orbital. Figure 5.19

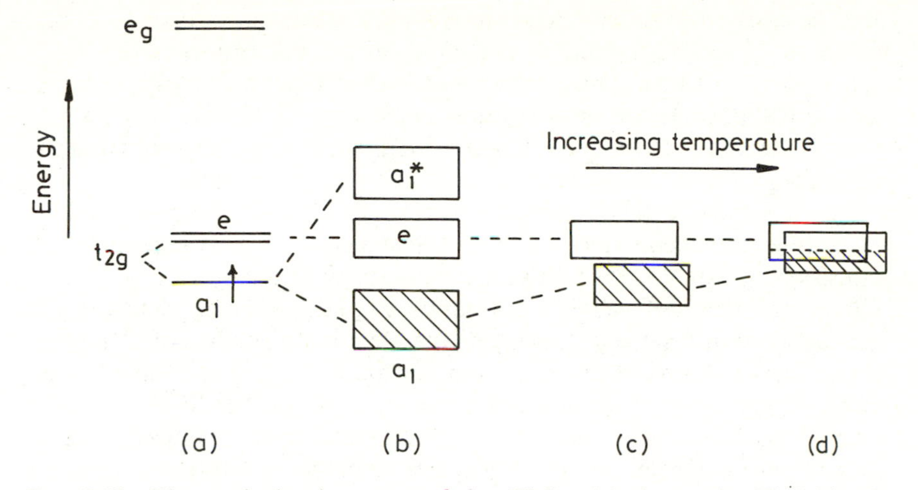

FIG. 5.19 Electronic levels proposed for Ti_2O_3: (a) shows the Ti 3*d* levels with a trigonal splitting and orbital occupancy appropriate to a single Ti^{3+} ion, (b) shows the formation of bonding and antibonding levels accompanying metal–metal bonding (with occupied levels shaded), and (c)–(d) show the evolution of energies as a function of temperature.

(c)–(d) shows what is proposed as the temperature increases. Lattice expansion weakens the Ti–Ti overlap, and the small gap between the a_1 and *e* bands decreases, giving eventually a *band-crossing* transition at 390 K. As the crossing occurs, some electrons are transfered from the a_1 into the *e* band. The Ti(1)–Ti(2) bonding therefore starts to weaken even more sharply, and the distance increases. The *e* orbitals are oriented to overlap best between the in-plane Ti(1)–Ti(3) pairs. Although the distance is longer and true metal–metal bonds cannot be formed, some favourable interaction does nevertheless occur as the *e* band is populated, and the Ti(1)–Ti(3) distance decreases slightly.

Although the scheme just outlined provides a straightforward explanation of many features of Ti_2O_3, it is doubtful whether any band-crossing transition of this type is really very simple.[46] The electrons and holes formed in low concentrations close to the crossing temperature probably behave as polarons, with a considerably enhanced effective mass. By replacing some Ti^{3+} with 1–4 per cent V^{3+} it is possible to obtain a compound that remains metallic to low temperatures. A large electronic specific heat is found, indicating a high effective mass for carriers. Hall measurements show that the conductivity of $Ti_{2-x}V_xO_3$ is *p*-type, whereas the simplest arguments would predict that replacing Ti by V should *add* an electron and give *n*-type behaviour. Thus a rigid-band model for the substituted compound is not appropriate; it is likely that the more contracted V 3*d* orbitals form a localized state, so that their electrons do not contribute to the band. Some disruption of the Ti–Ti pairing on substitution may lead to both electrons and holes being present, and it has been proposed that it is the electrons in the very narrow *e* band that have the enhanced mass measured from the specific heat, while the holes in the broader a_1 band carry most current.

Rather similar behaviour is shown with another d^1 compound, VO_2. At high temperatures this has the rutile structure, and is metallic.[3,60] A transition at 340 K leads to a semiconductor with a distorted structure. The even spacing of vanadium atoms along edge-sharing chains in the regular structure (see Fig. 1.8) is replaced by an alternation of distances, with pairs of atoms at 265 pm, comparable to that in the metallic element. The simplest interpretation is the same as that for Ti_2O_3, with the formation of a metal–metal bonding band, occupied at low temperatures with two electrons per pair of atoms. However, there are several features of VO_2 which seem to be more complicated than Ti_2O_3, and which are still controversial. The transition is first order, whereas that in Ti_2O_3 occurs gradually. This must be partly related to the structural change, which involves a lowering of symmetry from tetragonal to monoclinic in the low-temperature phase of VO_2. By contrast, Ti_2O_3 retains the corundum structure throughout. In VO_2 a significant distortion of the VO_6 octahedra also occurs, which must change the V–O bonding interactions, and may alter the orientation of the lowest-lying *d* orbitals in such a way as to favour the metal–metal bonding.[63] Other properties of VO_2 suggest that electron correlation is probably important. Thus the metallic phase has a magnetic susceptibility much enhanced from the expected Pauli value, and quite inconsistent with the very small density of states at the Fermi level seen in the photoelectron spectrum.[64] This enhanced susceptibility has allowed measurements by magnetic neutron diffraction, which show that, as expected, the states at the Fermi level are t_{2g}-like V 3*d* orbitals.[65] Although the low temperature form is diamagnetic, some anomalous effects are observed on doping with magnetic impurities, which suggest that it is on the verge of being a magnetic insulator.[66] These magnetic properties indicate that electron correlation is probably very important in VO_2, and that one-electron chemical bonding or band-structure arguments are inadequate. VO_2 may be close to the Mott-transition borderline like V_2O_3 (see Section 5.2.3).

In spite of these complications, it does seem that the formation of metal–metal bonds can provide a first approximation to the behaviour of VO_2. It is tempting to relate the pairing of metal atoms accompanying the MNM transition to the type of Fermi-surface instability described in Section 5.3.1. The closest approach between metal atoms in the regular rutile structure occurs within the edge-sharing chains of octahedra along the *c* axis. One could imagine a quasi-one-dimensional band formed by *d* orbitals oriented for overlap in this direction. In the metallic d^1 system this band would be half-full, and the predicted Peierls distortion would indeed lead to an alternation of V–V distances, giving a lower band filled by two electrons per vanadium pair.[2] Although this argument is qualitatively attractive, it is probably misleading. Band-structure calculations in metallic VO_2 do *not* indicate any appreciable one dimensionality.[67] There is some degree of Fermi-surface nesting, not along *c*, however, but in the [101] lattice direction (ΓR in the Brillouin-zone notation shown for RuO_2 in Fig. 2.14). It has been claimed that the rather complex distortion occuring in VO_2 can be related to a nesting vector in this direction. There is, however, another argument against the Fermi-surface interpretation of the MNM transition in VO_2. The sudden, *first-order* nature of this transition contrasts with the gradual structural and electronic transitions more characteristic of Fermi-surface instabilities.

It is interesting to look at some other dioxides in the light of the properties of VO_2. Among 3*d* compounds, only the d^1 VO_2 shows the distorted structure; ferromagnetic CrO_2 retains the regular form. In the 4*d* series, however, both NbO_2 (d^1) and MoO_2 (d^2) have distortions involving metal–metal bonding.[3] NbO_2 is distorted and semiconducting up to much higher temperatures than VO_2, consistent with the stronger M–M bonding expected with 4*d* elements. MoO_2 is metallic in spite of the distortion. This is easy to understand in the model discussed above, as there is one electron per Mo more than required to fill the Mo–Mo bonding band. Thus the Fermi level falls within the band of t_{2g} levels that are unaffected by the bond formation. Such an interpretation is supported both by band-structure calculations and the photoelectron spectra shown in Fig. 5.20.[68] In contrast to the spectra of d^1 compounds or of the later dioxides such as RuO_2 which have the regular structure, there is a pronounced splitting of the 4*d* band in MoO_2: the peak at higher binding energy presumably comes from the metal–metal bonding level.

Metal–metal bonding is not found in oxides of later transition elements. This may be partly due to the contraction of the *d* orbitals, which weakens any direct overlap between them. But it is also a natural consequence of increasing the number of *d* electrons. A distorted structure with a metal–metal bonding level will also have antibonding orbitals at higher energy within the *d* band. Adding *d* electrons will eventually start to populate this antibonding level, and the distortion will no longer be favourable. Metal–metal bonding sometimes occurs with the d^2 and even d^3 electron configurations in the 4*d* series, but only with d^1 for 3*d* elements. The more contracted radial distributions of 3*d* orbitals give weaker overlap than in the later series, so that bonds are weaker; as the

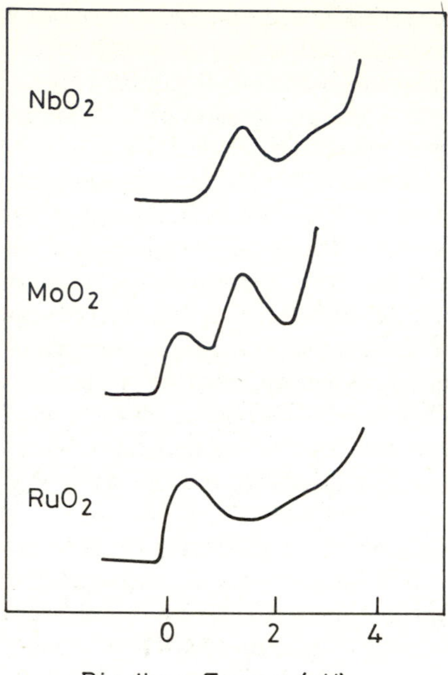

Fig. 5.20 Photoelectron spectra showing the 4*d* ionization in NbO_2, MoO_2, and RuO_2. The Fermi edge in metallic MoO_2 can be seen, ánd also a peak at higher binding energy, coming from the Mo–Mo bonding levels, which is absent in the undistorted RuO_2.[68]

examples of VO_2 and CrO_2 show, however, electron correlation and magnetic anomalies are also more important in the 3*d* series.

5.3.3 Mixed-valency compounds: charge ordering and disproportionation

Many metal oxides are mixed-valency compounds, containing a transition element that cannot be assigned a unique integral oxidation state. This category includes not only the non-stoichiometric and doped compounds discussed in Chapter 4, but also many stoichiometric ones such as Fe_3O_4 and Ti_4O_7. An interesting question is whether one should assign different *integral* oxidation states to the element concerned (Fe^{2+} and Fe^{3+}, or Ti^{3+} and Ti^{4+} in the examples quoted), or rather an equal, but *fractional* oxidation state (for example $Fe^{2.67+}$ or $Ti^{3.5+}$). Such a distinction forms the basis of the classification scheme for mixed-valence compounds proposed by Day and Robin, and widely used by chemists.[69] **Class I** compounds have distinguishable oxidation states in very different environments, so that essentially no electronic phenomena can be associated with charge transfer between them. This situation is rather uncommon in transition metal oxides, but an example might be KCr_3O_8, containing octahedral Cr^{3+} and tetrahedral CrO_4^{2-} ions. **Class II** compounds are ones where different oxidation states can be identified by structural or electronic studies, but where the energy required for electron transfer is small, so that semiconduction occurs. In **class III** compounds on the other hand, metal atoms have equal fractional oxidation states, normally associated with metallic conduction. Thus the metallic tungsten bronzes should be assigned to class III, whereas many semiconducting oxides such as $Li_xMn_{1-x}O$ and $Fe_xCo_{3-x}O_4$ fall in class II. The type of MNM transition which can occur as the doping level increases (see Section 4.4) could be regarded, according to this classification, as a transition from class II to class III behaviour.

The electron localization associated with class II character in doped and non-stoichiometric oxide semiconductors is caused by a combination of the potential field of defects, structural disorder, and lattice interactions forming polarons. Class II behaviour can also be found sometimes in stoichiometric compounds, where it is the lattice interactions—essentially the different atomic geometries appropriate to different oxidation states—which cause electronic trapping. We have already seen an unusual example of this in the non-metallic compound AgO, formulated on the basis of its crystal structure and properties as $Ag^+Ag^{3+}O_2$ (Section 3.2). There are also some stoichiometric oxides that are borderline in this respect, and which show MNM transitions associated with 'valence trapping', or a change to class II behaviour, at low temperatures.

The magnetic properties of Fe_3O_4 were described in Section 5.2.2, and we noted that the ferromagnetic interactions between the octahedral Fe sites are associated with the facile electron transfer between Fe^{2+} and Fe^{3+}.[3,70] Although the conductivity is low for a metal (around 200 $(\Omega\ cm)^{-1}$) and tends to rise slightly with increasing temperature, Fe_3O_4 can probably be formulated as a class III compound, with octahedral iron behaving as $Fe^{2.5+}$ (the magnetic and other properties suggest that the tetrahedral iron is not involved in the conduction process, and is best regarded as localized Fe^{3+}). However, as Fig. 5.21 shows, Fe_3O_4 undergoes a transition at about 120 K, where the conductivity falls into the semiconducting region.[71] In spite of many decades of investigation, the nature and cause of this so-called **Verwey transition** remain something of a puzzle. The clearest evidence that valence trapping occurs comes from Mössbauer spectroscopy, examples of spectra taken above and

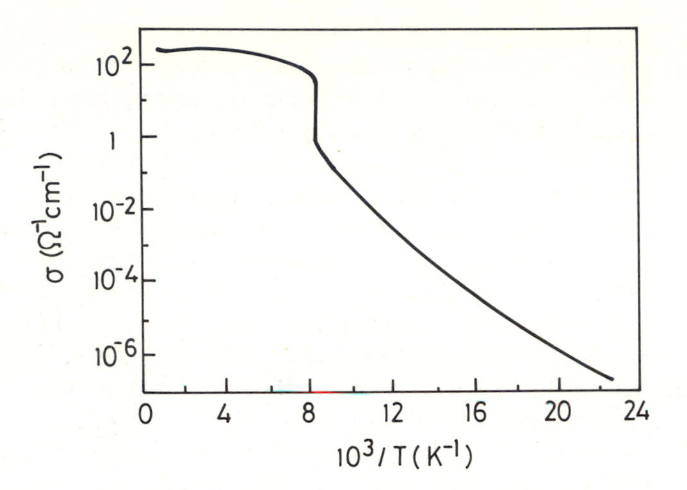

FIG. 5.21 Conductivity as a function of inverse temperature for Fe_3O_4, showing the Verwey transition at 120 K.[71]

below the transition temperature being illustrated in Fig. 5.22. At 298 K the spectrum shows one tetrahedral Fe^{3+} and two octahedral Fe ions with an average oxidation state of $Fe^{2.5+}$; the six-line spectrum from each type of atom is caused by hyperfine interactions between the nuclear magnetic moment and the ordered electron spins of the ferrimagnetic state.[72] Below 120 K the spectra are much more complex, and depend on the polarization of the observed transitions with respect to the domain magnetization controlled by an applied magnetic field.[73] The details of the assignment have been disputed, but the octahedral sites now seem to contain *four* different types of Fe atom, with equal numbers present as Fe^{2+} and Fe^{3+}.

Associated with the Verwey transition is a complex and still controversial structural distortion. The suggestion of early workers that there is a straightforward alternation of Fe^{2+} and Fe^{3+} sites along chains of octahedra in the spinel structure does not seem correct, but no clear alternative has been proposed.[74] However, it does seem quite clear that some charge ordering must exist: i.e. that we now have a class II compound associated with distinguishable oxidation states. The major interaction giving the transition is also controversial. On the one hand, one might expect that the difference in ionic radii of Fe^{2+} and Fe^{3+} would help to give a localization of these states. This is similar to the idea of small-polaron formation, although the extension of the small-polaron concept to compounds with such a high concentration of carriers has been criticized.[46] On the other hand, it may be that the long-range electrostatic

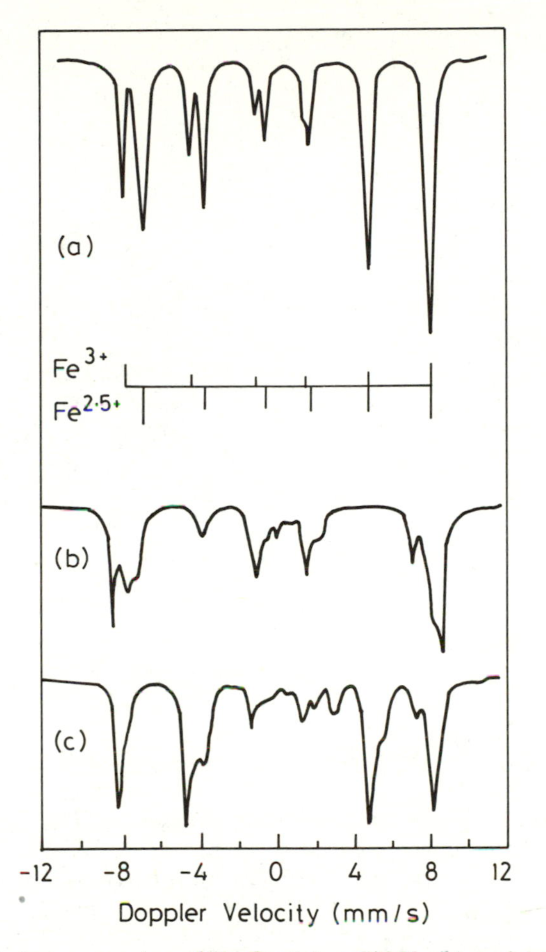

FIG. 5.22 Mössbauer spectra of Fe_3O_4: (a) at 298 K; (b) and (c) in two polarizations below 120 K. Each type of Fe gives six lines as shown in the analysis of the 298 K spectrum. (Based on ref. 73.)

repulsion between electronic carriers is important: thus the Verwey transition has sometimes been thought of as a **Wigner transition**, following the suggestion that long-range electrostatic interactions may cause a breakdown in the metallic behaviour of a low-density electron gas.[75] Possibly both types of interaction are important; and their influence may extend to temperatures above the transition, where many features of the transport properties, such as thermopower and the variation of conductivity with temperature, cannot be explained with any simple model.

Another compound showing a 'Verwey' type of transition is Ti_4O_7,

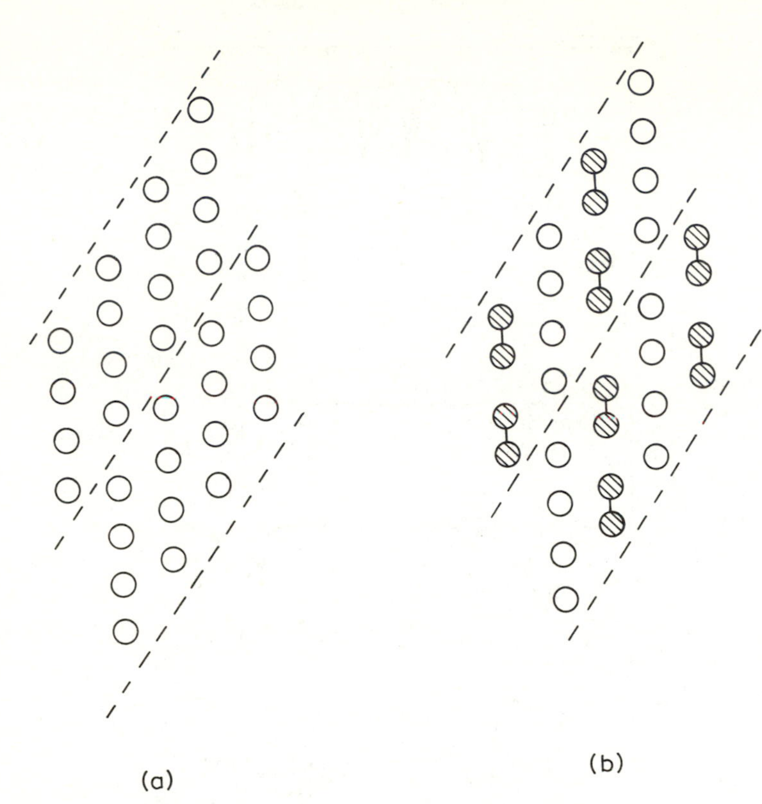

FIG. 5.23 Schematic representation of the structure of Ti_4O_7, showing the Ti positions in chains interrupted by shear planes. (a) Above 150 K showing equivalent $Ti^{3.5+}$ ions; (b) below 130 K, with ordered Ti^{3+} pairs.[76]

although the details of this are very different from Fe_3O_4, and show some of the features found in other reduced titanium oxides.[76] The low-temperature structure (see Fig. 5.23) shows the presence of clearly distinguishable Ti^{3+} and Ti^{4+} sites. Ti^{3+} pairs can be seen, probably with metal–metal bonds. As expected, this phase is non-metallic. Above 150 K, Ti_4O_7 is metallic, and the structure indicates equivalent $Ti^{3.5+}$ sites. Thus it seems that lattice expansion causes the Ti–Ti bonds to dissociate as in Ti_2O_3, the electrons localized in these bonds at low temperatures becoming delocalized in a band extending over all titanium sites. Between 130 and 150 K it is found that semiconducting behaviour coexists with a structure where all titanium atoms appear to be equivalent. The interpretation which has been given is that localized Ti^{3+} pairs exist on a short time-scale, but that these are disordered and *mobile* in the structure. Thus electrons move by hopping, as in a small-polaron conductor, but between paired states known as **bipolarons.**

As we shall see later (Section 5.4) bipolarons have been suggested as a possible mechanism for superconductivity, since the paired state of two electrons can behave as a boson. The example of Ti_4O_7 shows one problem with this idea: in order for bipolarons to produce superconductivity they would have to remain mobile at low enough temperatures to undergo Bose–Einstein condensation to form a **quantum fluid**, with very different properties from the 'classical fluid' proposed for mobile bipolarons in Ti_4O_7. Most classical fluids do not survive at low temperatures, but freeze into a solid. This is essentially what happens to the bipolarons in the low-temperature ordered from of Ti_4O_7.

As with Fe_3O_4, the forces responsible for the charge ordering in low-temperature Ti_4O_7 are uncertain. Metal–metal bonding (not present in Fe_3O_4) is probably involved in the pairing of Ti^{3+} ions to form the 'bipolarons' themselves. The ordering of bipolarons into a regular structure could be a result of lattice interactions similar to those responsible for shear-plane ordering in the CS structures discussed in Section 1.3.3. But electrostatic interactions could also be important.

Fe_3O_4 and Ti_4O_7 provide examples of compounds where the chemical formula itself indicates mixed valency. The case of AgO illustrates a different situation, of 'hidden' mixed valency where a study of the structure and the physical properties is necessary to show that different oxidation states are present. An interesting borderline example in this class is $CaFeO_3$, a perovskite which is a paramagnetic metal at room temperature. A transition to semiconducting behaviour below 115 K is again accompanied by a change in the Mössbauer spectrum, shown in Fig. 5.24.[77] The low-temperature spectrum indicates a **disproportionation** of iron, according to the equation

$$2Fe^{4+} = Fe^{3+} + Fe^{5+}. \tag{5.17}$$

In view of our previous discussion of magnetic insulators and the Hubbard model, this behaviour is quite surprising. For gas-phase ions, the reaction (5.17) is highly endothermic, requiring an energy input of about 20 eV. As we have seen, the *Hubbard U* in solids is much smaller, being screened by lattice polarization and covalency. Nevertheless, values of at least 3 eV are typical of magnetic insulators such as Fe_2O_3. The fact that reaction (5.17) is energetically favourable indicates a **negative** U, where the lattice screening more than compensates the Coulomb repulsion.

Negative U values are known in other situations, most commonly with post-transition elements. Examples are Sb_2O_4 and Cs_2SbCl_6, which are both class II mixed-valency compounds containing distinguishable Sb^{3+} and Sb^{5+}.[69] A similar type of disproportionation is thought to occur at

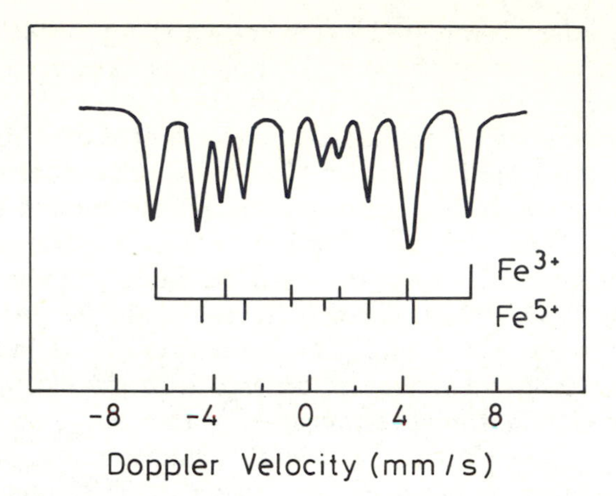

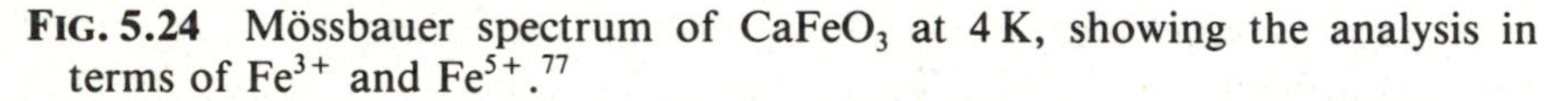

FIG. 5.24 Mössbauer spectrum of $CaFeO_3$ at 4 K, showing the analysis in terms of Fe^{3+} and Fe^{5+}.[77]

defect sites in chalcogenide semiconductors such as As_2Se_3.[78] The negative U in transition metal compounds is quite rare, and is associated with oxidation states that have an odd d electron of the e_g type. This is the case with AgO, where the hypothetical Ag^{2+} precursor (which does not exist in oxides) has a $4d^9$ configuration. The metallic state of $CaFeO_3$ has been interpreted as containing low-spin Fe^{4+}, but magnetic observations suggest that high spin ($t_2^3 e_g^1$) is more likely.[79] The absence of a Jahn–Teller distortion is a sign that the e_g electron is in a delocalized band-like state (as with $LaNiO_3$). In localized electron configurations, the marked Jahn–Teller distortions result from the highly antibonding nature of the e_g orbitals, which give a strong interaction with the lattice (see Section 2.1.1). This strong interaction may also be involved in the tendency to disproportionation. The negative U would thus be related to an especially large difference in radii of the Fe^{5+} (t_{2g}^3) and Fe^{3+} ($t_{2g}^3 e_g^2$) states, giving a large lattice relaxation term accompanying the disproportionation. There does, however, appear to be a mystery here. From the properties of $CaFeO_3$ we might expect to see a superlattice structure at low temperatures with an ordering of Fe^{3+} and Fe^{5+} states. Although no low-temperature structural studies appear to have been made on this compound, some slightly more complex examples such as $Sr_2LaFe_3O_9$ have been studied. Mössbauer spectra and neutron diffraction show a clear ordering of Fe^{3+} and Fe^{5+} at low temperatures, but apparently *no* structural distortion of the kind that might be expected.[80]

5.4 Superconductivity

Prior to the discovery of 'high-temperature' superconductors based on copper oxides, the highest known superconducting transition temperature was that of Nb_3Ge, with $T_c = 24$ K. In 1986 Bednorz and Muller reported superconduction in the layer perovskite phase $La_{2-x}Ba_xCuO_4$. The analagous strontium compound $La_{2-x}Sr_xCuO_4$ with $x \sim 0.15$ was then found to have T_c values up to 40 K, and much higher temperatures were soon to follow. Best known is the so-called '1:2:3' compound $YBa_2Cu_3O_{7-x}$, with a $T_c \sim 95$ K, making it the first solid discovered to be superconducting at liquid-nitrogen temperature.[81] Many other superconducting copper-oxide phases are now known, with T_c values up to about 125 K.

The copper compounds are by no means the only superconducting oxides, and some other examples are listed in Table 5.1. They include NbO, a metallic oxide with a defect-NaCl structure in which metal–metal bonding seems to be important, reduced $SrTiO_3$ (which was discussed from a different point of view in Section 4.4.2) and the mixed-valency titanium-containing spinel $Li_{1+x}Ti_{2-x}O_4$. The table also shows some non-transition metal perovskites containing lead and bismuth in mixed oxidation states. But because of their especially high T_c values, it is the copper compounds that have attracted most attention. Possibly more papers and review volumes have been published on these compounds than on all other transition metal oxides together. Only a small selection can be mentioned here[47, 81–97] in which many further references can be found.

Table 5.1 *Some superconducting oxides*

Compound	Transition temperature (K)
NbO	11
$SrTiO_{3-x}$	~1
$BaPb_{1-x}Bi_xO_3$	20
$Ba_{1-x}K_xBiO_3$	30
$Li_{1+x}Ti_{2-x}O_4$	15
$La_{2-x}Sr_xCuO_{4-y}$('2:1:4')	40
$YBa_2Cu_3O_{7-x}$('1;2:3')	95
$Bi_2Sr_2CaCu_2O_8$('2:2:1:2')	90
$Tl_2Ba_2Ca_2Cu_3O_{10}$('2:2:2:3')	125
$Nd_{2-x}Ce_xCuO_{4-y}$	25

5.4.1 *Characteristics of 'high T_c' superconductors*

Figure 5.25 illustrates the structures of a few of the most widely studied compounds. The structural characteristics can be seen most simply in $La_{1-x}Sr_xCuO_4$. The layer perovskite structure has a two-dimensional network of corner-sharing metal–oxygen octahedra (see also Figs 1.6 and 1.9). These are strongly distorted, with the four in-plane Cu–O distances being shorter than the two others. Such a geometry is characteristic of Cu^{2+}, and as explained in Section 2.1.1 is generally attributed to a Jahn–Teller distortion caused by the odd e_g electron of the $3d^9$ configuration. Replacing La^{3+} by the smaller Nd^{3+} leads to the structure in Fig. 5.25 (b), where the arrangement of ions between the copper–oxygen layers is different. Cu^{2+} now has square planar coordination, but the squares are linked together to form a corner-sharing planar network similar to that in the layer perovskite. The '1:2:3' compound shown in Fig. 5.25 (c) and (d) has two different copper positions, labelled Cu(1) and Cu (2). In $YBa_2Cu_3O_7$, Cu(1) has a square planar geometry, these squares forming chains in the structure, whereas in the oxygen-deficient and non-superconducting phase $YBa_2Cu_3O_6$ the Cu(1) positions show the *linear* coordination expected for Cu^+. Throughout this composition range the Cu(2) ions retain five-coordinate square-pyramidal coordination, connected to form the same types of plane as in the 2:1:4 structures.

Although the coordination of copper may vary from square planar to pyramidal or distorted octahedral, the corner-connected planar network shown by the compounds in Fig. 5.25 seems to be characteristic of high T_c oxides. In structures such as the bismuth- or thallium-containing phases (2:2:1:2 and 2:2:2:3) these copper–oxygen planes may be more widely spaced, with several intervening layers containing the other metal atoms. There is a sense in which all these compounds can be regarded as *intergrowths* between different structures. For example the layer perovskite structure can be thought of as perovskite layers with $ACuO_3$ composition alternating with AO rock-salt layers. The different layers are not electrically neutral, and the internal electric fields may have some influence on the relative energy levels of different orbitals, such as Cu $3d$ and oxygen $2p$. There may also be a slight structural *mismatch* between the different layers. Some $(La,Sr)_2CuO_4$ compositions show a distortion which lowers the overall symmetry from the ideal tetragonal form to orthorhombic. This results from a 'buckling' of the perovskite layers so that the important Cu–O framework is no longer precisely planar, and may be due to a compression of this layer imposed by other parts of the structure. Other compounds show different structural effects that could be attributed to structural mismatches; but since these differ very much from compound to compound it is doubtful whether they are crucial to the phenomenon of superconductivity.

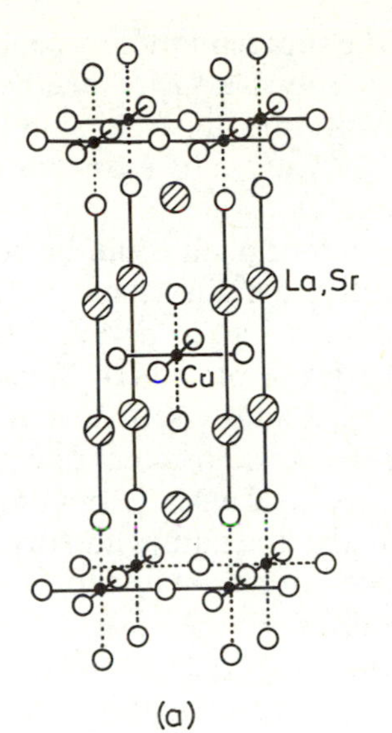

Nd,Ce
Cu
(b)

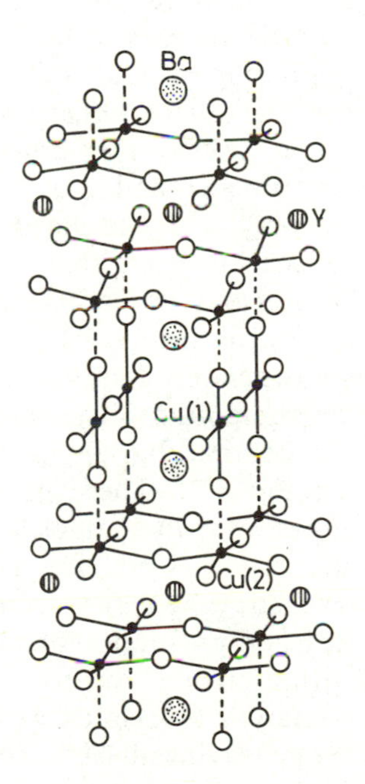

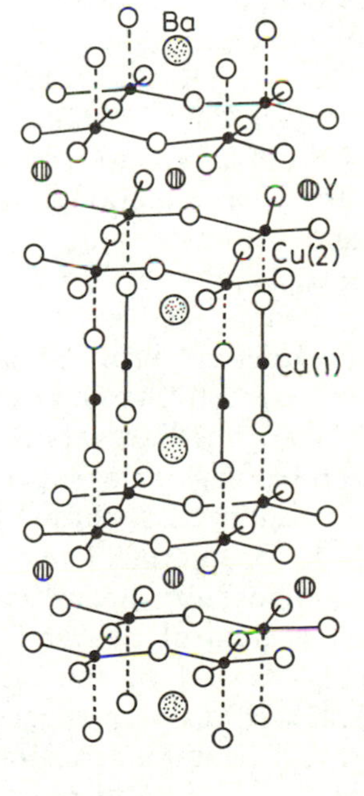

Fig. 5.25 Structures of some superconducting oxides containing copper: (a) $La_{2-x}Sr_xCuO_4$; $Nd_{2-x}Ce_xCuO_4$ (c) $YBa_2Cu_3O_7$; (d) non-superconducting $YBaCu_3O_6$.[87]

Another apparently essential feature of all the superconducting copper oxides is *mixed valency*. Oxides containing only Cu^{2+} are magnetic insulators, with a local moment compatible with one unpaired spin per copper atom. This is the case in CuO and La_2CuO_4. In most known superconductors the formal copper oxidation state is greater than two. In compounds such as $La_{2-x}Sr_xCuO_{4-y}$ the oxidation state can be controlled by altering the proportion of the other metallic elements. As indicated by the formula, however, oxygen contents also vary, and this is one reason why the superconducting properties of most high T_c compounds are extremely sensitive to preparative conditions. Under high oxygen pressures, some compounds can take up *excess* oxygen, and it is possible to make superconducting compositions La_2CuO_{4+y}, containing no strontium. Not only the stoichiometry, but also the nature and ordering of defects such as oxygen vacancies or interstitials, may depend on thermal treatment and other processing variables.

The 1:2:3 compounds normally vary only in the oxygen content, and the synthesis of superconducting compositions requires annealing in oxygen or air at well below the normal reaction temperatures. The control of copper oxidation state in these compounds appears to be quite subtle, and depends largely on changes taking place in the Cu(1) sites of Fig. 5.25. In $YBaCu_3O_7$ all the Cu(1) atoms are square planar with rather short Cu–O distances appropriate to Cu^{3+}. The other limiting composition $YBa_2Cu_3O_6$ has these atoms in linear coordination as Cu^+. To a first approximation the removal of oxygen only alters the coordination and oxidation state at these sites. It seems, however, that some charge transfer occurs between the Cu(1) chain sites and the Cu(2) layers, giving a slight increase of Cu oxidation state in the layers. Although superconductivity probably involves the layers, the chains therefore provide an essential 'charge reservoir' in the 1:2:3 compounds. The charge in the layers does not vary linearly with oxygen content, but apparently in a step-like fashion as suggested by the data in Fig. 5.26. There are two plateau regions of T_c: around 95 K for the highest oxygen compositions and 60 K for intermediate ones. Superconductivity does not disappear until compositions near $YBa_2Cu_3O_{6.4}$ are reached, where the *average* oxidation state of copper is below 2. At this composition a marked anomaly is seen in the *c*-axis parameter (the vertical direction in Fig. 5.25). It has been suggested that the anomaly is associated with a charge redistribution from the chains to the layers, leaving all Cu(2) layer sites as Cu^{2+} for lower oxygen contents.[86]

It is also possible to induce superconductivity by *decreasing* the oxidation state, doping with electrons rather than holes. This is the situation in $Nd_{2-x}Ce_xCuO_{4-y}$ which contains Ce^{4+}, and has the structure illustrated in Fig. 5.25 (b). Electron doping seems to be impossible in

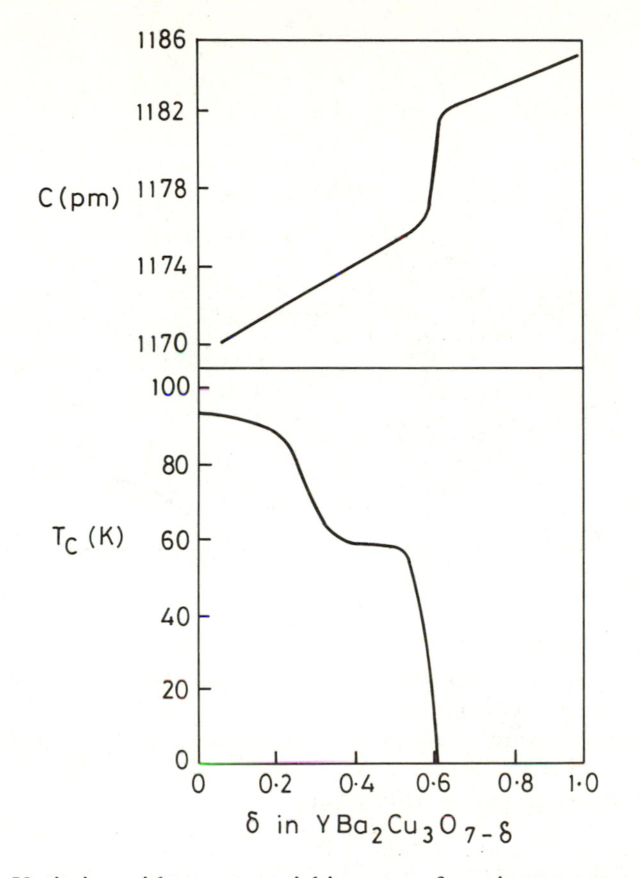

Fig. 5.26 Variation with oxygen stoichiometry of *c*-axis parameter and superconducting transition temperature for 1:2:3 compounds $YBa_2Cu_3O_{7-\delta}$, showing the anomaly associated with the disappearance of superconductivity.[86]

La_2CuO_4, and the different packing arrangements of the La–O and Nd–O layers may be crucial here.[87] If the Cu–O layers are under compressive stress, then hole doping, which increases the Cu oxidation state and shortens the bonds, will be favoured. It appears that this stress is relieved in the Nd_2CuO_4 structure, so that it is now possible to increase the Cu–O distance by adding electrons.

As expected from the discussion of Sections 3.4.3 and 3.4.4, the undoped copper oxides are antiferromagnetic, ordering of the unpaired Cu^{2+} spins taking place by Cu–O–Cu superexchange interactions. The strongly two-dimensional nature of these interactions gives rise to some unusual properties. Neutron diffraction studies show that a high degree of short-range magnetic order persists above the Néel temperature (326 K

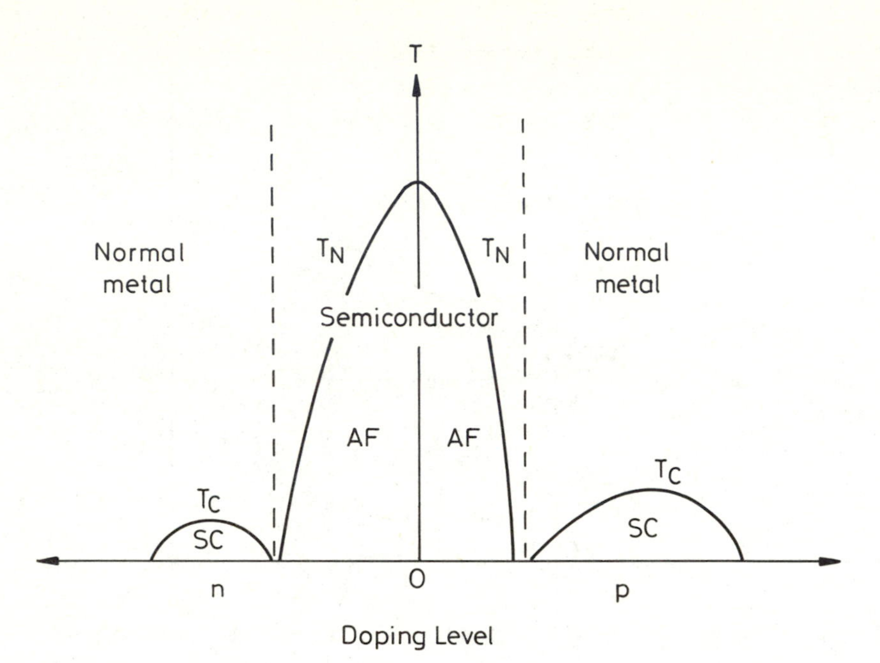

FIG. 5.27 Composite (and very schematic) phase diagram for copper oxide superconductors, showing the antiferromagnetic (AF) and superconducting (SC) phases, the semiconductor–metal transition, and the variation of Néel temperature (T_N) and superconducting transition temperature T_c as a function of electron (n) or hole (p) doping.

in La_2CuO_4). Doping with holes or electrons tends to destroy the antiferromagnetic order, although some short-range order may persist in the superconducting region. It is possible to construct the generalized phase diagram of Fig. 5.27, which shows the disappearance of the antiferromagnetic phase (decreasing T_N) and its replacement by a superconducting phase (increasing T_c) as doping is increased. A semiconductor to metal transition takes place near to the critical concentration for superconductivity, and the low-temperature magnetic properties in this region suggest a 'spin glass' phase with local moments frozen in random orientations. There is only a qualitative symmetry with respect to electrons and holes: for example in $La_{2-x}Sr_xCuO_4$ the Néel temperature falls extremely sharply with x, reaching zero at a value around 0.02. Superconducting compositions fall in the range $x = 0.07–0.25$. In the electron-doped compound $Nd_{2-x}Ce_xCuO_4$ on the other hand antiferromagnetism does not disappear completely until $x = 0.14$, and superconductivity is confined to a narrow range up to $x = 0.18$.

Figure 5.28 shows some of the properties associated with superconductivity in $YBa_2Cu_3O_7$. Most obvious is the drop in conductivity to zero, which occurs over a narrow temperature range in well annealed samples, although some compounds may have a more gradual transition. The very nearly linear variation of resistivity above T_c is interesting, and has been found in many high-T_c superconductors. The similar temperature coefficients found do not correlate at all with the T_c value, which suggests that the phonons which determine resistivity in the normal state cannot be responsible for the superconductivity.

An important feature of all superconductors in the **Meissner effect**: this implies the total expulsion of a magnetic field by the superconducting phase, and is the basis for the 'magnetic levitation' experiments popular in simple classroom demonstrations of high-T_c superconductivity. Figure 5.28 (b) shows the sharp onset of diamagnetic susceptibility at T_c and its evolution to lower temperatures. Meissner-effect measurements provide a useful way of showing what fraction of a given sample is really superconducting. Both the sharpness of the transition, measured by conductivity or the Meissner effect, and the superconducting fraction, are very sensitive to preparation conditions. This may be partly due to the difficulty in obtaining genuinely single-phase materials. We noted earlier, however, (Section 4.4.2), that the superconducting fraction in $Li_{1+x}Ti_{2-x}O_4$ varies with x in a way that is consistent with the existence of a percolating network of Ti^{3+}ions. This suggests that a *structurally* single-phase solid may behave electronically as if two different but intimately mixed phases were present. Similar behaviour may be seen in high-T_c compounds with compositions close to the superconducting limit, although it is difficult to establish whether these samples are genuinely homogeneous in composition.

Magnetic fields higher than a critical value penetrate a superconducting sample. In so-called **type I** materials, superconductivity is destroyed quite sharply by a magnetic field exceeding a critical value H_c. In **type II** superconductors, however, magnetic fields start to penetrate the sample above a lower critical value H_{c1}, and do not destroy the superconducting properties completely until an upper critical value H_{c2}. All high-T_c superconductors are strongly type II in character, with rather low H_{c1} and exceptionally high H_{c2} values. Figure 5.25 (c) shows the evolution of the lower critical field below the transition temperature. These magnetic properties are quite anisotropic, and for magnetic fields perpendicular to the Cu–O planes (which is the polarization required to interact strongly with electrons in the planes) H_{c2} values in excess of 100 T are estimated for low temperatures; these cannot be measured directly but must be extrapolated from the lower critical fields measured close to T_c.

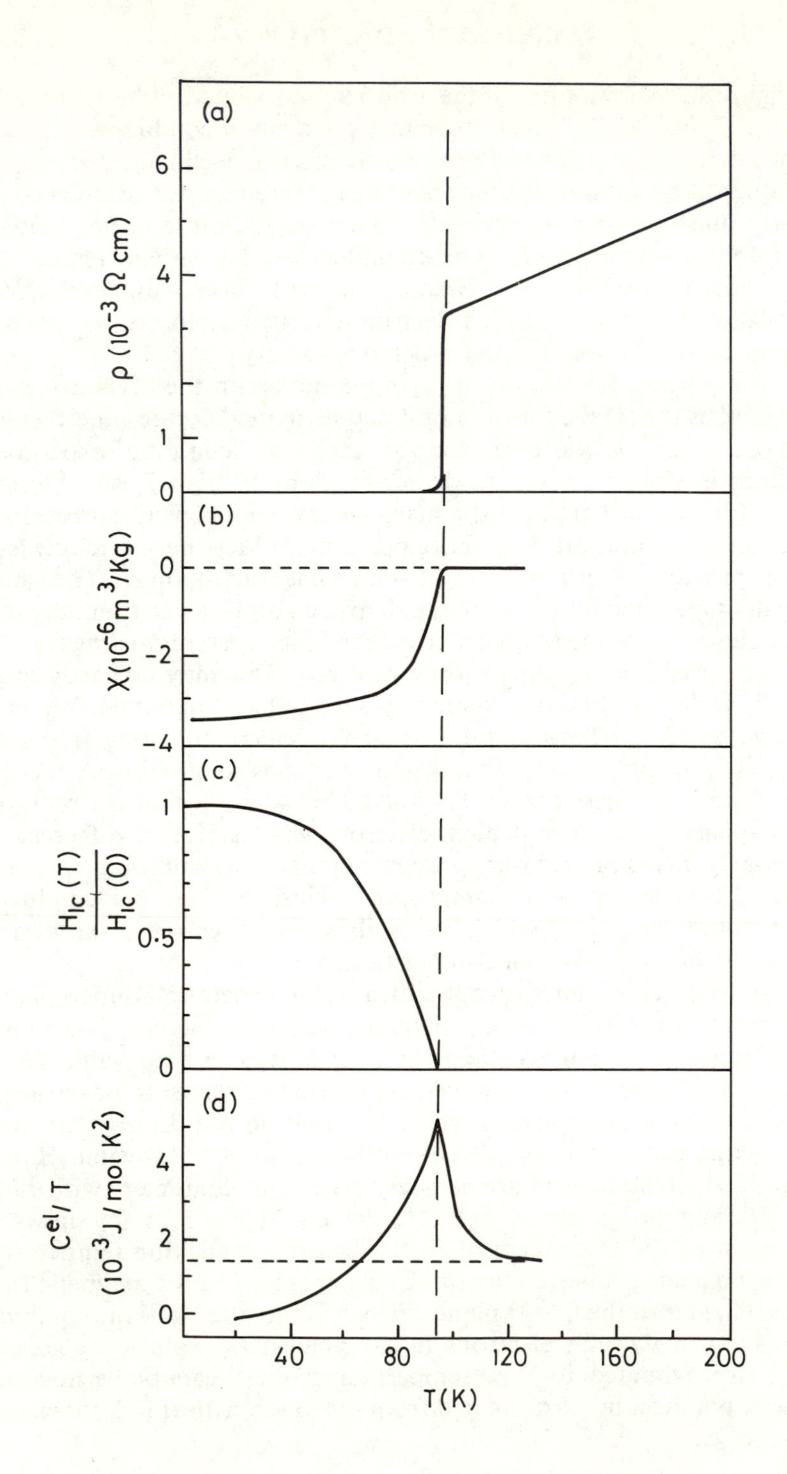

(a)
6
4
1
0
ρ (10⁻³ Ω cm)
(b)
0
-2
-4
X(10⁻⁶ m³/Kg)
(c)
1
0·5
0
H_Ic (T) / H_Ic (O)
(d)
4
2
0
C^el / T
(10⁻³ J / mol K²)
40
80
120
160
200
T(K)

Magnetic fields penetrating a superconducting sample, or passing through a superconducting ring, show the phenomeon of **flux quantization**. This is exploited in superconducting quantum-interference devices (SQUIDs) used in sensitive magnetometers. SQUID-type measurements on high T_c compounds have established that, as in other superconductors, the magnetic flux is quantized in units of $\hbar/2e$. This implies that superconducting current is carried by *pairs* of carriers with a charge $\pm 2e$. In most previously known superconductors, the 'Cooper pairs' are highly extended, giving a large **coherence length** for the superconducting wavefunction. Coherence lengths (ξ) may be estimated in various ways, for example from the upper critical field according to

$$H_{c2} = \Phi_0/(2\pi\mu_0\xi^2) \tag{5.18}$$

where Φ_0 is the flux quantum $\hbar/2e$. Thus high H_{c2} values are associated with abnormally small coherence lengths in high-T_c superconductors: around 1 nm (10 Å) compared with 100 nm found for superconducting Pb. Coherence lengths are also highly anisotropic, being much smaller in directions perpendicular to the Cu–O planes. Even within the planes, the values correspond to only a few atomic diameters, and suggest that the interactions responsible for the pairing of carriers have a much shorter range than the long-wavelength phonons involved in the conventional BCS mechanism (see below).

Small coherence lengths are partly responsible for the disappointing technological properties of the high-T_c oxides, as the superconducting wavefunction decays rapidly at the surface, making good superconducting contacts between grains of a ceramic material difficult. Combined with severe processing problems, this difficulty seems to limit the superconducting current that can be carried.

The weak attractive interactions that lead to carrier pairing in a superconductor show their influence most strongly on electrons close to the Fermi level, giving rise to a **gap** in the one-electron density of states as shown in Fig. 5.29. This is rather similar to the gaps caused by some types of static lattice distortion and discussed in Section 5.3, although its effect on the conductivity is very different! The change predicted by the BCS theory is shown in Fig 5.29 (b). In this theory the gap parameter Δ should be related to T_c according to

---

Fig. 5.28 Properties associated with superconductivity in $YBa_2Cu_3O_7$: (a) resistivity showing sharp drop to zero at $T_c \sim 95$ K; (b) magnetic susceptibility illustrating the Meissner effect; (c) lower critical field H_{c1} normalized to low-temperature value;[90] (d) anomaly in the conduction-electron specific heat, plotted as C^{el}/T.[91]

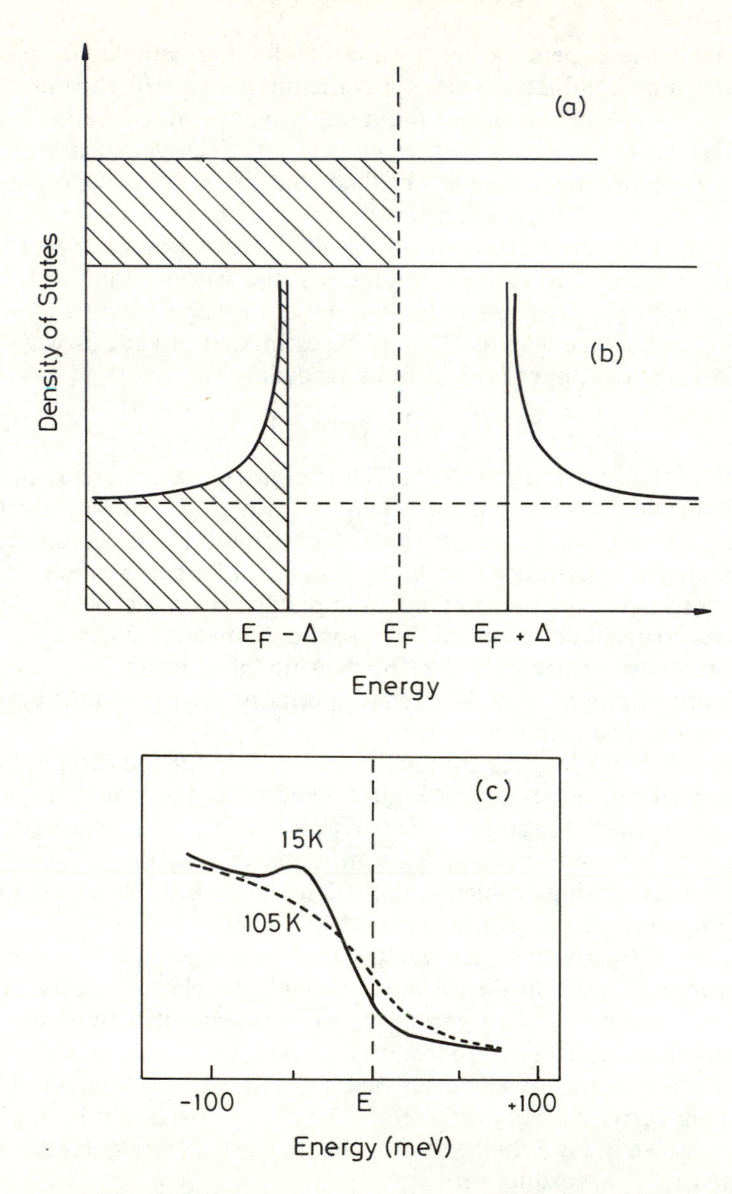

FIG. 5.29 Density of states close to the Fermi level, showing the predicted change between (a) normal and (b) superconducting states at $T = 0$ (with filled levels shaded); (c) photoelectron spectra of '2:2:1:2' Bi–Sr–Ca–Cu–O superconductor above and below T_c, showing the formation of a gap.[92]

$$2\Delta/kT_c \sim 3.5 \tag{5.19}$$

One manifestation of the formation of a gap is a strong anomaly in the electronic specific heat, as illustrated in Fig. 5.28 (d). The peak close to T_c is associated with the decrease in energy of electrons close to the Fermi level as the gap forms. At lower temperature the specific heat falls below the value expected for a normal metal, because a minimum energy 2Δ is now required to excite electrons. The magnitude of the anomaly allows the density of carrier pairs to be estimated. In compounds such as 1:2:3 this is around 10^{21} per cm^3, a significant fraction (more than one tenth) of the total carrier density.[91] Again, this is exceptional, as in most superconductors only a very small fraction of carriers close to the Fermi level is paired. Specific-heat measurements on compounds in their normal state can be used to estimate the density of states at the Fermi level, as discussed earlier (Section 5.1.1). The values found in many compounds appear to be considerably larger than predicted by band theory, showing that electron correlation is important.

Superconducting gaps in high-T_c compounds have been measured by a variety of techniques. Because the value of Δ is larger than in conventional superconductors, it is possible to observe the change in the density of states close to E_F directly by photoemission.[92] Figure 5.29 (c) shows the result of measurements above and below the transition temperature (88 K) for a 2:2:1:2 bismuth compound. When the predictions are corrected for both instrumental and thermal broadening, the low-temperature spectrum is consistent with $\Delta \sim 30$ meV. Far-IR spectroscopy and tunnelling experiments with junctions can also be used. Values obtained with different techniques are not entirely consistent. There are difficulties in interpretation of some experiments because the compounds have vibrational frequencies with energies similar to that of the gap. As with other properties, the gaps also appear to be highly anisotropic (for example, depending on the direction of motion of electrons close to the Fermi surface, and on the polarization of IR radiation used for reflectivity measurements). It seems, however, that the in-plane gap values give $2\Delta/kT_c$ around 7, about twice the BCS prediction.

One of the original observations which suggested the phonon mechanism of the BCS theory is the **isotope effect** on T_c. BCS theory predicts that

$$T_c \propto M^{-\alpha} \tag{5.20}$$

where M is the isotopic mass and a is an exponent equal to 0.5 in the ideal theory. Values rather less than this are often found in practice, and in some transition metal elements the effect is very small. There was originally some controversy about the isotope effects of high-T_c

superconductors; it seems to be established, however, that significant effects are seen on ^{18}O substitution of $La_{2-x}Sr_xCuO_4$, but very little in the 1:2:3 compound, or others with T_c in the higher temperature ranges. Although this has been taken as evidence that phonons are not involved in the mechanism of high-T_c superconduction, this is not necessarily true, as some theories which go beyond the BCS model do not require an isotope effect.

More direct observations on phonons can be made by Raman and IR spectroscopy. Measurements on 1:2:3 and 2:2:1:2 compounds show that changes occur around T_c in the energies and intensities of a number of phonon modes.[85] Although this may suggest that such phonons play a role in superconductivity, it is also possible that the anomalies are the result of a change in electronic screening accompanying the formation of a superconducting gap, and therefore show nothing about the mechanism involved.

5.4.2 Electronic structure of doped copper oxides

Before a convincing mechanism for high-T_c superconductivity can be established, it is essential to understand the electronic levels in copper oxides, and in particular the nature of the charge carriers introduced by doping. Undoped Cu^{2+} oxides are magnetic insulators, with antiferromagnetic ordering of the single spin coming from strong in-plane superexchange. Figure 5.30 shows the important orbitals within a Cu–O plane. The singly occupied orbital in Cu^{2+} is expected to be the $3d_{x^2-y^2}$ according to the energy diagram for tetragonal coordination shown in Fig. 2.2. Superexchange interactions, as discussed in Section 3.4.4, involve the in-plane oxygen $2p_x$ and $2p_y$ orbitals, shown doubly occupied in Fig. 5.30. Magnetic neutron-diffraction experiments have shown that the moment on each copper in the antiferromagnetic materials is about 0.5 μ_B, considerably reduced from the value of 1 μ_B expected from simple arguments. The lower moment results from two effects. Superexchange interactions partly 'delocalize' the unpaired spin, the opposed values for two adjacent coppers cancelling on the shared oxygen. But also the classical picture of the Néel state is rather misleading, and lower moments are expected from quantum 'zero-point' fluctuations in the spin direction. These are particularly important in low-dimensional systems.

Conventional band-theory approaches cannot be expected to give a good description of these materials, and indeed all first-principles local-density band-structure calculations predict that undoped oxides such as La_2CuO_4 should be metallic.[84] It was suggested in Sections 2.3.3 and 3.4.2 that band theory can in principle overcome this difficulty by using a spin-density wave formalism. Unlike the situation in MnO and NiO,

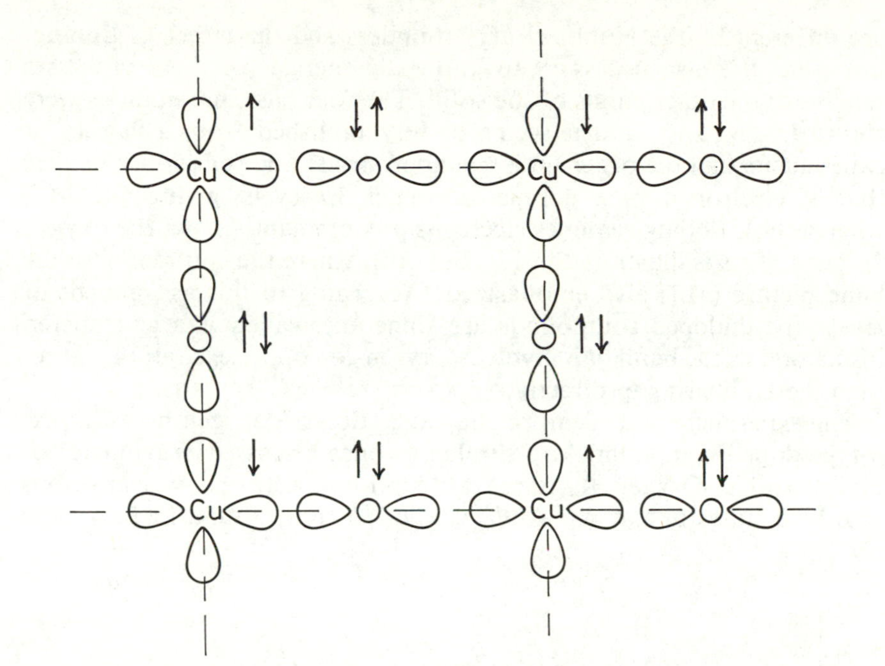

Fig. 5.30 Structure of copper–oxygen plane in high-T_c superconductors, showing the in-plane p_x and p_x oxygen orbitals (filled) and the singly occupied copper $d_{x^2-y^2}$.

however, calculations using the local spin-density formalism do not predict an antiferromagnetic insulating state for copper oxides. But although these calculations are unsatisfactory in many respects, they do confirm some chemical expectations about the Cu^{2+} state. The important levels are found to involve the copper and oxygen orbitals illustrated in Fig. 5.30, with the Fermi level lying in the middle of a band formed from antibonding combinations of these.

It was argued in Section 3.4.2 that the best simple interpretation of magnetic insulators is provided by the Hubbard model, which incorporates the strong Coulomb repulsion of electrons on the same metal atom. Energies associated with the removal and addition of electrons, that is to the processes

$$Cu^{2+} = Cu^{3+} + e^-$$

and

$$Cu^{2+} + e^- = Cu^+$$

are different by the Hubbard U. To understand the effect of doping, however, it is also necessary to know the energies of these processes relative to the other bands of the solid. The four basic possibilities were shown in Fig. 3.15, and it seems to be established from a variety of experiments that the order (c) is the appropriate one for copper oxides: that is, electron doping populates copper $3d$ levels, giving $Cu^{+}3d^{10}$, whereas hole doping removes electrons predominantly from the oxygen $2p$ band. This is illustrated in Fig. 5.31 (b), where the contrast with the band picture (a) is also emphasized. According to this assignment of levels, the undoped compounds are sometimes called 'charge-transfer' insulators, as the band gap involves oxygen-to-copper excitation, rather than the Hubbard gap directly.

The experiments that lead to the suggestion of 'oxygen holes' in the p-type doped compounds are similar to ones discussed previously for nickel oxides. Oxygen K_{α} X-ray absorption spectra show transitions into hole states in $La_{2-x}Sr_{x}CuO_{4}$ similar to those found in $Li_{1-x}Ni_{x}O$

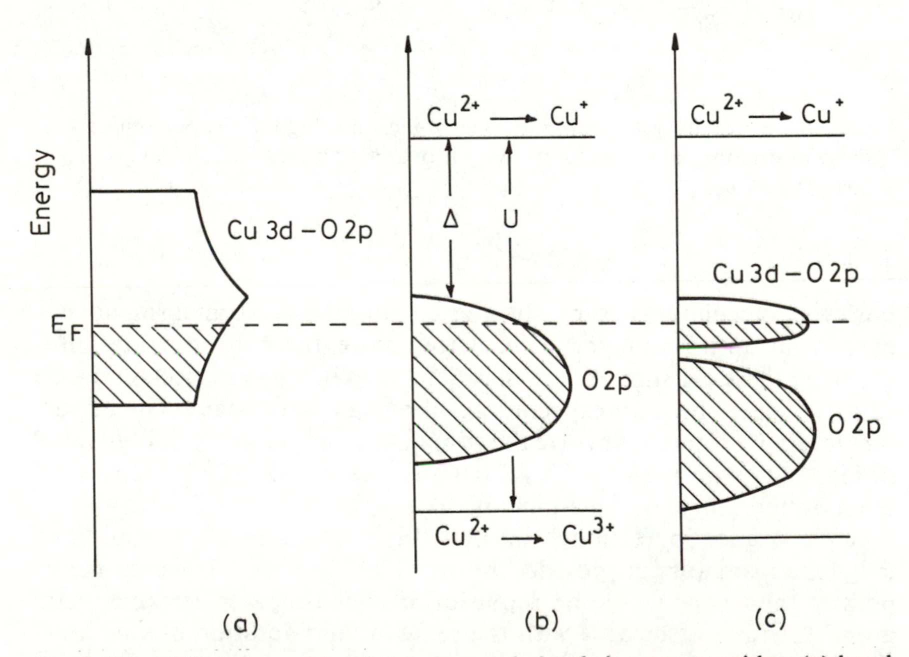

FIG. 5.31 Different models of the electronic levels in copper oxides: (a) band formed by strong mixing between Cu $3d$ and in-plane O $2p$ orbitals; (b) ionic levels with strong correlation, giving a Hubbard gap (U) larger than the charge-transfer gap (Δ); (c) result of Cu $3d$–O $2p$ mixing, to form an 'impurity state' above the main valence band. The position of the Fermi level is appropriate to hole-doped compounds.

(see Section 4.3.1). Corresponding absorptions on Cu core spectra are much weaker. Results such as this should not always be accepted at face value, as the creation of a core hole in X-ray absorption or photoemission experiments produces a strong perturbation which can redistribute valence electrons. Thus the strongest peaks appearing in the XPS of all copper oxides correspond to final states with the $3d^{10}$ configuration, whatever the oxidation state of copper. This comes about because of the strong Coulomb interaction of the core hole with the copper $3d$ electrons. In order to reconstruct the *ground state* electronic structure it is necessary to use model calculations of the cluster or impurity type (see Section 2.2.2 and 2.4.3) in which the important Coulomb interactions are included. Calculations based on PES data suggest Hubbard U values for the copper $3d$ orbitals of around 9 eV, which is much larger than the band gap of about 1.5. eV in the undoped oxides.

Given the predominantly 'oxygen hole' nature of carriers, it is still important to know the orientation of the relevant oxygen $2p$ orbitals with respect to the Cu–O planes. Holes could be formed either in the in-plane p_x and p_y orbitals able to mix directly with the copper $3d_{x^2-y^2}$ (shown in Fig. 5.30), or in ones orthogonal to them. It is tempting to refer to molecular orbital diagrams such as that of Fig. 2.8, which would suggest that the top-filled oxygen orbitals should be *non-bonding* $2p$ combinations. But because of the strong correlation this picture is very misleading. Calculations show that the orbitals in Fig. 5.30 are pushed *up* in energy by interaction with the lower-Hubbard sub-band of copper. This is suggested as the formation of a type of 'impurity state' in Fig. 5.31 (c), and Fig. 5.32 shows a more detailed picture of the state found with a single hole.[93] Located on oxygen $2p$ orbitals surrounding one copper, it is coupled in spin and symmetry to the singly occupied $3d_{x^2-y^2}$ orbital, giving a singlet state $^1A_1$. A number of experiments seem to confirm this model, including polarized X-ray absorption measurements which show in-plane holes, and NMR results which are interpreted in terms of singlet carriers.[94] With this model it is also easy to understand the rapid disappearance of antiferromagnetism on hole doping. In Fig. 5.32 the oxygen hole is shown coupling to a single copper, but of course it will interact with neighbouring ones as well. This gives a strong ferromagnetic contribution to the net coupling between copper spins, probably much larger than the normal antiferromagnetic superexchange. The interactions of carriers with spins are important in some theories of superconductivity, which are discussed in the following section.

The formation of oxygen holes in high-T_c oxides (and in some other compounds such as doped NiO) leads to a dilemma over how to represent the oxidation states in these compounds. Conventional chemical assignments (in the absence of structural features associated with O_2^{2-}

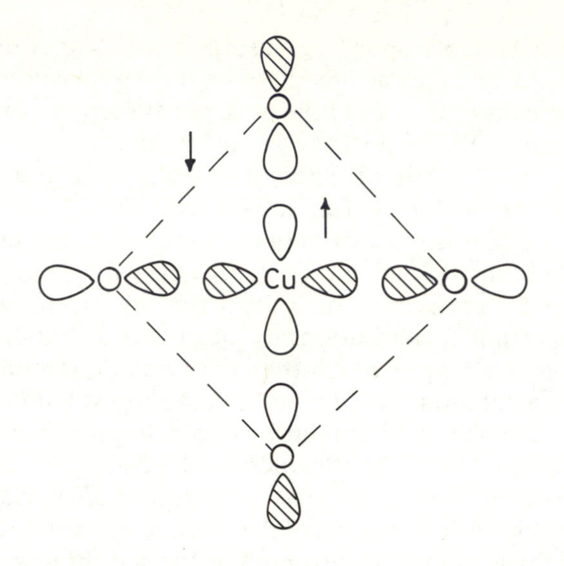

FIG. 5.32 Representation of the singlet state presumed to be formed by strong antiferromagnetic coupling of an oxygen hole with that on Cu^{2+}. The relative signs of the different orbital lobes is indicated by the shading.[93]

groups—see Section 1.2.1) require some Cu^{3+} to be present in hole-doped materials. On the other hand, the physical location of the charge seems to correspond more closely to O^-, with all copper remaining in the Cu^{2+} state. There may be no entirely satisfactory solution to this problem, but one way out is to recognize that chemical oxidation states are determined by certain *formal* rules, and are not intended to indicate a *real* charge distribution: for example one cannot realistically claim that W^{6+} represents the 'real' charge on tungsten atoms in WO_3. In this view, the designation Cu^{3+} remains the most useful way of describing the chemistry of hole-doped copper oxides, and should not be abandoned just because of physical measurements showing oxygen holes.

Although this is a reasonable 'formal' resolution of the difficulty, it does leave unanswered some interesting questions which might be important in understanding high-T_c superconductivity. Why is something resembling O^- found in copper oxides, when normally it is much less stable, either existing as a defect at low concentrations, combining to form peroxide O_2^{2-}, or disproportionating to O^{2-} and O_2? And why does the discrepancy between conventional oxidation states and 'real' charges seem to be more apparent here than in many other situations? These questions are certainly related, and the answers must be connected with the way in which O^- is stabilized by mixing with other configurations. In the models just discussed the ground state would be described as a mixture of ionic states such as [Cu^{2+}–O^-] and [Cu^{3+}–O^{2-}]. The accessibility of 'real' (as opposed to 'formal') Cu^{3+} must therefore be essential in giving O^- a stability that it does not have, for example, with Mg^{2+} or Zn^{2+}. (Even so, one should be a little suspicious of these 'real' charges, as the models do not take explicit account of the Cu 4*s* and 4*p* orbitals which are certainly involved in bonding.) But the position of copper late in the 3*d* series is probably crucial in another way, as the 3*d* valence orbitals are quite contracted and almost 'core-like' in size. The overlap interaction necessary for the mixing of states is therefore quite weak, and significant covalent interaction occurs only because the copper 3*d* and oxygen 2*p* states are similar in energy. Furthermore, the strong Coulomb interactions involved in electronic transitions can strongly perturb this mixing: the final states observed in high-energy X-ray spectroscopies correspond closely to the pure 'unmixed' ionic states. This contrasts with the situation in the early early transition elements, where overlap between metal *d* and oxygen *p* orbitals is much larger. It is therefore much harder to make either an experimental or a theoretical distinction between oxygen-and metal-based orbitals.

5.4.3 *Problems in understanding high-T_c superconductivity*

An enormous variety of 'explanations' of high-T_c superconductivity has been proposed, but at the time of writing there is no sign of an emerging consensus as to which one might be correct. There are also serious disputes about the nature of the normal (non-superconducting) metallic state. The aim of this final section is to summarize some of the important issues, without proposing any definite conclusions.

The discussion above concentrated on the nature of individual carriers, especially the 'oxygen holes' of the *p*-type compounds. The semiconductor-to-metal transition that occurs as the doping level increases is probably similar to that described for some other oxides in Section 4.4, but the nature of the metallic state formed is more controversial. Among the questions that have been asked are: should one describe this state as a Fermi liquid? Is the strongly correlated energy-level structure shown in Fig. 5.31 (c) appropriate in the metallic state? Do the unfilled copper and oxygen levels behave as two separate bands, or do they couple strongly enough to make a single-band description more appropriate?

The term *Fermi liquid* implies a metallic state with a well defined Fermi surface in $\boldsymbol{k}$-space, and single-particle levels which can be described by $E(\boldsymbol{k})$ curves, at least in the vicinity of the Fermi surface. It is a generalization of the independent-electron model discussed in Sections

2.2 and 5.1, in that the appropriate 'single-particle' levels may not be those of individual electrons, but of 'quasi-particles' strongly influenced by electron–electron interactions. There are a number of theorems that seem to guarantee the existence of a Fermi surface in a metal even when strong correlation is present, but their applicability to high-T_c copper oxides has been disputed.[95] One particular difficulty associated with holes moving in an antiferromagnetically coupled spin system is illustrated in Fig. 5.33. When the hole moves from site to site by hopping between near neighbours, it leaves a 'trail' of reversed spins behind. This picture may be highly oversimplified, as it treats the holes as if they were on copper rather than oxygen (but see below), and also neglects the possibility of motion between next-nearest neighbours where the spins are parallel. Also one might dispute its applicability to the metallic state where long-range antiferromagnetic order no longer exists: against this

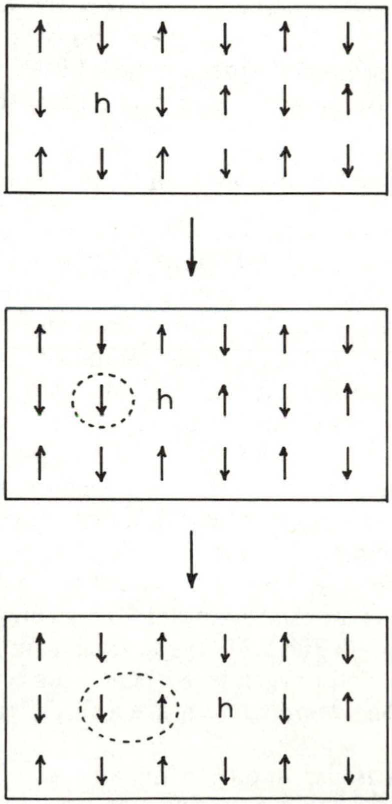

FIG. 5.33 Showing how the motion of a hole (h) by nearest-neighbour hopping in an antiferromagnetic array leaves behind a 'trail' of reversed spins (ringed).

argument, however, is the fact that *short-range* antiferromagnetic correlations seem to persist into the superconducting regime. But in spite of its shortcomings, the model suggests why it may be impossible to separate the motion of the carriers from the spin degrees of freedom, and that the very notion of a 'band structure' for carriers has some serious difficulties.

Experimentally, it is established from photoelectron spectra of the normal state, such as that of the 2:2:1:2 compound at 105 K shown in Fig. 5.29, that a sharp Fermi *energy* exists the metallic copper oxides. Whether there is a true Fermi *surface* in **k**-space is harder to say. Angular resolved photoelectron measurements give peaks which seem to disperse with wavevector and thus suggest that well defined $E(\boldsymbol{k})$ excitations exist.[96] But an alternative interpretation is that such peaks merely show the maximum in a largely *incoherent* spectrum, and not true $E(\boldsymbol{k})$ curves.

Implicit in the picture of Fig. 5.33 is that strong correlation remains important in the metallic state, preventing double occupancy of copper sites. Some theories have supposed that screening by metallic electrons reduces or even removes the correlation gap. Spectroscopic measurements on doped compounds do not support this idea, but rather show that the correlated energy levels shown in Fig. 5.31 persist. One must be careful, however, to distinguish between the screening mechanisms important in different situations. The models used to interpret high-energy spectroscopies give values for the Hubbard U and the charge-transfer gap that exclude near-neighbour screening interactions such as metal–oxygen covalency. They are appropriate values to use in theories which incorporate such interactions in an explicit way. It is quite understandable that any additional effect of metallic screening is unlikely to show up in these parameters, given the very short effective time-scale of the experiments, and the nature of the model used to interpret them. Some theories, on the other hand, use the concept of a correlation gap in a quite different way, and include *implicitly* the effects of all possible screening mechanisms. It is plausible that such a gap might be significantly reduced in the metallic state. If this happens, strong charge fluctuations between copper and oxygen could occur, although on a much slower time-scale than that probed by spectroscopy.

Another problem is whether it is better to describe the metallic state in terms of two bands or one band. It has been proposed that oxygen holes form a band largely independent of the copper levels, so that there are two weakly coupled systems: the charge carriers on oxygen, and the spins on the copper. But the theories developed to describe isolated holes suggest something quite different. According to the model shown in Fig. 5.32 the overlap interaction between oxygen and copper orbitals gives a very strong antiferromagnetic coupling between the spins on oxygen- and copper-based electrons. The state illustrated has the spin and symmetry of a low-spin Cu^{3+}ion, and in some ways has the same effect as a hole in the singly occupied copper 3*d* orbital. This idea forms the basis for models which describe spins and holes in a *single* band. One such is the ***t–J* model**, where *t* is the effective transfer integral for the motion of a hole, and *J* the exchange parameter describing the anti-ferromagnetic interaction of spins.[97]

Given the problems just discussed, it is hardly surprising that very divergent views have been expressed concerning the most appropriate theory for describing high-T_c superconductivity. Whatever the interaction involved, most theories of superconductivity require two related phenomena.

1. There is an attractive interaction between carriers giving rise to the formation of *pairs*. Normally the spins in a pair are assumed to be coupled to form a singlet, but 'triplet pairing' is also possible.

2. Electron pairs behave as bosons and undergo a type of Bose–Einstein condensation, forming a macroscopically coherent state similar to that occurring in superfluid liquid helium.

In BCS and similar theories these two processes are inseparable. The effective size of 'Cooper pairs' is the coherence length discussed in Section 5.4.1. In most superconductors this is very large, so that when pairs form, the strong overlap of their wavefunctions is sufficient to guarantee that Bose–Einstein condensation happens at the same time: indeed the coherence of the condensed-state wavefunction itself assists the formation of pairs. These theories predict a superconducting transition temperature given by[14]

$$kT_c \sim \hbar\omega \exp(-1/\lambda) \tag{5.21}$$

where $\hbar\omega$ is the excitation energy associated with the interactions that cause pairing, and λ an effective 'coupling strength'.

In the traditional version of the BCS theory, the interaction is provided by long-wavelength acoustic phonons: one can think of one electron polarizing the lattice, leaving a transient 'energy trough' which attracts a second electron. Then, in eqn (5.21), ω is the Debye frequency appropriate to acoustic phonons, and λ is related to an electron–phonon coupling parameter V_{ep}

$$\lambda = V_{ep}N(\mathrm{E_F}). \tag{5.22}$$

The isotope effect in this theory arises because

$$\omega \propto M^{-1/2}. \tag{5.23}$$

However, this prediction is modified by more elaborate theories, which go beyond the 'weak-coupling' approximation and also incorporate the effect of Coulomb repulsion which opposes the pairing. A small or even negligible isotope effect is not incompatible with a phonon mechanism for superconduction. The short coherence length in high-T_c compounds suggests that long-wavelength phonons cannot be involved here, but a BCS-type theory could incorporate other types of interaction, for example metal–oxygen (optic) vibrations, or electronic polarization. Various other observations seem to contradict the simplest BCS predictions: for example the relation between the gap magnitude and T_c discussed earlier. As with the isotope effect, however, these could be consistent with modified 'strong-coupling' versions of the theory.

If the interaction that forms pairs is sufficiently strong, a different type of strong-coupling theory may be appropriate. In *bipolaron theories*, carrier pairs ('bipolarons') are present even in the normal metallic state.[89] The superconducting transition is then associated with the Bose–Einstein condensation of the pairs. The temperature at which this happens is given by

$$kT_c = 3.3\,\hbar^2 n^{2/3}/m^* \qquad (5.24)$$

where n is the density of pairs, and m^* their effective mass. The variation of T_c with coupling strength is quite different in the two theories: in the BCS limit T_c increases strongly with coupling, but in bipolaron theories a stronger coupling provides a tighter binding of the pairs and increases their effective mass, so that T_c is *reduced*.

Bipolaron theories are sometimes described in terms of pairing in 'real space', as opposed to $\boldsymbol{k}$-space for BCS theories. This distinction emphasizes that the physical proximity of the carriers making up the pair is important in bipolaron formation. In the BCS model on the other hand pairing takes place between carriers close to the Fermi surface and with opposite wavevectors, their location in real space being of secondary importance. Bipolaron models are an attractive alternative for high-T_c superconductors because of the short coherence lengths found experimentally, but it is not clear how far other observations are compatible with these theories. Specific-heat measurements (see Fig. 5.28 (d)) show that anomalies extend somewhat above T_c, suggesting that some carrier pairs may be forming in the normal state. But the magnitude of the anomalies indicates that only a fraction of carriers (perhaps ten per cent) can be involved in pair formation. Other properties such as the superconducting gap seem to be more in line with BCS-type predictions than with bipolaron theories. The distinction between these two models is not absolute, however: they can be thought of as opposite limits, with the real situation lying perhaps somewhere between. It is even possible that a series of compounds, in which the carrier concentration or some other parameter is varied, may pass from one limit to the other without any change in the microscopic mechanism involved in superconductivity.

Arguments concerning 'strong' or 'weak' coupling are important as they determine the *formal* nature of a theory of superconductivity, but they do not directly say anything about the type of interaction responsible for pairing the carriers. Three different phenomena that could be involved are:

(1) interaction through phonons: i.e. by lattice polarization;

(2) interaction through 'excitons', or electronic polarization;

(3) interaction through the unpaired copper spins.

Although many theories concentrate on one of these factors at the expense of others, they are not necessarily mutually exclusive. Indeed one of the difficulties here is that plausible arguments exist for the importance of all three types of interaction in copper oxides. The role played by lattice polarization, giving rise to polarons and other effects, has been emphasized in many previous sections. It can be argued that some polaronic influence is inevitable because of the small spatial extent of the pairs, even if it is not basically responsible for pair formation. On the other hand, the copper oxide planes have a high electronic polarizability, because of the relatively facile excitation of electrons between oxygen and copper. And finally, any complete theory must take account of the localized spins which are certainly present in the undoped compounds, and which may persist in some form in the superconducting regime.

As already emphasized, the short coherence lengths rule out the involvement of the long-wavelength acoustic phonons important in the traditional BCS theory. Many suggestions have been made, however, about different types of short-wavelength mode that could couple strongly with electronic carriers. A potential difficulty with all these theories is that a sufficiently strong electron–phonon coupling will generally cause static distortions which can localize the carriers, rather than provide a dynamic pairing mechanism. For example, the bipolarons present in Ti_4O_7 (discussed in Section 5.3.3) lead to activated conduction, and 'freeze out' into an ordered static structure at low temperature. It has been suggested that the dielectric properties of copper oxides may have some unusual features that allow the interaction with optic phonons to form 'large' rather than 'small' bipolarons.

One of the original motivations for studying copper oxides was the importance of the Jahn–Teller effect in these compounds. The fact that strong *static* distortions are found with Cu^{2+} seems to argue against their involvement in dynamic processes. But it is possible that the electronic energy differences associated with tetragonal distortions are quite small, and in fact some X-ray absorption measurements on *p*-type compounds suggest that a proportion of holes may occupy the copper $3d_{z^2}$ levels. If this is so, the Jahn–Teller active vibrations could provide some kind of coupling mechanism. Another idea that is chemically attractive suggests that individual oxygen holes might be only marginally stable with respect to a dimerization to peroxide O_2^{2-}. Unusual vibrations in the near-neighbour O–O distances could give a dynamic pairing of holes, as opposed to the static pairing normally found in peroxides.

The idea of replacing the phonons of the BCS theory by excitons has a history which long predates the discovery of superconducting copper oxides. The motivation comes from looking at the $\hbar\omega$ prefactor in the BCS equation (5.21). A much higher T_c could be predicted if this term represents an electronic rather than a vibrational energy quantum. As a guide to the prediction of new superconductors this idea has never been fruitful, but the discovery of high-T_c superconductors revived suggestions of an excitonic mechanism, where carrier pairing is produced by electronic instead of lattice polarization. Various electronic modes have been proposed, including *d–d* types (which might also involve Jahn–Teller vibrations), copper–oxygen charge transfer, and plasma excitations which could have some unusual features because of the two-dimensional electronic structure.

A rather different type of theory is that which starts from the idea of a 'collapse' of the correlation gap when metallic screening is taken into account. Significant charge fluctuations could then take place, either between copper and oxygen, or of the type

$$2Cu^{2+} = Cu^{+} + Cu^{3+}.$$

In silver compounds such as AgO this process leads to a chemical disproportionation of Ag^{2+} (see Sections 3.2 and 5.3.3). Although Cu^{2+} does not show static disproportionation, it may be that the energetics are finely balanced, and that dynamic fluctuations of this type could occur. An alternative idea is that there is a fine balance between a localized, correlated state, and an itinerant state in which correlation is less important.[87] If such a balance occurs, it may be tipped in one direction or the other by small changes in lattice spacing. Short-wavelength acoustic phonons, corresponding to local dilatations and compressions of the lattice, could then interact with carriers and form a pairing mechanism.

Even more radical theories of pair formation in high-T_c superconductors are those which involve interactions between mobile carriers and localized spins. This type of interaction has already been discussed in Section 5.2.2, in connection with the double-exchange mechanism for ferromagnetic coupling in compounds such as Fe_3O_4 and

$La_{1-x}Pb_xMnO_3$. The double-exchange process depends on the exchange interaction between unpaired electrons on the same ion, and cannot operate in compounds based on Cu^{2+} where there is only one unpaired electron. But carriers undoubtedly do couple strongly with the spins in these compounds, as is shown for example by the rapid disappearance of long-range magnetic ordering as doping is introduced. One approach to this problem is to think of carriers moving in the oxygen 2*p* band and having a strong antiferromagnetic coupling with neighbouring coppers. This will induce an effectively ferromagnetic coupling between Cu^{2+} spins in the vicinity of the hole. At small doping levels the existing antiferromagnetic order will provide a localizing influence on the carriers. It is likely that they will form *spin polarons*, each carrier being surrounded by a small region of ferromagnetically coupled spins. There have been suggestions that spin polarons have an attractive interaction at short range, so forming bipolarons which could be the carrier pairs responsible for superconductivity.

Some other spin-based models are based on the one-band picture, assuming that the coupling between oxygen holes and copper spins is so strong that they cannot be separated. As suggested above, this coupling may generate a state with the appropriate spin and symmetry to behave a a 'copper hole', that is a low-spin (singlet) Cu^{3+}ion. This picture was used in Fig. 5.33, where it was shown how the motion of a carrier interferes with the antiferromagnetic ordering of spins. The 'trail' of reversed spins gives an increase of energy, and provides a possible mechanism for carrier pairing: if a second carrier follows in the 'wake' of the first it may restore the low-energy antiferromagnetic state.

Theories based on spin interactions have many attractive features. They suggest that high-T_c superconductivity is a consequence of doping a system with antiferromagnetically coupled spin-$\frac{1}{2}$ ions, and thus provide some 'explanation' of why copper may be peculiar. Some of these theories can also account naturally for the approximate symmetry found between hole and electron doping, as Cu^+ ions also give 'spin-zero' carriers. At the same time, this type of theory emphasizes the delicate balance between the different interactions involved. If super-exchange is sufficiently weak, introducing carriers will lead to a ferro-magnetic state where they can move without the problems of interference shown in Fig. 5.33. On the other hand, a very strong superexchange interaction would maintain the long-range antiferromagnetic order and tend to keep carriers localized. If the pairing of mobile carriers does arise through the spins, the relative magnitudes of superexchange and the carrier band width are crucial.

It must be clear that one of the difficulties with high-T_c superconductivity is not the absence of a possible theory, but rather the embarrassing superabundance of candidates. Many of the ideas discussed above are no more than plausible suggestions, and in order to elevate one of them into a convincing theory and to discount the others, progress is necessary in two directions. In the first place, it is essential to have a better idea of the parameters which govern the strengths of the different possible interactions. When this is achieved, the parameters need to be put together into a properly formulated many-body theory which does not neglect anything important. One motivation for this endeavour is to understand how to optimize the properties of known compounds, and possibly even to predict and synthesize new superconductors. But there is another reason why work in this area is important. The reader who has followed this book from the beginning will realize that although the properties of copper oxide superconductors are unusual, there is nothing very peculiar about the difficulties in understanding them. The complex range of chemical, structural, and physical interactions discussed in this section has also formed a common theme through much of the book. Even if the applications of high-T_c superconductors should prove ultimately disappointing, one of the lasting benefits of the immense effort put into studying them may be a better general understanding of transition metal oxides.

References

1. Kittel, C. (1974). *Introduction to solid state physics*, (5th edn). Wiley, New York.
2. Cox, P.A. (1987). *The electronic structure and chemistry of solids*. Oxford University Press.
3. Goodenough, J.B. (1971). *Prog. Solid State Chem.*, **5**, 145.
4. Mattheis, L.F. (1969). *Phys. Rev.*, **181**, 987; (1970). Ibid. B, **2**, 3918.
5. Marcus, S.M. (1968) *Phys. Lett.*, **27A**, 584.
6. Ferretti, A., Rodgers, D.B., and Goodenough, J.B. (1965). *J. Phys. Chem. Solids*, **26**, 2007.
7. Dickens P.G. and Whittingham, M.S. (1968). *Quart. Rev. Chem. Soc.*, **22**, 30.
8. Bullett, D.W. (1983). *J. Phys. C: Solid State Phys.*, **16**, 2197.
9. Tunstall, D.P. and Ramage, W. (1980). *J. Phys. C: Solid State Phys.*, **13**, 725.
10. Hill, M.D. and Egdell, R.G. (1983). *J. Phys. C: Solid State Phys.*, **16**, 6205.
11. Chazalviel, J.N., Campagna, M., Wertheim, G.K., and Shanks, H.R. (1978). *Phys. Rev.* B, **16**, 697.
12. Vest, R.M., Griffel, M., and Smith, J.F. (1958). *J. Chem. Phys.*, **28**, 293.
13. Greiner, J.D., Shanks, H.R., and Wallace, D.C. (1962). *J. Chem. Phys.*, **36**, 772.
14. Ziman, J.M. (1972). *Principles of the theory of solids*, (2nd edn). Cambridge University Press.

15. Zumsteg, F.C. (1976). *Phys. Rev.* B, **14**, 1406.
16. Muhlestein, L.D. and Danielson, G.C. (1967) *Phys. Rev.*, **158**, 825; Ibid. **160**, 562.
17. Wiseman, P.J. and Dickens, P.G. (1976). *J. Solid State Chem.*, **17**, 91.
18. Feinleib, J., Scouler, W.J., and Ferretti, A. (1968). *Phys. Rev.*, **165**, 765.
19. Ibach, H. and Mills, D.L. (1982). *Electron energy loss spectroscopy and surface vibrations*. Academic, New York.
20. Owen, J.F., Teegarden, K.J., and Shanks, H.R. (1978). *Phys. Rev.* B, **18**, 3827.
21. Wertheim, G.K., Mattheis, L.F., Campagna, M., and Pearsall, T.P. (1974). *Phys. Rev. Lett.*, **32**, 997.
22. Wertheim, G.K., Campagna, M., Chalzaviel, J.N., and Shanks, H.R (1976). *Chem. Phys. Lett.*, **44**, 50.
23. Cox, P.A., Goodenough, J.B., Tavener, P., Telles, D., and Egdell, R.G. (1986). *J. Solid State Chem.*, **62**, 360.
24. Beatham, N., Cox, P.A., Egdell, R.G., and Orchard, A.F. (1981). *Chem. Phys. Lett.*, **69**, 479.
25. de Angelis, B.A. and Schiavello, M. (1976). *Chem. Phys. Lett.*, **58**, 249.
26. Kotz, R., Lewerenz, H.J., and Stucki, S. (1983). *J. Electrochem. Soc.*, **130**, 825.
27. Campagna, M., Wertheim, G.K., Shanks, H.R, Zumsteg, F., and Banks, E. (1975). *Phys. Rev. Lett.*, **34**, 738.
28. Kotani, A. and Toyazawa, Y. (1974). *J. Phys. Soc. Japan*, **37**, 912.
29. Steiner, P., Hochst, H., and Hufner, S. (1979). In *Photoemission in solids II* (ed. L. Ley and M. Cardona), Topics in Applied Physics, Vol. 27. Springer, Berlin.
30. Kemp, J.P., Beal, D.J., and Cox, P.A. (1990). *J. Phys.: Condensed Matter*, **2**, 3767.
31. Cox, P.A., Egdell, R.G., and Orchard, A.F. (1980). In *Mixed-valence compounds*, (ed. Brown D.B.), NATO ASI Series. Reidel, D. Dordrecht.
32. Cox, P.A. (1981). In *Proc. X-80 Conf. on inner shell and X-ray physics of atoms and solids*, (ed. D. Fabian). Plenum, New York.
33. Goodenough, J.B., Mott, N.F., Pouchard, M., Demazeau, G., and Hagenmuller, P. (1973). *Mater. Res. Bull.*, **8**, 547.
34. Kemp J.P. and Cox, P.A (1990). *Solid State Commun.*, **75**, 731.
35. Figgis, B.N. (1966). *Introduction to ligand fields*. Wiley, New York.
36. Cox, P.A., Egdell, R.G., Goodenough, J.B., Hamnett, A., and Naish, C.C. (1983). *J. Phys. C: Solid State Phys.*, **16**, 6221.
37. Schwarz, K.H. (1986). *J. Phys. F: Met. Phys.*, **16**, L221.
38. Kuhota, B. (1960). *J. Phys. Soc. Japan*, **15**, 1706.
39. Zener, C. (1951). *Phys. Rev.*, **81**, 440.
40. Blakemore, R. (1975). *Science*, **190**, 377.
41. Anderson, P.W. and Hasegawa, H. (1955). *Phys. Rev.*, **100**, 675.
42. Cox, P.A. (1980). *Chem. Phys. Lett.*, **69**, 340.
43. Goodenough, J.B. (1963). *Magnetism and the chemical bond*. Wiley, New York.
44. Goodenough, J.B. and Longo, J.M. (1970). *Crystallographic and Magne-*

tic properties of perovskites and related compounds, In *Magnetic and other properties of oxides*, (ed. K.-H. Hellwege). Landolt-Bornstein New Series III, Vol. 4. Springer, Berlin.
45. Searle, C.W. and Wang, S.T. (1970). *Can. J. Phys.*, **48**, 2023.
46. Mott, N.F. (1974). *Metal–insulator transitions*. Taylor and Francis, London.
47. Fukuyama, H., Maekawa, S., and Malezemoff A.P. (eds) (1989). *Strong correlation and superconductivity*, Springer Series in Solid State Science, Vol. 89. Springer, Berlin.
48. McWhan, B.D. and Remeika, J.P. (1970). *Phys. Rev.* B, **2**, 3734; McWhan, D.B., Menth, A., Remeika, J.P., Brinkman, W.F., and Rice, T.M. (1973). *Phys. Rev.* B, **7**, 1920.
49. Ganguly, P., Parkash, D., and Rao, C.N.R. (1976). *Phys. Status Solidi* (a), **36**, 669; McLean D.A. and Greedon, J.E. (1981). *Inorg. Chem.*, **20**, 1025.
50. Brinkman, W.F. and Rice, T.M. (1970). *Phys. Rev.* B, **2**, 4302.
51. Ram, R.A.M., Ganapathi, L., Ganguley P., and Rao, C.N.R. (1986). *J. Solid State Chem.*, **63**, 139.
52. Ram, R.A.M., Ganguley P., and Rao, C.N.R. (1987). *J. Solid State Chem.*, **70**, 82.
53. Wilson, A.J. (1972). *Adv. Phys.* **21**, 143.
54. Zaanen, J., Sawatzky, G.A., and Allen, J.W. (1985). *Phys. Rev. Lett.*, **55**, 418.
55. Schlenker, C., Dumas, J., Escribe-Filippini, E., Guyot, H., Marcus, J., and Fourcaudot, G. (1985). *Phil. Mag.* B, **52**, 643.
56. Fogle, W. and Perlstein, J.H. (1972). *Phys. Rev.* B, **6**, 1402.
57. Travaligni, G., Wachter, P., Marcus, J., and Schlenker, C. (1981). *Solid State Commun.*, **37**, 599.
58. Pouget, J.P., Kagoshima, S., Schlenker, C., and Marcus, J. (1983). *J. Phys. Paris*, **44**, L113.
59. Dumas, J., Schlenker, C., Marcus, J., and Bader, R. (1983). *Phys. Rev. Lett.*, **50**, 757.
60. Hamnett, A. and Goodenough, J.B. (1984). *Binary transition metal oxides*. In *Physics of non-tetrahedrally bonded compounds*, Landolt-Bornstein New Series III, Vol. 17g. Springer, Berlin.
61. Rice, C.E. and Robinson, W.R. (1977). *Acta Cryst.* B, **33**, 1342.
62. Ashkenazi, J. and Chuchem, T. (1975). *Phil. Mag.*, **32**, 763.
63. Cox, P.A. and Burton, A.D. (1985). *Phil. Mag.* B, **51**, 255.
64. Beatham, N., Fragala, I.L., Orchard, A.F., and Thornton, G. (1980). *J. Chem. Soc. Faraday Trans. II*, **76**, 929.
65. Brown, P.J. and Ziebeck, K.R.A. (1978). *J. Phys. C: Solid State Phys.*, **10**, 2791.
66. Fabritchinyi, P.B., Bayard, M., Pouchard, M., and Hagenmuller, P. (1974). *Solid State Commun.*, **14**, 603.
67. Gupta, M., Freeman, A.J., and Ellis, D. (1977). *Phys. Rev.* B, **16**, 3338.
68. Beatham, N. and Orchard, A.F. (1979). *J. Electron Spectrosc. Relat. Phenom.*, **26**, 77.
69. Robin, M. and Day, P. (1967). *Adv. Inorg. Chem. Radiochem.*, **10**, 247.

70. Gleitzer, C. and Goodenough, J.B. (1985) *Structure and Bonding*, **61**, 1.
71. Miles, P.A., Westphal, W.B., and von Hippel, A. (1957) *Rev. Mod. Phys.*, **29**, 279.
72. Greenwood, N.N. and Gibb, T.C. (1971). *Mössbauer spectroscopy*. Chapman and Hall, London.
73. Umemura, S. and Iida, S. (1979). *J. Phys. Soc. Japan*, **45**, 458.
74. Iizumi, M., Koetzle, T.F., Shirane, G., Chikazumi, S., Matsui, M., and Todo, S. (1982). *Acta Cryst.* B, **38**, 2121.
75. Care, C.M. and March, N.H. (1975). *Adv. Phys.*, **24**, 101.
76. Lakkis, S., Schlenker, C., Chakravaty, B.K., Ruder, R., and Marezio, M. (1979). *Phys. Rev.* B, **14**, 1429.
77. Takeda, Y., Naka, S., Takano, M., Shinjo, T., Takada, T., and Shimada, M. (1978). *Mater. Res. Bull.*, **13**, 61.
78. Mott, N.F., Davis, E.A., and Street, R.A. (1975). *Phil. Mag.*, **32**, 961.
79. Battle, P.D., Gibb, T.C., and Nixon, S. (1988). *J. Solid State Chem.*, **77**, 124.
80. Battle, P.D., Gibb, T.C., Lightfoot, P., and Matsuo, M. (1990). *J. Solid State Chem.*, **85**, 38.
81. Bednorz, J.G. and Muller, K.A. (1988). *Angew. Chem. Int. Ed.*, **100**, 757.
82. Hatfield, W.E. and Miller J.H. (eds) (1988) *High temperature superconducting materials*. Marcel Dekker, New York.
83. Bednorz, J.G. and Muller K.A. (eds) (1990). *Superconductivity*, Springer Series in Solid State Science, Vol. 90. Springer, Berlin.
84. Picket, W.E. (1989). *Rev. Mod. Phys.*, **61**, 433.
85. Kuzmany, K., Mehring, M., and Fink J. (eds) (1990). *Electronic properties of high-*T_c *superconductors and related compounds*, Springer Series in Solid State Science, Vol. 99. Springer, Berlin.
86. Cava, R.J., Batlogg, B., Rabe, K.M., Rietman, E.A., Gallagher, P.K., and Rupp, L.W. (1988). *Physica* C, **156**, 523.
87. Goodenough, J.B. (1990) *Supercond. Sci. Technol.*, **3**, 26.
88. Bedell, K., Coffey, D., Maltzer, D., Pines, D., and Schrieffer J.R. (eds) (1990). *High temperature superconductivity—the Los Alamos Symposium 1989*. Addison-Wesley, Redwood City, Calif.
89. Mott, N.F. (1990). *Adv. Phys.*, **39**, 55.
90. Batlogg, B. (1990). In ref. 85, p. 1 and ref. 88, p. 37.
91. Loram, J.W. and Mirza, K.A. (1990). In ref. 85, p. 92.
92. Imer, J.M., Patthey, F., Dardel, B., Schneider, W.D., Baer, Y., Petroff, Y., and Zettl, A. (1989). *Phys. Rev. Lett.*, **63**, 102.
93. Zhang, F.C. and Rice, T.M. (1988). *Phys. Rev.* B, **37**, 3754; Eskes H. and Sawatzky, G.A. (1990). In ref. 85, p. 127.
94. Rice, T.M., Mila, F., and Zhang, F.C. (1991). *Phil, Trans. Roy. Soc. Lond.* A, **334**, 459.
95. Anderson, P.W. (1991). *Phil. Trans. Roy. Soc. Lond.* A, **334**, 473.
96. Claessen, R., Buslaps, T., Fink, J., Mante, G., Manzke, R., and Skibowski, M. (1990). In ref. 85, p. 120.
97. Zhang, F.C. and Rice, T.M. (1990). *Phys. Rev.* B, **41**, 7243.

COMPOUND INDEX

Compounds discussed in the text are listed here according to the transition metal(s) they contain (ordered alphabetically). The notation MgO:Co denotes Co as an impurity in MgO.

Ag_2O_2 115, 117, 243

CoO 11, 21, 78, 134, 138, 146, 151
$Co_{1-x}O$ 176, 180, 183
Co_3O_4 11, 24
$CoFe_2O_4$ 1, 167
$Co_{3-x}Fe_xO_4$ 164, 167, 243
$LaCoO_3$ 115, 116, 233
$La_{1-x}Sr_xCoO_3$ 116
$LiCoO_2$ 14, 116
$Li_xCo_{1-x}O$ 162, 177, 188, 191
Al_2O_3:Co 46, 52
MgO:Co 46, 121
$SrTiO_3$:Co 121
TiO_2:Co 128

CrO_2 32, 76, 221, 241
CrO_3 105
Cr_2O_3 31, 46, 134, 140, 146, 232
K_3CrO_8 6
KCr_3O_8 243
$LaCrO_3$ 134, 146
$La_{1-x}Sr_xCrO_3$ 162, 166, 169
$SrCrO_3$ 233
Al_2O_3:Cr 45, 46, 52, 118, 130, 170
MgO:Cr 46, 120, 121, 122, 170
$SrTiO_3$:Cr 90, 121
TiO_3:Cr 120, 128, 170
V_2O_3:Cr 225–7

CuO 134, 142
Cu_2O 30, 88, 115
$Bi_2Sr_2CaCu_2O_8$ 248, 259
$LaCuO_3$ 220, 233
La_2CuO_4 19, 225
$La_{2-x}Sr_xCuO_4$ 18, 23, 32, 249–54, 206
$Nd_{2-x}Ce_xCuO_4$ 249–54
$Tl_2Ba_2Ca_2Cu_3O_{10}$ 249
$YBa_2Cu_3O_{7-x}$ 1, 3, 7, 249–55, 260
Al_2O_3:Cu 52
TiO_2:Cu 128

$Fe_{1-x}O$ 1, 12, 14, 21, 26, 78, 134, 144, 151, 182
Fe_3O_4 1, 6, 12, 24, 32, 153, 222, 242–5
Fe_2O_3 134, 140, 148
Fe_2CoO_4 1, 167
$Fe_xCo_{3-x}O_4$ 164, 167, 243
$FeTiO_3$ 5, 21
$BaFeO_4$ 7
$CaFeO_3$ 247–8
$LaFeO_3$ 134, 146, 233
$SrFeO_{2.5}$ 15, 27
$SrFeO_3$ 27, 233
$Sr_2LaFe_3O_9$ 248
$Y_3Fe_5O_{12}$ 24, 31, 134, 148
Al_2O_3:Fe 46, 52, 131
$LiNbO_3$:Fe 57, 123
MgO:Fe 46, 121, 123, 127
$SrTiO_3$:Fe 121, 127
TiO_2:Fe 128

IrO_2 233

MnO 10, 21, 78, 134, 138, 140, 144, 147, 151, 232
$Mn_{1-x}O$ 162, 165, 176, 180
MnO_2 134, 146, 232
$KMnO_4$ 107
$LaMnO_3$ 134, 152, 233
$La_{1-x}Pb_xMnO_3$ 223–4
$La_{1-x}Sr_xMnO_3$ 153, 195, 223–4, 232
$Li_xMn_{1-x}O$ 31, 162, 177, 191, 243
$SrMnO_3$ 233
Al_2O_3:Mn 46, 52
MgO:Mn 121
$SrTiO_3$:Mn 121
TiO_2:Mn 128

MoO_2 21, 241
MoO_3 18, 105, 109, 234
Mo_4O_{11} 237
$K_{0.3}MoO_3$ 3, 32, 80, 195, 235–7
$K_{0.33}MoO_3$ 3
$K_{0.9}Mo_6O_{17}$ 237
$NaMo_4O_8$ 19

NbO 21, 32, 249

NbO_2 21, 241
Nb_2O_5 105
$Nb_{18-n}W_{8+n}O_{69}$ 193
$LiNbO_3$ 5, 21, 22, 57, 105, 106, 114
$LiNbO_{3-x}$ 162, 168, 176
$KNbO_3$ 105, 112
$NaNbO_3$ 109
$Ti_{1-x}Nb_xO_2$ 158

NiO 1, 5, 21, 31, 50, 59, 61, 78, 83, 133–44, 151
$Ni_{1-x}O$ 31, 89, 159, 165, 168, 174, 176
$LaNiO_3$ 1, 32, 41, 220, 231, 233
$LiNiO_2$ 148, 152
$Li_xNi_{1-x}O$ 159, 162, 177, 188–91, 193
Al_2O_3:Ni 46, 52
MgO:Ni 46, 121, 123
$SrTiO_3$:Ni 121
TiO_2:Ni 128

OsO_4 1

PdO 30, 43, 115, 117

ReO_3 1, 21, 32, 205–6, 213–16

$LaRhO_3$ 30, 115, 116

RuO_2 32, 68–74, 204, 216, 221, 241
$Bi_2Ru_2O_7$ 228–31
$CaRuO_3$ 221, 228–31
$Pb_2Ru_2O_{7-x}$ 228–31
$SrRuO_3$ 22, 221, 228–31
$Y_2Ru_2O_7$ 228–31

Ta_2O_5 105
$LiTaO_3$ 105
$KTaO_3$ 22, 30, 109
$KTaO_{3-x}$ 162, 185, 191, 195–6, 199
$NaTaO_3$ 105, 106
$Na_xW_{1-y}Ta_yO_3$ 195–6

TiO_x 21, 26, 232
TiO_2 1, 3, 5, 14, 21, 29, 50, 101, 105, 106, 109, 128
TiO_{2-x} 31, 158, 162, 168, 174, 176–7, 182–3
Ti_2O_3 21, 22, 30, 32, 42, 237–40
Ti_4O_7 245–7
$Ti_{2-x}Nb_xO_2$ 158, 162
$Ti_{2-x}V_xO_3$ 239
$BaTiO_3$ 105, 110, 113
$BaTiO_{3-x}$ 162, 169, 176
$CaTiO_3$ 109
$FeTiO_3$ 5, 21
$KTiO(PO_4)$ 114
$LaTiO_3$ 227, 231, 233
$La_2Ti_2O_7$ 105
$Li_{1-x}Ti_{2+x}O_4$ 195, 200, 249
$MgTiO_3$ 105, 106
$PbTiO_3$ 1, 23, 109, 110, 113
$SrTiO_3$ 105, 106, 109, 112, 121
$SrTiO_{3-x}$ 32, 162, 169, 176, 185, 191, 195–6, 199
Al_2O_3:Ti 46, 52, 131
MgO:Ti 121

VO_x 21, 26, 232
VO_2 1, 21, 30, 32, 80, 240
V_2O_3 1, 32, 225–7
V_2O_5 105
$V_{2-x}Cr_xO_3$ 225–7
$V_{2-x}Ti_xO_3$ 226
$Cu_xV_2O_5$ 166
$LaVO_3$ 22, 134, 146, 233
$La_{1-x}Sr_xVO_3$ 162, 169, 193, 195
$Li_xV_2O_5$ 158, 162, 166, 195
$Li_xMg_{1-x}V_2O_4$ 195, 200
$SrVO_3$ 233
Al_2O_3:V 46, 52
MgO:V 121
$SrTiO_3$:V 121
TiO_2:V 128
Ti_2O_3:V 239

WO_3 21, 105, 109
WO_{3-x} 4, 29, 185
$CaWO_4$ 30
Na_xWO_3 1, 3, 4, 6, 23, 32, 187, 194–6, 206–19
$Na_xW_{1-y}Ta_yO_3$ 195–6
$Nb_{18-n}W_{8+n}O_{69}$ 193–5

ZrO_2 110
$CaZrO_3$ 109
$SrZrO_3$ 105

SUBJECT INDEX

absorption spectra 5, 45, 94, 118, 123, 184–5, 195; *see also* crystal field spectra; charge transfer
acceptor level 126, 159; *see also* defect level; impurity state
activation energy 95, 127, 132, 161–9, 191, 192–3, 236
activity 172, 180, 183
Anderson localization 91, 198, 200
angular overlap 57
antibonding orbital 40, 54–7, 66, 114, 118, 205, 238, 248
antiferromagnetism 31, 76–8, 130, 134–8, 142–51, 227, 253–4, 260, 263–7, 272
Arrhenius equation 161, 193; *see also* activation energy
atomic orbital, *see* crystal field theory; *d* band; electron configuration; linear combination of atomic orbitals; orbital overlap; oxygen 2*p* band; partial density of states
augmented plane wave 69

band, *see* conduction band; valence band
  gap 29–31, 47–53, 76–80, 83, 101–7, 115–17, 126, 131–4, 136–42, 158, 214, 239, 257–8
  structure 32, 66–9, 101, 205–10, 235–41, 267
  theory 36, 64–81, 116, 136–8, 205, 261
  width 48–50, 83–6, 88, 106–7, 116, 169, 205, 209, 218–19, 227–34, 272
BCS theory 259, 268–71
binding energy 49; *see also* ionization energy; photoelectron spectra
bipolaron 186, 247, 269, 272
Bloch's theorem 64
Bohr radius 89, 191
bonding orbital 54–6, 66, 205, 238, 241
bond lengths 15–19, 46, 57, 111, 114, 120, 145, 234, 238–40; *see also* ionic radii
Bose–Einstein condensation 247, 268–9
bremsstrahlung, *see* inverse photoelectron spectra
Brillouin zone 66–9, 78, 80
bronze 4, 23, 32, 81, 89, 166, 195, 206–16, 234–7
brownmillerite structure 27

carrier 157–77, 210–13, 257–9
  scattering 73, 95, 168, 185, 210–11, 224–5
charge density wave 78–80, 234–7
charge transfer 45, 47–53, 59, 91, 122–31, 150, 271
charge transfer insulator 142, 262
chemical
  analysis 3–4
  equilibrium 10, 172–6
  potential 157, 180; *see also* Fermi level
cluster model 36, 53–63, 129, 263
coherence length 257, 269
compensated semiconductor 160, 192, 198
conduction band 30, 32, 101, 106–7, 123–8, 157, 185, 205, 207–10, 228, 242
conductivity 73, 160–9, 192–200, 206, 222, 226, 235–8, 243–6, 255
configuration coordinate model 93–5, 186
configuration interaction 60–3, 91, 141–2
coordination 15–19
  distorted 18–19, 40–2, 110, 114, 234, 250
  linear 115, 117, 250
  number 16
  octahedral 17–18, 21–5, 30, 36–40, 45, 110, 115–16; *see also* corner-sharing, edge-sharing, *and* face-sharing octahedra
  square planar 16–19, 30, 42, 115, 117, 250
  tetrahedral 16–19, 42
corner-sharing octahedra 21–4, 106, 228, 235, 250
correlation, *see* electron correlation
corrosion 178
corundum structure 21, 45, 106, 115,

corundum structure (*cont.*)
118–20, 226, 238
Coulomb force 19, 26, 48, 112, 183, 196–8; *see also* electron correlation; electron repulsion; Hubbard model
covalent bonding 19, 47, 54–7, 63, 64–6, 129, 148–52, 231, 233, 265
critical field 257–8
crystal field
model 17, 19, 37–47, 118–20, 228
spectra 31, 61, 88, 118–20, 123, 132–5, 138
splitting 30, 37–46, 57, 61, 71, 101, 115–17, 120, 137, 205, 234
stabilization energy 40
crystallographic shear plane 28–9, 183, 246
crystal structure, *see* coordination, structure
Curie temperature 110–13, 221, 224
Curie–Weiss law 112, 135–6

d^0 oxide 9, 19, 28, 50, 56, 100–15, 234
d band 30, 67–72, 140–2, 205; *see also* conduction band
d–d transitions, *see* crystal field spectra
defect 25–9, 36, 52, 157–200
calculations 26, 28
charge 170, 180, 182
clustering 26–7
level 31, 52; *see also* impurity state
ordering 26–9
degeneracy 40, 54–6, 121, 206
de Haas–van Alphen effect 206
density of states 70–2, 157–8, 205, 215
at Fermi level 74–6, 207–10, 228, 231, 240
diamagnetism 30, 74, 115, 255
dichroism 103
dielectric
constant 28, 49, 53, 87, 89, 92, 108–15, 190, 193
function 73, 213–14
loss 188
different orbitals for different spins 58, 76, 90, 124
diffusion 14, 178–82
coefficient 178–82
direct gap 104
disorder 36, 196–200, 210
dispersion 66–8, 139, 205
disproportionation 117, 247–8, 271
donor level 30, 126, 159–60, 196; *see also* defect level; impurity state
doping 31, 157–200, 221–25, 243, 252–4, 260–64
double exchange 153, 221–5, 271
Drude theory 73, 213–14

edge-sharing octahedra 21, 28, 106, 235, 240
EELS, *see* electron energy-loss spectrum
effective mass 72, 87, 89, 158, 162, 169, 190, 209, 214, 228–9, 239, 269
electron
affinity 53
configuration 9, 19, 37–43, 56, 67, 84–5, 115–17, 125, 138, 150–2, 167, 232–4, 248
correlation 32, 44, 60, 63, 82, 219–34, 240, 259, 261–3, 271–2; *see also* Hubbard model
energy-loss spectrum 101, 135, 187, 194, 214, 228
microscopy 4, 26
repulsion 39, 43–4, 47, 51, 57, 59, 82–7, 124, 130
electron–phonon coupling, *see* lattice interactions
emission spectra 5, 118–20, 126
entropy 136, 227
EPR, *see* ESR
ESR 5, 113, 120–2, 129, 183, 186
exchange
energy 39, 47, 53, 58–9, 75, 84, 128, 137, 233
interaction 130, 148–53, 220–4, 268
exciton 87–8, 103, 107, 138, 140, 270
exhaustion 160

face-sharing octahedra 21–2
Fermi
energy, *see* Fermi level
level 70–2, 92, 157–60, 194, 198, 205, 212, 240, 259, 267
liquid 265
surface 73–5, 206, 209, 213, 265–7, 269
nesting 76, 80, 234–7, 240
Fermi–Dirac distribution 157–60
ferrimagnetism 134, 148, 223
ferrite 24, 31, 148
ferroelectric 110–14, 191
ferromagnetism 76, 148–9, 151–3, 220–25, 263
Fick's law 179
free-carrier absorption 184–5
free-electron model 32, 72, 207, 211–14
fluorescence 106
Franck–Condon principle 53, 93–5, 107, 120, 126, 186, 188, 228

Frenkel
  defect 25
  exciton 88, 138

garnet structure 24, 31, 148
Gibbs free energy 10
Green's function 90
g value 46, 121

Hall effect 163, 167–8, 211–12
Hartree–Fock method 59, 138
heat capacity, *see* specific heat
Heikes formula 166
Heisenberg Hamiltonian 147, 149
high spin 10, 39–42, 59, 124, 128, 232–3, 248
hopping
  conduction, *see* small polaron; variable-range hopping
  integral 149–50, 219
hole, *see* acceptor level, semiconductor
Hubbard model 52, 82–7, 88, 117, 128, 130, 137–43, 149, 189, 198, 227, 247, 261–3
Hubbard *U* 83–6, 128, 137–9, 233, 261–3, 267
Hund's rule 9, 39, 117, 149, 151–2, 247
hydrogenic model
  for defects 89, 190
  for excitons 87, 103
hyperfine splitting 121, 244, 248

ilmenite 5, 21
impurity 30, 52, 57, 90, 113, 118–31
  band 196–8
  state 89–91, 122–9, 142, 159, 217, 263; *see also* defect level
indirect gap 104
infrared spectrum 259–60
insulator 29–31, 110–153; *see also* d^0 oxide; magnetic insulator
interstitial 25–7, 170–6, 182
intervalence transition 186
intrinsic semiconductor 133, 158, 160, 164
inverse photoelectron spectra 133, 140
inverse spinel, *see* spinel structure
ionic
  conduction 178
  model 5, 8, 28, 36–53, 89, 105, 127, 129, 140, 191, 265
  radius 15–19, 23, 231, 248
ionization energy 50, 53, 84, 105, 159–60
isotope effect 259, 268

irreducible representation 37, 40

Jahn–Teller distortion 19, 40–2, 57, 152, 224, 248, 270

Knight shift 207
Kröger-Vink notation 170–8, 180

lanthanide 17, 85, 136, 227
  contraction 17
large polaron 95, 169
laser 119
lattice
  distortion 32, 78–81, 110–14
  interaction 32, 36, 234–48, 270–1; *see also* Peierls distortion; polaron; vibronic transition
  vibration, *see* vibration
layer perovskite structure 23, 231, 250
law of mass action 172, 183
ligand field theory 39; *see also* crystal field theory
linear combination of atomic orbitals 54–6, 64–7
line phase 4, 12–14
local density method 57, 59, 137–8, 261
localization 36, 81–95, 106, 116, 193, 198, 270; *see also* metal/non-metal transition; magnetic insulator; small polaron
low-dimensional properties 32, 78–81, 231, 234–7, 240, 253, 260
low spin 10, 39–42, 115–7, 152, 221, 228, 232–4
luminescence 106, 118–20

Madelung potential 48, 51
magnetic
  insulator 31, 50–2, 76–8, 131–53, 226–34, 252, 260
  moment 136, 146, 221, 223, 252, 260
  ordering 135–6, 142–53, 220–25
  properties 9, 30–2, 46, 75–8, 135–6, 142–35, 219–34, 240, 255–6; *see also* antiferromagnetism; diamagnetism; ferrimagnetism; ferromagnetism; Meissner effect; paramagnetism
  unit cell 144
magnetite 222–3, 243–5
magnetostriction 144
magnon 147
Marcus–Hush theory 167
mean free path 206, 223–5

Meissner effect 255
metal ion, *see* ionic model
metallic oxide 29, 31-2, 36, 68-75, 191, 192-200, 204-73
metal-metal bonding 19, 21, 29, 32, 237-42, 246
metal/non-metal transition 31-2, 192-200, 225-48, 254, 265
minimum metallic conductivity 200
mixed valency 6, 87, 117, 131, 164-7, 186, 217-19, 221-5, 242-8, 252-4, 260-5
mobility 95, 161-2, 167-9, 191, 223-5; *see also* conductivity
  edge 92, 197-8
molecular orbital 53-60, 63, 81, 130
Mössbauer spectrum 5, 144, 243-5, 247-8
Mott
  criterion 199
  insulator 131, 142; *see also* magnetic insulator
  transition 225-34, 240; *see also* metal/non-metal transition
Mott-Hubbard splitting 131; *see also* Hubbard *U*
muffin-tin approximation 68

Néel temperature 134, 143-6, 151, 168, 227, 253
neutron
  diffraction 26, 144, 211, 223, 240, 248, 253, 260
  scattering 147
negative *U* 117, 247
nickel arsenide structure 22
NMR 5, 207, 263
non-degenerate semiconductor 158
non-stoichiometry 4, 11-15, 21, 25-9, 117, 157-200
n-type, *see* donor level; doping; semiconductor

octahedra, *see* coordination; corner-sharing; edge-sharing; face-sharing
optical properties 29, 31, 46, 57, 73, 102-7, 109, 118-19, 122-35, 213-14
  non-linear 114-15
  see also absorption spectra, emission spectra, reflectivity
orbital, *see* atomic orbital; *d* orbital; molecular orbital
  energy 54, 58
  overlap 47, 53-6, 65-8, 86, 149-52, 196-9, 235
overlap integral 54, 61, 149
oxidation state 5-8, 15-16, 32, 120, 165, 217, 233, 242-8, 252-4, 263-5
oxide ion 6-7, 16, 48, 190
oxygen
  hole 190, 218, 233, 262-5
  2*p* band 30, 67-72, 101, 262-5; *see also* valence band
  pressure 10-14, 172-8, 180, 252

paramagnetism 74-6, 135-6, 209, 220, 226, 247
partial density of states 70, 215
Pauli paramagnetism 74-5, 209, 228
Peierls distortion 78-80, 234-7, 241
percolation 199-200
periodic lattice distortion 78, 234-7
perovskite structure 22, 27, 102-6, 109-13, 116, 145, 206, 223, 231-3, 247, 249
peroxide 6-7, 264, 271
phase
  diagram 11-14, 226, 254
  rule 12
phonon, *see* vibration
photoconductivity 103
photoelectron spectrum 6, 49, 61-2, 71-2, 116, 133, 140-2, 187-8, 194, 207-8, 215-19, 228-30, 240-2, 259, 263
  satellites 63, 141, 216-19
photoemission, *see* photoelectron spectrum
piezoelectricity 110
plasma frequency 73, 75, 194-5, 213-14, 229, 271
platinum-group element 9
point defect 25-7, 169-84
polarization 109-14, 269-70
polarization energy 28, 49, 51-3, 60, 85-6, 92, 129, 191
polaron 89, 92-5, 186, 196, 243, 270; *see also* bipolaron; large polaron; small polaron
post-transition element 115, 227-31, 247
p-type, *see* acceptor level; doping; semiconductor
pyrochlore structure 228-31

quenching
  chemical 15, 27
  orbital 136

Racah parameter 43-4, 47

Raman spectrum 187, 260
redox potential 53, 106
reflectivity 73, 110, 140, 195, 213, 259
refractive index 109, 110, 213
resistivity, *see* conductivity
resonance integral 149–50
rhenium trioxide structure 21, 205–6
rocksalt structure 21, 144, 151, 249
ruby 45, 118–20
rutile structure 21, 68, 101–6, 120, 127, 221, 240–2

sapphire 131
Schottky defect 25
scattering, *see* carrier scattering
screening 75, 82, 211, 260, 267
Seebeck effect 163–6, 177, 194, 200, 212
selection rule
  atomic 101
  Laporte 46
  spin 45, 118
  wavevector 104
semiconductor 29, 31, 117, 157–200, 237, 240, 241, 243, 246
semiconductor/metal transition, *see* metal/non-metal transition
shear plane 28–9, 183, 246
shell model 49, 127, 191
small polaron 92–5, 131, 163–9, 188, 191, 244
soft mode 112
specific heat 74, 207–8, 227, 239, 259, 269
spectroscopy 53; 101–8, 118–42, 213–19; *see also* absorption spectrum; emission spectrum; ESR; inverse photoelectron spectrum; Mossbauer spectrum; NMR; photoelectron spectrum; X-ray spectrum
spin, *see* electron configuration; ESR; exchange energy; high spin; low spin; magnetic moment; magnetic properties
spin density wave 77, 220, 260
spinel structure 24, 42, 148, 163, 167, 222, 243–5, 249
spin-only formula 136; *see also* magnetic moment
spin–orbit coupling 43, 136, 206, 216, 220
spin polaron 225, 272
spin wave 138, 147
SQUID 257
Stokes shift 107, 126–7
stoichiometry 3–6, 253; *see also* line phase, non-stoichiometry
Stoner
  model 75–6
  enhancement 75, 220
structural distortion 21, 196, 240, 250; *see also* ferroelectricity; Jahn–Teller distortion; metal–metal bonding; Peierls distortion
structure 20–8, 235, 250–3; *see also* coordination; structural distortion
sub-band 83
sub-lattice magnetization 144, 146
superconductor 4, 7, 18, 32, 200, 225, 249–73
superexchange 147–53, 253, 260, 263, 272
superoxide 6
susceptibility; *see* magnetic properties
symmetry 37, 40–2, 54–6, 64, 104, 114, 118, 205–6

Tanabe–Sugano diagram 44
term symbol 39, 118–20
thermodynamics 10–14, 25, 53, 129, 172, 183, 227
thermopower 163–6, 194, 200, 212, 245
tight-binding method 54, 64–7
transport property 160–9, 210–13, 224, 226; *see also* conductivity; Hall effect; Seebeck effect

unpaired electron 9, 31, 131–2
UPS, *see* photoelectron spectrum

vacancy 25–8, 122, 170–84
valence band 30, 101, 125, 128, 139, 177; *see also* oxygen 2*p* band
valence trapping 93; *see also* mixed valency
variable-range hopping 92, 193, 198–9
Verwey transition 243–6
vibration 104, 106, 112–3, 186–7, 210, 268
vibronic transition 46, 104; *see also* Franck–Condon principle; selection rule

Wannier exciton 87–8, 103
wavevector 64–9, 73, 79, 104
Weiss constant 134, 136
Wigner transition 245

XPS, *see* photoelectron spectrum

X-ray
  diffraction 4, 14, 29
  spectrum 49, 101–2, 133, 190, 218, 262–3, 271
$X\alpha$ method 57–9

Zeeman splitting 75, 121
zero-field splitting 121

π bonding 54–7, 114, 205, 231, 233
σ bonding 54–7, 233